Trademarks

The following are registered trademarks of Autodesk, Inc.: AutoCAD, AutoCAD Architectural Desktop, Revit, Autodesk, AutoCAD Design Center, Autodesk Device Interface, VizRender, and HEIDI

Microsoft, Windows, Word, and Excel are either registered trademarks or trademarks of Microsoft Corporation.

All other trademarks are trademarks of their respective holders.

Copyright 2014 by Elise Moss

All rights reserved. No part of this book may be reproduced, stored in a retrieval system, or transcribed in any form or by any means – electronic, mechanical, photocopying, recording, or otherwise – without the prior written permission of Schroff Development Corporation.

Moss, Elise
 The Unofficial Revit 2015 Certification Exam Guide
 Elise Moss
ISBN-13: 978-1503158382
ISBN-10: 1503158381

Examination Copies:

Books received as examination copies are for review purposes only and may not be made available for student use. Resale of examination copies is prohibited.

Electronic Files:

Any electronic files associated with this book are licensed to the original user only. These files may not be transferred to any other party.

The author and publisher of this book have used their best efforts in preparing this book. These efforts include the development, research, and testing of material presented. The author and publisher shall not be held liable in any event for incidental or consequential damages with, or arising out of, the furnishing, performance, or use of the material herein.

Printed and bound in the United States of America.

Preface

This book is geared towards users who have been using Revit for at least six months and are ready to pursue their official Autodesk certification. Autodesk offers two certification exams for Revit: a User certificate and a Professional certificate. It is advisable to take both exams. You can locate the closest testing center on Autodesk's website.

I wrote this book because I taught a certification preparation class at SFSU. I also teach an introductory Revit class using my Revit Basics text. I heard many complaints from students who had taken my Revit Basics class when they had to switch to the Autodesk AOTC courseware for the exam preparation class. Students preferred the step by step easily accessible instruction I use in my texts. When my support staff advised me that Autodesk had informed them that there would not be an exam guide available for Revit 2010, I decided to tackle my own guide. Because of the interface changes between Revit 2008 (the last year an AOTC guide is available) and 2010, I did not feel comfortable using a 2008 textbook to prepare students for a 2010 certification test.

This textbook includes exercises which simulate the knowledge users should have in order to pass the certification exam. I advise my students to do each exercise two or three times to ensure that they understand the user interface and can perform the task with ease.

I have endeavored to make this text as easy to understand and as error-free as possible…however, errors may be present. Please feel free to email me if you have any problems with any of the exercises or questions about Revit in general.

I have posted videos for some of the lessons from this text on my website at www.mossdesigns.com on the Tutorials page. The videos are free to view. Videos are also posted on Youtube – search for Moss Designs and you will find any videos for my book on that channel.

Exercise files can be accessed and downloaded from onedrive at:

http://1drv.ms/1ByXaSF

If you need access to the link or have problems accessing the files, email me at elise_moss@mossdesigns.com and I will send you a personal invite to the folder. There is no charge to you.

I no longer use dropbox because too many users were deleting and modifying files. With onedrive, you have to download files before you can use them.

Acknowledgements

A special thanks to Rick Rundell, Gary Hercules, Christie Landry, and Steve Burri, as well as numerous Autodesk employees who are tasked with supporting and promoting Revit.

Additional thanks to Gerry Ramsey, Will Harris, Scott Davis, James Cowan, James Balding, Rob Starz, and all the other Revit users out there who provided me with valuable insights into the way they use Revit.

My eternal gratitude to my life partner, Ari, my biggest cheerleader throughout our years together.

Elise Moss
Elise_moss@mossdesigns.com

TABLE OF CONTENTS

Preface i

Acknowledgements ii

Table of Contents iii

Introduction - FAQs

Lesson One
Building Information Modeling and Revit Basics
    Exercise 1-1
    Quick Access Toolbar 1-3
    Exercise 1-2
    Exploring the User Interface 1-7
    Exercise 1-3
    Recover and Use Backup Files 1-8
    Exercise 1-4
    Design Options 1-12
    Exercise 1-5
    Phases 1-27

    Practice Exam 1-41

Lesson Two
The Basics of Building a Model
    Exercise 2-1
    Wall Options 2-2
    Exercise 2-2
    Placing a Wall Sweep 2-6
    Exercise 2-3
    Create a Wall Sweep Style 2-10
    Exercise 2-4
    Create a Custom Profile 2-12
    Exercise 2-5
    Stacked Walls 2-15

Exercise 2-6
Dividing a Wall into Parts — 2-23

Exercise 2-7
Creating an In-Place Mass — 2-31

Exercise 2-8
Editing an In-Place Mass — 2-45

Exercise 2-9
Mass Properties — 2-52

Practice Exam — 2-56

Lesson Three
Component Families

Exercise 3-1
Level-Based Component — 3-2

Exercise 3-2
Creating a Family — 3-4

Exercise 3-3
Creating a Door Panel — 3-19

Exercise 3-4
Create a Nested Door Family — 3-39

Exercise 3-5
Identifying a Family — 3-50

Exercise 3-6
Assigning OMNI Class Numbers — 3-51

Exercise 3-7
Room Calculation Point — 3-53

Practice Exam — 3-55

Lesson Four
View Properties

Exercise 4-1
Creating a Level — 4-2

Exercise 4-2
Story vs. Non-Story Levels — 4-4

Exercise 4-3
Creating Column Grids — 4-7

Exercise 4-4
Setting View Depth — 4-14

Exercise 4-5
Create a Cropped View — 4-15

Exercise 4-6
Change View Display — 4-17

Exercise 4-7
Reveal Hidden Elements — 4-21

Exercise 4-8
Create a View Template — 4-22

Exercise 4-9
Create a Scope Box ... 4-29
Exercise 4-10
Duplicating Views ... 4-33
Exercise 4-11
Segmented Views ... 4-38

Practice Exam ... 4-40

Lesson Five
Dimensions and Constraints
Exercise 5-1
Placing Permanent Dimensions ... 5-3
Exercise 5-2
Modifying Dimension Text ... 5-9
Exercise 5-3
Converting Temporary Dimensions to Permanent Dimensions ... 5-12
Exercise 5-4
Applying Constraints ... 5-14
Exercise 5-5
Equality Formula ... 5-19
Exercise 5-6
Alternate Dimensions ... 5-21
Exercise 5-7
Multi-Segmented Dimensions ... 5-22

Practice User Exam ... 5-25

Lesson Six
Developing the Building Model
Exercise 6-1
Modifying a Floor Perimeter ... 6-2
Exercise 6-2
Modifying a Ceiling ... 6-5
Exercise 6-3
Creating Stairs by Sketch ... 6-9
Exercise 6-4
Modifying Assembled Stairs ... 6-14
Exercise 6-5
Creating a Stair by Component ... 6-19
Exercise 6-6
Railings ... 6-25
Exercise 6-7
Creating a Roof by Footprint ... 6-30
Exercise 6-8
Creating a Roof by Extrusion ... 6-32
Exercise 6-9
Modifying a Roof Join ... 6-38

Exercise 6-10
Creating a Sloped Ceiling   6-43

Practice Exam   6-49

Lesson Seven
Detailing and Drafting
Exercise 7-1
Creating Drafting Views   7-2
Exercise 7-2
Reassociate a Callout   7-7
Exercise 7-3
Filled Regions   7-10
Exercise 7-4
Callout from Sketch   7-15
Exercise 7-5
Save and Re-Use a Drafting View   7-20
Exercise 7-6
Adding Tags   7-22
Exercise 7-7
Creating a Detail View   7-25
Exercise 7-8
Creating a Detail Group   7-33
Exercise 7-9
Creating a Drafting View   7-37
Exercise 7-10
Revision Control   7-44
Exercise 7-11
Modify a Revision Schedule   7-47
Exercise 7-12
Add Revision Clouds   7-51
Exercise 7-13
Aligning Views Between Sheets   7-56

Practice Exam   7-61

Lesson Eight
Construction Documentation
Exercise 8-1
Creating a Door Schedule   8-2
Exercise 8-2
Creating a Legend   8-5
Exercise 8-3
Adding Rooms to a Floor Plan   8-12
Exercise 8-4
Creating an Area Scheme   8-15
Exercise 8-5
Creating an Area Plan   8-18
Exercise 8-6
Creating a Room Schedule   8-22
Exercise 8-7   8-25

Creating a Drawing List
Exercise 8-8
Create a Note Symbol 8-28
Exercise 8-9
Add Notes 8-33
Exercise 8-10
Create a Material TakeOff Schedule 8-35
Exercise 8-11
Creating a Mass Floor Schedule 8-38

Practice Exam 8-40

Lesson Nine
Presenting the Building Model
Exercise 9-1
Creating a Toposurface 9-2
Exercise 9-2
Defining Camera Views 9-12
Exercise 9-3
Sun Settings 9-19
Exercise 9-4
Rendering Settings 9-22
Exercise 9-5
Render in Cloud 9-24
Exercise 9-6
Place a Decal 9-27
Exercise 9-7
Adding a Background 9-32
Exercise 9-8
Using Transparency Settings 9-37
Exercise 9-9
Custom Render Settings 9-38
Exercise 9-10
Using Element Graphic Overrides in a Hidden View ... 9-41

Practice Exam 9-44

Lesson Ten
Collaboration
Exercise 10-1
Monitoring a Linked File 10-3
Exercise 10-2
Interference Checking 10-9
Exercise 10-3
Using Shared Coordinates 10-14
Exercise 10-4
Worksets 10-21
Exercise 10-5
Import an IFC File 10-28

Practice Exam 10-30

Introduction
FAQs

The first day of class students are understandably nervous and they have a lot of questions about getting certification. Throughout the class, I am peppered with the same or similar questions.

Why is this textbook the "Unofficial" Guide?

Autodesk reserves the right to publish any official training courseware. I have used and continue to use Autodesk courseware when I teach, but I have found that it is easier for me to teach from my own books. I did contact Autodesk and check if there would be any objection to the publication of an unofficial guide and was told that as long as I did not violate any copyrights this book could be published.

What is the difference between the User Level and Professional Exam?

Both exams have you use the software now, but test on different areas of the software. Both exams use a "secure" browser, which means that you cannot cut and paste or copy from the browser. You should use the Alt-TAB shortcut key to flip from the secure browser to the software application to do each problem. If you go on Autodesk's site, there is a list of topics covered for each exam. For the user exam, expect questions on the user interface, navigation, and zooming as well as how to place doors and windows. For the professional exam, the topics include collaboration, creating Revit families, linking and monitoring files, as well as more complex wall families.

It is helpful during both exams to have a piece of paper and a pencil so you can make notes.

You do not need a calculator for either exam and calculators are against the rules anyway.

What are the exams like?

The first time you take the exam, you will create an account with a login. Be sure to write down your login name and password. You will need this regardless of whether you pass or fail.

If you fail and decide to retake the exam, you want to be able to log in to your account.

If you pass, you want to be able to log in to download your certificate and other data.

Both exams are timed. This means you only have one to two minutes for each question. You have the ability to "mark" a question to go back if you are unsure. This is a good idea because a question that comes later on in the exam may give you a clue or an idea on how to answer a question you weren't sure about.

Both exams pull from a question "bank" and no two exams are exactly alike. Two students sitting next to each other taking the same exam will have entirely different experiences and an entirely different set of questions. However, each exam covers specific topics. For example, you will get at least one question about BIM or Revit projects. You will probably not get the same question as your neighbor.

At the end of the exam, the browser will display a screen listing the question numbers, the answers you chose, and indicate any marked or incomplete questions. The questions themselves are not displayed on this page. Any questions where you forgot to select an answer will be marked incomplete with a capitol I. Questions that you marked will display the letter M. You can click on those answers and the browser will link you back directly to those questions so you can review them and modify your answers.

I advise my students to mark any questions where they are struggling and move forward, then use any remaining time to review those questions. A student could easily spend ten to fifteen minutes pondering a single question and lose valuable time on the exam.

Once you have completed your review, you will receive a prompt to END the exam. Some students find this confusing as they think they are quitting the exam and not receiving a score. Once you end the exam, you may not change any of your answers. There will be a brief pause and then you will see a screen where you will be notified whether you passed or failed. You will also see which questions you missed, listed by the question number. Again, you will not see any of the actual questions. However, you will see the topic, so you might see that you missed Question 4 and that question is on view properties.

You will have to rely on your memory to recall what question was about view properties. Some students might find this confusing because they didn't even know that particular question fit into the view properties category.

If you failed the test, you do want to note which categories you missed. For example, you might have missed a question on BIM, a question on view properties, and a question on constraints. Write these down. Then review those topics both in this guide and in the software to help prepare you to retake the exam.

How many times can I take the test?

You can take each test up to three times in a 12-month period. There is no waiting period between re-takes, so if you fail the exam on Tuesday, you can come back on Wednesday and try again. Of course, this depends on the testing center where you take the exam.

Do a lot of students have to re-take the test or do they pass on the first try?

About half of my students pass the exam on the first try, but it definitely depends on the student. Some people are better at tests than others. My youngest son excels at "multiple guess" style exams, which is what the user certification exam is. He can pretty much ace any multiple guess exam you give him regardless of the topic. Most students are not so fortunate. Some students find a timed test extremely stressful. For this reason, I have created a simulated version of both the user and the professional level exams for my students simply so they can practice taking an on-line timed exam. This has the effect of "conditioning" their responses so they are less stressed taking the actual exam.

Some students find the multiple choice style extremely confusing, especially if they are non-native English speakers. There are exams available in many languages, so if you are not a native English speaker, check with the testing center about the availability of an exam in your native language.

Why take a certification exam?

The competition for jobs is steep and employers can afford to be picky. Being certified provides employers with a sense of security knowing that you passed a difficult exam that requires a basic skill set. It is important to note that the certification exam does not test your ability as a designer or drafter. The certification exam tests your knowledge of the Revit software. This is a fine distinction, but it is an important one.

If you pass the exam, you have the option of having your name published on Autodesk's website as a Certified User or Certified Professional. Some employers and headhunters use that list to find a potential new hire. You do not have to have your name published if you don't want to be bothered.

How long is the certification good for?

Your certification is good for a particular year of software. It does not expire. If you pass and are certified for Revit 2015, you are always certified for Revit 2015. That said, you are not certified for Revit 2016 or Revit 2020. Most employers want you to be certified within a couple of years of the most current release, so if you wish to maintain your "competitive edge" in the employment pool, expect that you will have to take the certification exam every two to three years. I recommend students take an "update" class from an Autodesk Authorized Training Center before they take the exam to improve their chances of passing. Autodesk recently changed their requirements and if you have a prior certification, then you only need to re-take one exam to be re-certified. However, this may change, so check with your testing center about the current requirements for certification.

How much does it cost?

The cost for the exam is posted on Autodesk's website. Autodesk sometimes charges students less than professionals, so if you are a student or unemployed, check to see if you qualify for a discount or a free exam. Members of AUGI, Autodesk User Group International, may also qualify for a discount or a free exam. Membership in AUGI is free. If you attend Autodesk University, some years the conference will allow attendees to take a certification exam for free or at a discount. It is worth it to check with the testing center to find out what criteria is necessary to qualify for a discount or free exam. Some testing centers offer "certification days" where users can take the exam at a discount on specific dates. Check with your local reseller or ATC to find out if there are any upcoming certification days available.

Can I take the exam at home?

No. Autodesk requires users to take the exam in a proctored setting. When I proctor an exam, I ask my students to show valid ID, such as a driver's license. This is to ensure that the certification process maintains a certain level of integrity and meaningfulness. Otherwise, a user could pay someone to take the exam for him.

Do I need to be able to use the software to pass the exam?

This sounds like a worse question than intended. Some of my students have taken the Revit classes, but have not actually gotten a job using Revit yet. They are in that Catch-22 situation where an employer requires experience or certification to hire them but they can't pass the exam because they aren't using the software every day. For those students, I advise some self-discipline where they schedule at least six hours a week for a month where they use the software – even if it is on a "dummy" project – before they take the exam. That will boost the odds in their favor.

Is the professional exam a lot harder than the user exam?

Most of my students tell me that they find the user exam considerably harder than the professional exam. The user exam is a mix of thirty multiple choice questions with exercises. The professional exam is a set of thirty-five exercises. For each exercise, the user must open a Revit file, perform a specific task, and then answer a question. If they performed the task properly, the answer will be correct. It is a good idea if you have time to do each task twice just to verify your answer. You can take the professional exam first and then the user exam, if you prefer. You must pass **both** exams to be fully certified in Revit.

What happens if I pass?

Because you are taking the test in a testing center, you want to be sure you wrote down your login information. That way when you get back to your home or office, you can login to Autodesk's certification center and download your certificate. You also can download a logo which shows you are a Certified User or Certified Professional that you can post on your website or print on your business card.

How many times can I log into Autodesk's testing center?

You can log in as often as you like. The tests you have taken will be listed as well as whether you passed or failed.

Can everybody see that I failed the test?

Autodesk is kind enough to keep it a secret if you failed the exam. Nobody knows unless you tell them. If you passed the test, people only see that information if you selected the option to post that result. Regardless, only you and Autodesk will know whether you passed or failed unless *you* choose to share that information.

Can I change my mind? I said I didn't want it posted if I passed and now I am looking for a job.

Yes, you are allowed to change your mind. Log on to the Autodesk testing site and email them using the Contact Link. Provide them with your user name, email address, and the test results you want posted. They will verify that you passed and update their database listing. This may take a week or longer. So, if you change your mind, let them know right away. Do not go on a job interview and say you are certified without bringing along your certificate as proof. Many employers will check your claim against Autodesk's database. A lot of employers now require a Certificate Number as proof that you are actually certified.

What if I need to go to the bathroom during the exam or take a break?

You are allowed to "pause" the test. This will stop the clock. You then alert the proctor that you need to leave the room for a break. When you return, you will need the proctor to approve you to re-enter the test and start the clock again.

How many questions can I miss?

The User exam has 30 questions. The required passing score is 80%. Autodesk may change the passing score requirement, so check on this before you take the exam. This means you can miss as many as six questions and still pass.

The Professional exam has 35 questions. The required passing score is 80%. Autodesk may change the passing score requirement, so check on this before you take the exam. This means you can miss as many as four questions and still pass.

Do I have to take both tests? Can't I just take the Professional Exam?

In order to receive the Professional certificate, you must pass both the User and Professional exams. However, you can take the Professional exam before the User exam, if you prefer. If you have prior certification in Revit, you only need to take the Professional exam.

How much time do I have for each exam?

The user exam has a fifty minute time limit. That is roughly two minutes per question.

The professional exam has a two hour time limit. That is roughly three minutes per question.

Can I ask for more time?

You can make arrangements for more time if you are a non-English speaker or have problems with tests. Be sure to speak with the proctor about your concerns. Most proctors will provide more time if you truly need it. However, my experience has been that most students are able to complete the exam with time to spare. I have only had one or two students that felt they "ran out of time".

What happens if the computer crashes during the test?

Don't worry. Your answers will be saved and the clock will be stopped. Simply reboot your system. Get Revit launched again (if you are taking the professional exam) and let the proctor know when you are ready to start the exam again, so you can re-enter the testing area in your browser.

Do I have to take both exams if I want to be re-certified?

If you passed the user and professional level exams for a previous release, you only need to take the Professional exam again to be certified for the current release.

Can I have access to the practice exams you set up for your students?

I have placed a mock User exam on the internet at http://www.daypo.net/revit-user-certification-practice-exam.html. This exam has not been updated for the current release, but it will give you an idea of the type of questions to expect.

I have updated my practice exam for 2015 and have placed it here:

https://www.hightail.com/download/UlRTb3BORkVveE0xWjhUQw

If you have difficulty with the link, email me and I will send you the link electronically. This link takes you to a zipped/compressed file. In order to use the exam, you have to extract the four files (two Revit files and two quiz files). Then double left click on the quiz.html file to open a browser and start the test.

What sort of questions do you get in the exams?

Many students complain that the questions are all about Revit software and not about building design or the uniform building code. Keep in mind that this test is to determine your knowledge about Revit software. This is not an exam to see if you are a good architect or designer.

The exams have several question types:

One best answer – this is a multiple choice style question. You can usually arrive at the best answer by figuring out which answers do NOT apply.

Select all that apply – this can be a confusing question for some users because unless they know how *many* possible correct answers there are, they aren't sure. In some cases, you may be provided with the hint of selecting two or three out of five possible choices.

Point and click – this has a java-style interface. You will be presented with a picture, and then asked to pick a location on the picture to simulate a user selection. When you pick, a mark will be left on the image to indicate your selection. Each time you pick in a different area, the mark will shift to the new location. You do not have to pick an exact point…a general target area is all that is required.

Matching format – you probably are familiar with this style of question from elementary school. You will be presented with two columns. One column may have assorted terms and the second column the definitions. You then are expected to drag the terms to the correct definition to match the items.

What are the exercises like?

Each question follows the same process. You will be asked to open a file. You may be asked to open a specific view or a sheet. This tests your ability to navigate around Revit, so be familiar with the Project Browser and how it works. You may be asked to orient the model to a specific orientation or level. You will then be required to perform a specific task, either adding or modifying an element or determining an element's properties. You will then be asked to fill an answer into an input box. The answer will be either text or numeric. The answer must match exactly with what is presented on screen, so it is best to Copy and Paste between Revit and the secure browser. If you miss a punctuation mark, use the wrong case, or spell something wrong, your answer will be marked wrong.

If you are unsure about a question in the exam and want to mark it for review, perform a SaveAs Copy of the file, so you don't lose any work you did.

Any tips?

I suggest you read every question at least twice. Some of the wording on the questions is tricky.

Be well rested and be sure to eat before the exam. Most testing centers do not allow food, but they may allow water. Keep in mind that you can take a break if you need one.

Relax. Maintain perspective. This is a test. It is not fatal. If you fail, you will not be the first person to have failed this exam. Failing does not mean you are a bad designer or architect or even a bad person. It just means you need to study the software more.

Have Revit open and ready to go before you start the test. Verify that you know where the drawing files are located so you don't have to search for them every time you need to open a file. Write down the file path on your scratch paper just in case you panic. Don't close the file. Occasionally, different questions will use the same file, so if you already have the file open that saves time. You will not open and use every file in the data set, so do not open every file in anticipation of using it. Practice switching windows before the exam. Open Internet Explorer and open a session of Revit. Then use the Alt-Tab to switch between the windows. Practice this until you are comfortable.

Remember to write down your login name and password for your account. The proctor will not be able to help you if you forget.

Lesson One

Building Information Modeling and Revit Basics

This lesson addresses the following User level exam questions:

- Building Information Modeling
- User Interface
- Building Elements
- Revit Projects

There will be at least one question on the User exam regarding Building Information Modeling. You will be expected to understand what BIM means and how it works. Autodesk is extremely proud that Revit is BIM software.

BIM means that Revit uses intelligent objects to create and manage a building model. In AutoCAD, you draw a set of lines to symbolize a door. In Revit, you place a door object which has parameters embedded in the object. These parameters contain data concerning the door: everything from the material, cost, and size to function and manufacturer information. This information can be leveraged to be used in schedules and in Excel spreadsheets. You can create an unlimited number of views for your building model and they all reside in a single file.

Revit boasts "bidirectional associativity," which means that if you make a change in one view, all related views also update.

Revit has parametric relationships within the model. For example, floors are constrained to walls, so if a wall is shifted in any direction, the floor will automatically update.

When you first launch Revit, a startup window named Recent Files is displayed.

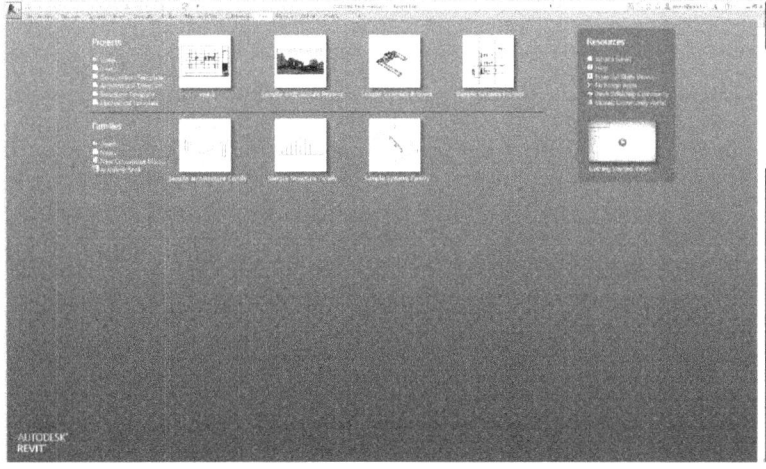

The Unofficial Revit 2015 Certification Exam Guide

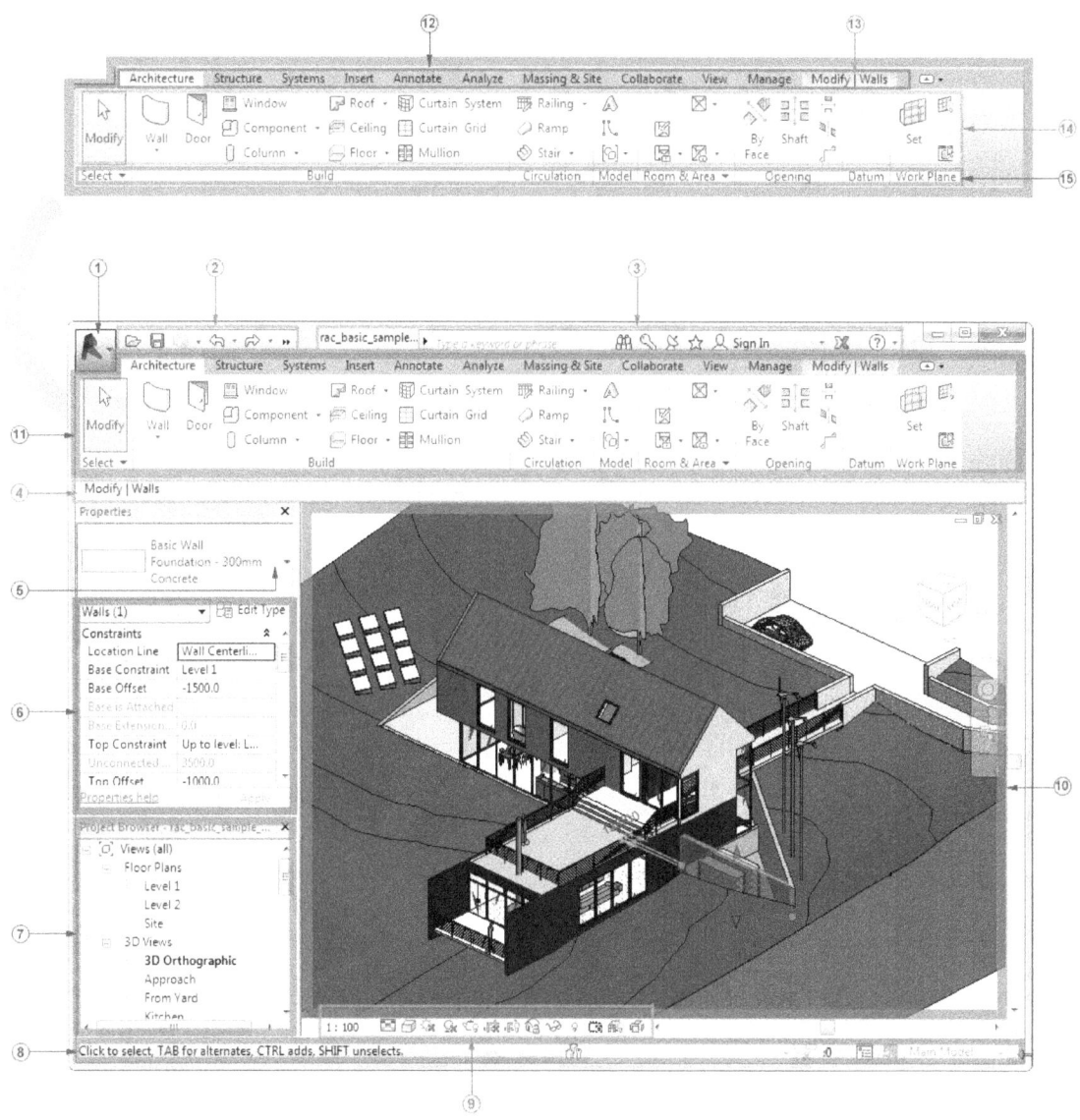

1	Application Menu
2	Quick Access Toolbar
3	InfoCenter
4	Options Bar
5	Type Selector
6	Properties Palette
7	Project Browser
8	Status Bar
9	View Control Bar
10	Drawing Area
11	Ribbon
12	Tabs on the ribbon
13	A contextual tab on the ribbon, providing tools relevant to the selected object or current action
14	Tools on the current tab of the ribbon
15	Panels on the ribbon

You will be expected to identify the different areas of the Revit User Interface in the exam.

For example, you may have a question asking you to indicate where the View Control Bar is located.

Building Information Modeling and Revit Basics

Command Exercise
Exercise 1-1 – Quick Access Toolbar

Drawing Name: **(none, start from scratch)**
Estimated Time to Completion: 10 Minutes

Scope

Learn how to add and remove tools from the Quick Access Toolbar.

Solution

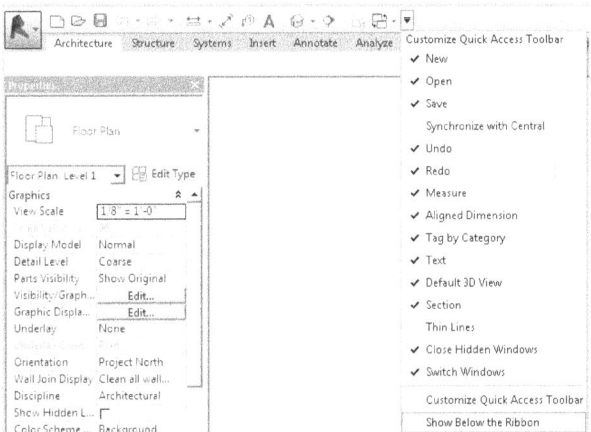

1. Select the drop-down arrow on the Quick Access toolbar.

Enable **New**.
Disable **Synchronize with Central**.
Disable **Thin Lines**.

The Quick Access toolbar updates with the new settings.

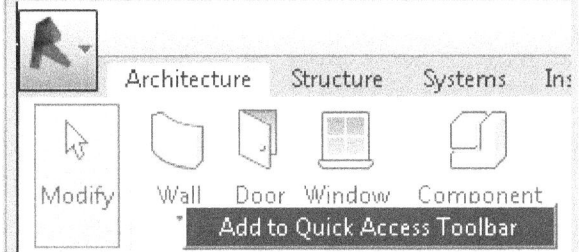

2. Place your mouse over the Wall tool on the Architecture ribbon.

Right click and select **Add to Quick Access Toolbar**.

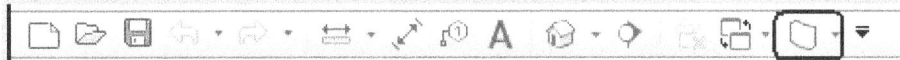

The Wall tool is added to the Quick Access toolbar (QAT).

3. Select the **Wall** tool on the QAT and place a wall in the drawing area.
To place the wall, just select two points like you are drawing a line.
Then press ESC to release the command.
Select the Wall and note that the ribbon changes to Modify mode.

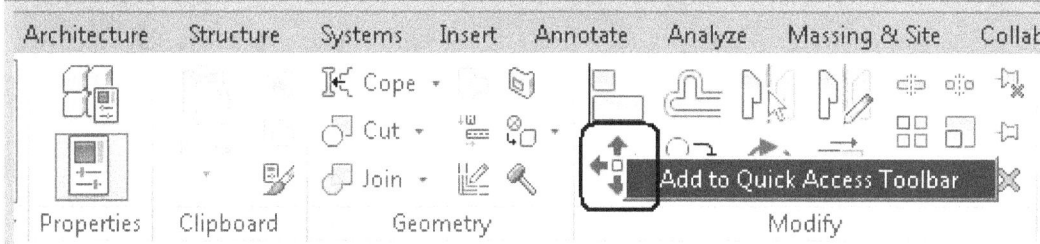

4. Activate the **Modify** ribbon.
Right click on the **Move** tool.
Select **Add to Quick Access Toolbar**.

The Quick Access toolbar now displays the Move tool.
5. Left click anywhere in the drawing area. This releases the wall from selection.

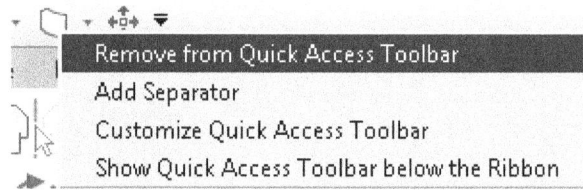

6. Right click on the **Wall** tool on the Quick Access Toolbar.
Select **Remove from Quick Access Ribbon**.
7. The Wall tool is removed.

 The Quick Access toolbar behaves like the ribbon as some tools may become disabled depending on the mode you are in.

Building Elements are used to create a building design. There are five classes of building elements: host, component, datum, annotation, and view. Building elements fall into three categories: Model, View, and Annotation. To pass the User exam, users need to identify which category a building element falls in.

Each element falls into a category, such as wall, column, door, window, furniture, etc. Each category contains different families. Each family can have more than one type. The type is usually determined by the size or parameters assigned to that family.

These are very difficult concepts for many students, especially if they have been used to dealing with lines, circles, and arcs.

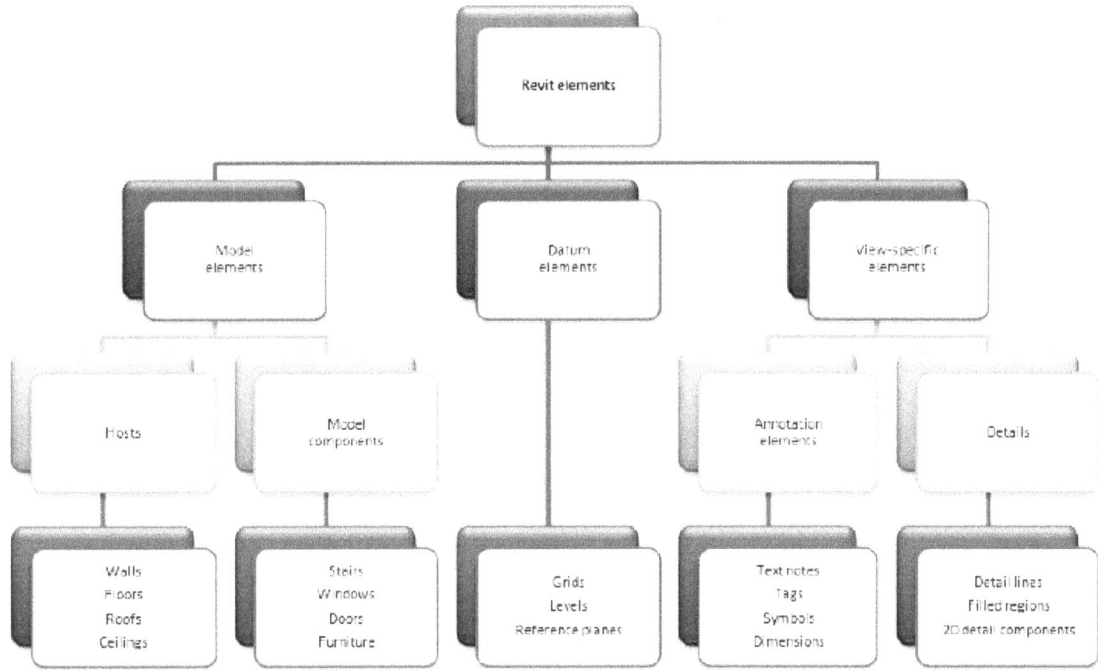

Revit elements are separated into three different types of elements: Model, Datum and View-specific. Users are expected to know if an element is model, datum or view-specific.

Model elements are broken down into categories. A category might be a wall, window, door, or floor. If you look in the Project Browser, you will see a category called Families. If you expand the category, you will see the families for each category in the current project. Each family may contain multiple types.

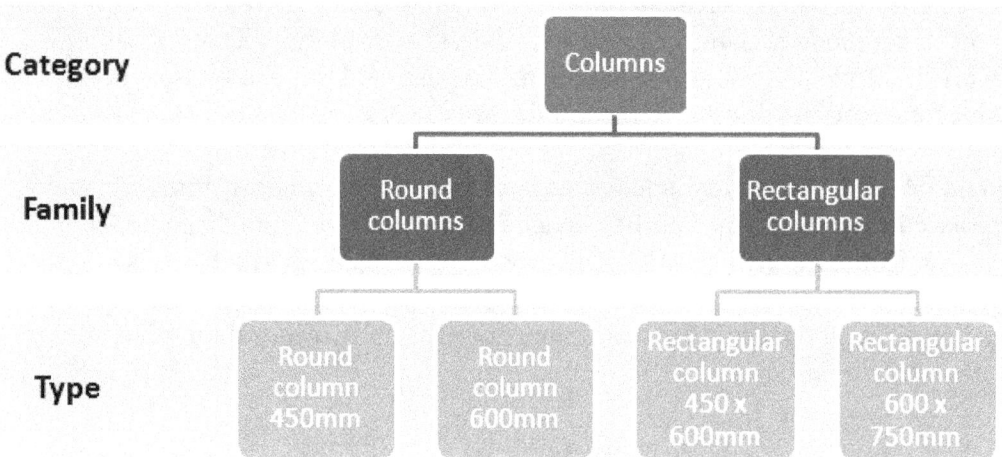

Every Revit file is considered a Project. A Revit project consists of the Project Environment, components, and views. The Project Environment is managed in the Project Browser.

Command Exercise

Exercise 1-2 – Exploring the User Interface

Drawing Name: **c_user interface.rvt**
Estimated Time to Completion: 5 Minutes

Scope

Review the user interface to prepare for the exam.

Solution

1. The file will open in a 3D view. Note that there is a ViewCube in the upper left corner.

2. Open the Level 1 Floor Plan view.

Double left click on Level 1 listed in the Project Browser.

The ViewCube is only visible in 3D views. This is a possible question on the exam.

3. Note that the ViewCube is no longer visible and has been replaced with the Navigation Bar.

Command Exercise

Exercise 1-3 – Recover and Use Backup Files

Drawing Name: **new**
Estimated Time to Completion: 5 Minutes

Scope
Recover and Use Backup Files

Solution

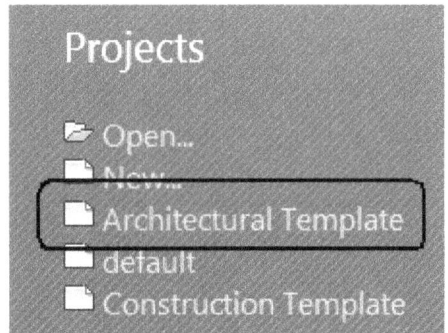

1. Close any open projects.

Select **Architectural Template** under Projects.

This starts a new project using the Architectural template.

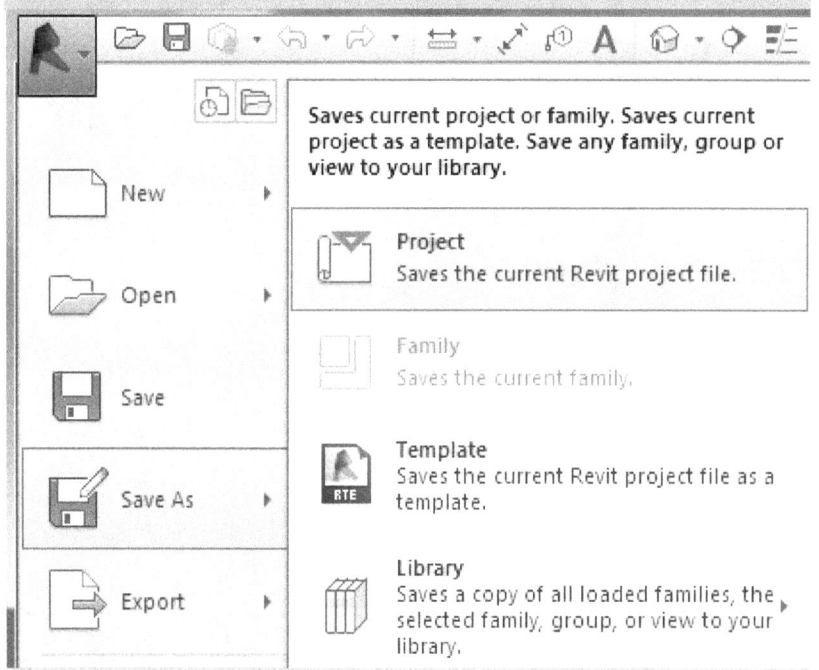

3. Go to **File→Save As→ Project**.

Building Information Modeling and Revit Basics

4. Select the **Options** button next to the file name.

5. Set the Maximum backups: to **5**.

Some students prefer not to save any backups so that their flash drive doesn't fill up. Those students set the number of backups to 0.

Press **OK**.

6. Save as *ex1-3.rvt*.

7. Draw four walls.

8. Press **Save**.

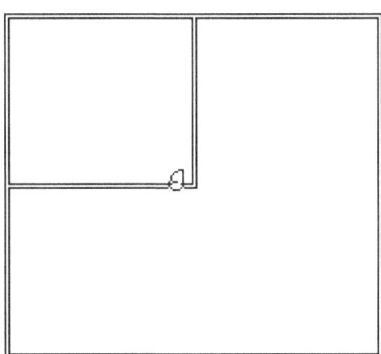

9. Add two more walls.
Add a door.

10. Press **Save**.

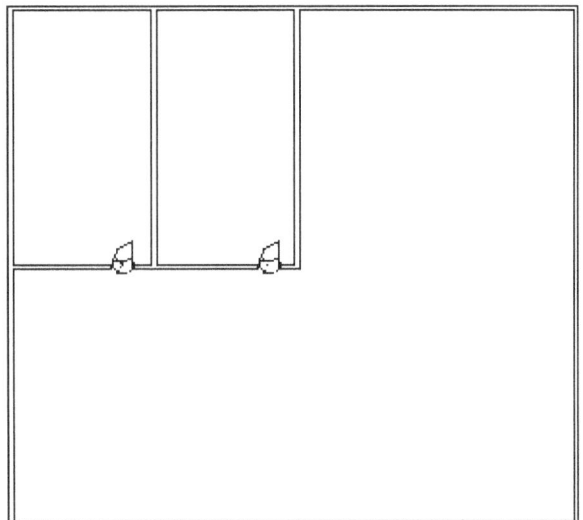

11. Add one wall.
12. Add one door.

 13. Press **Save**.

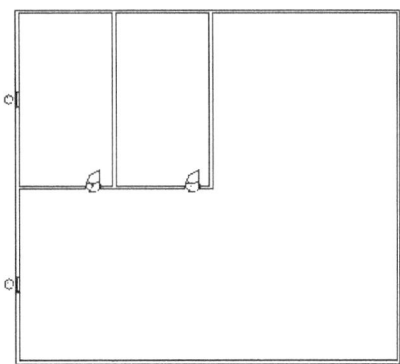

14. Add two windows.

 15. Press **Save**.

 16. Select **Open**.

 ex1-3.0001
ex1-3.0002
ex1-3.0003
ex1-3.0004
ex1-3

17. Note that you have several versions of ex1-3.

The .0000x indicates the backup number.

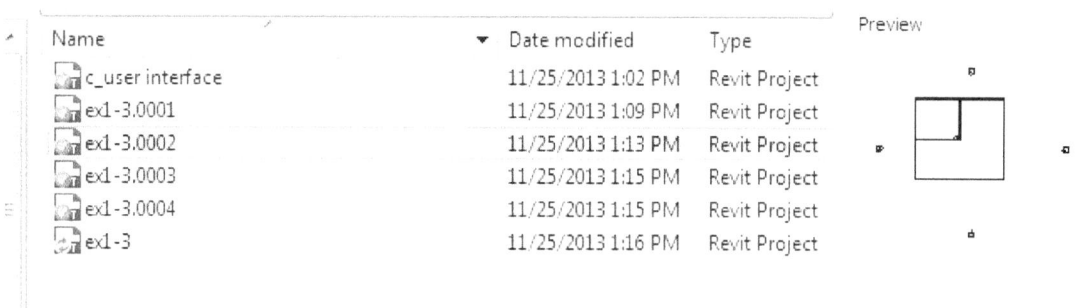

Note that you can highlight a version and check in the preview window which backup you want to select.

18. Open *ex1-3.0001.rvt*.
This is the first save you did.

ex1-3.0001 - Floor Plan: Level 1

19. *Note the file name at the top of the screen.*
Close all files without saving.

Command Exercise

Exercise 1-4 – Design Options

Drawing Name: **i_Design_Options**
Estimated Time to Completion: 90 Minutes

Scope

Use of Design Options

Solution

1. Activate the **Manage** ribbon.

Select **Design Options** under the Design Options panel.

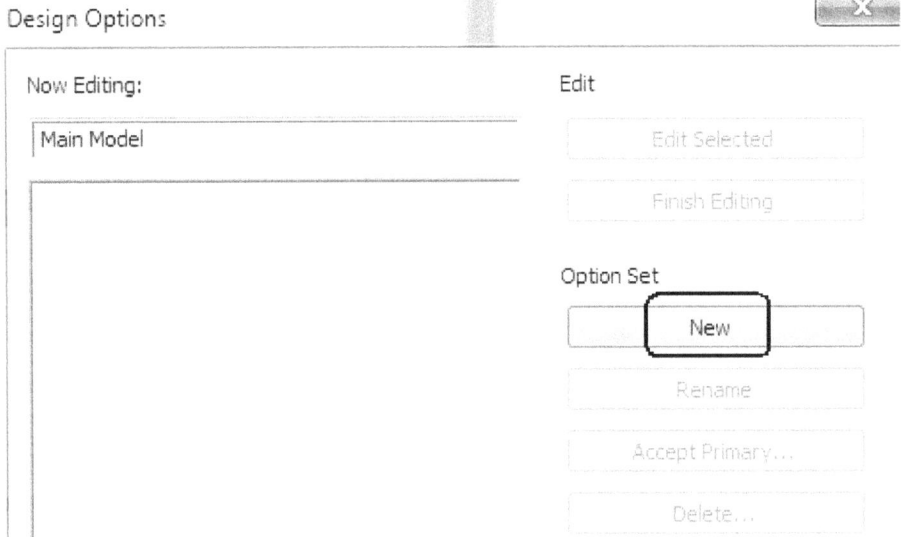

2. Select **New** under Option Set.

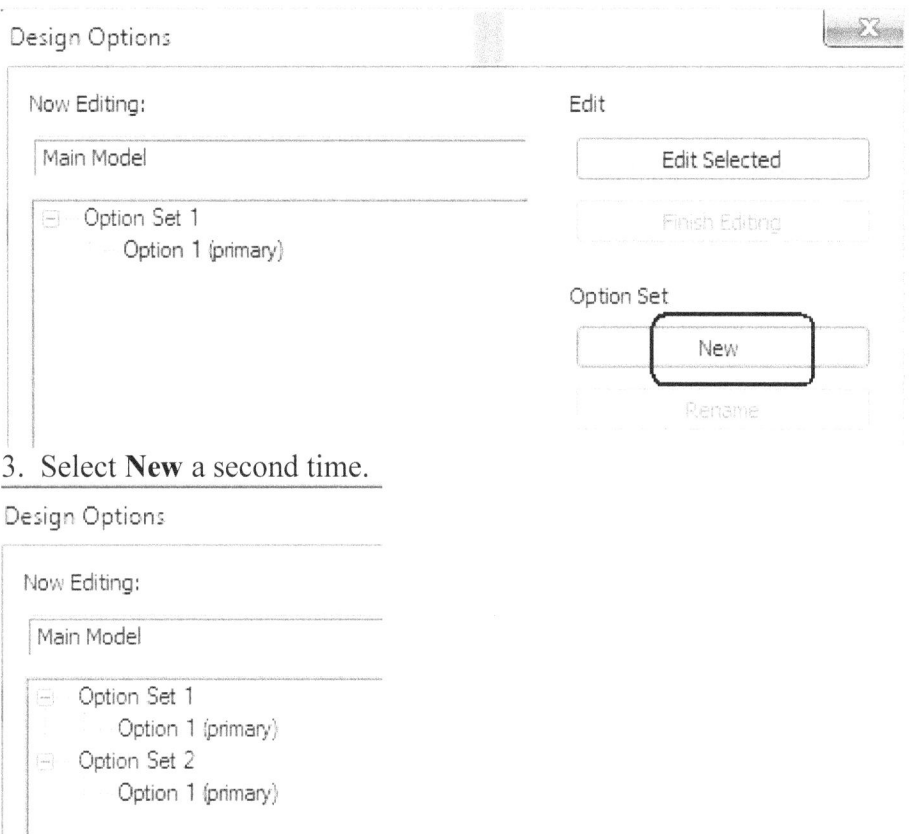

3. Select **New** a second time.

Two Option Sets have been created. Each Option set represents a design choice group. The Option set can have as many options as needed. The more options, the larger your file size will become.

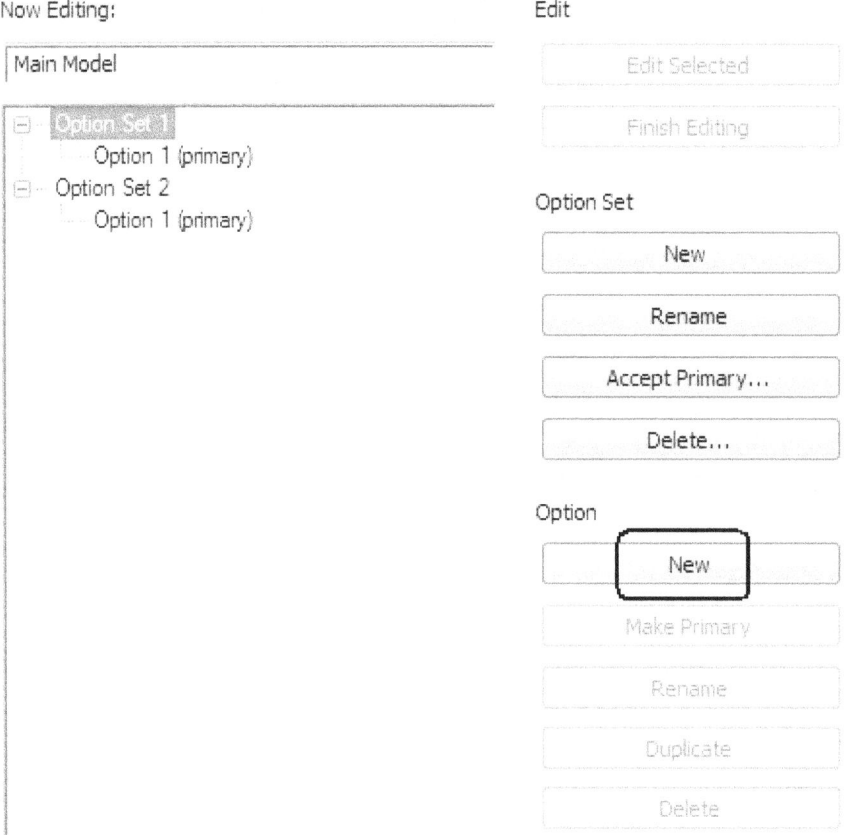

4. Highlight the **Option Set 1**.

Select the **New** button under Option.

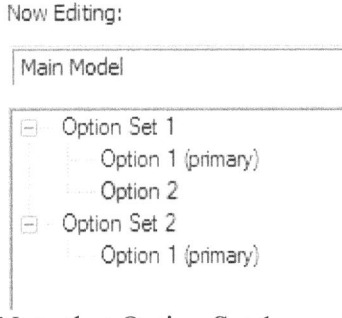

Note that Option Set 1 now has two sub-options.

Building Information Modeling and Revit Basics

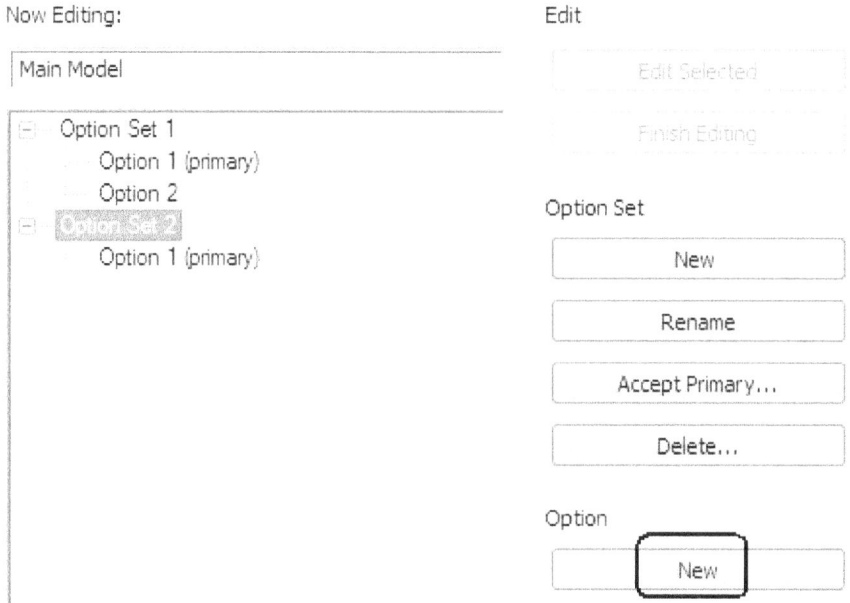

5. Highlight the **Option Set 2**.

Select the **New** button under Option.

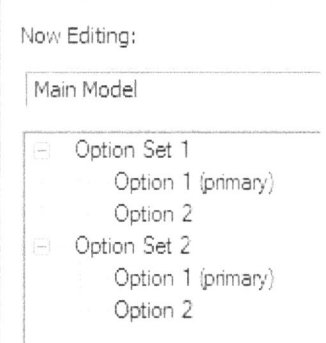

Note that Option Set 2 now has two sub-options.

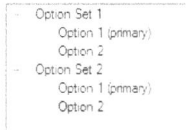

1. Highlight **Option Set 1**.

Select **Rename**.

2. Rename Option Set 1 **South Entry Door Options**.

   Press **OK**.

3. Highlight **Option 1 (primary)** under the South Entry Door Options.

4. Select **Rename**.

5. Rename to **Dbl Glass Door - No Trim**.

   Press **OK**.

6. Highlight **Option 2** under the South Entry Door Options.

7. Select **Rename**.

8. Rename to **Dbl Glass Door with Sidelights**.

   Press **OK**.

9. Highlight **Option Set 2**.

   Select **Rename**.

10. Rename Option Set 2 **Office Layout Design Options**.

    Press **OK**.

11. Highlight **Option 1 (primary)** under the **Office Layout Design Options**.

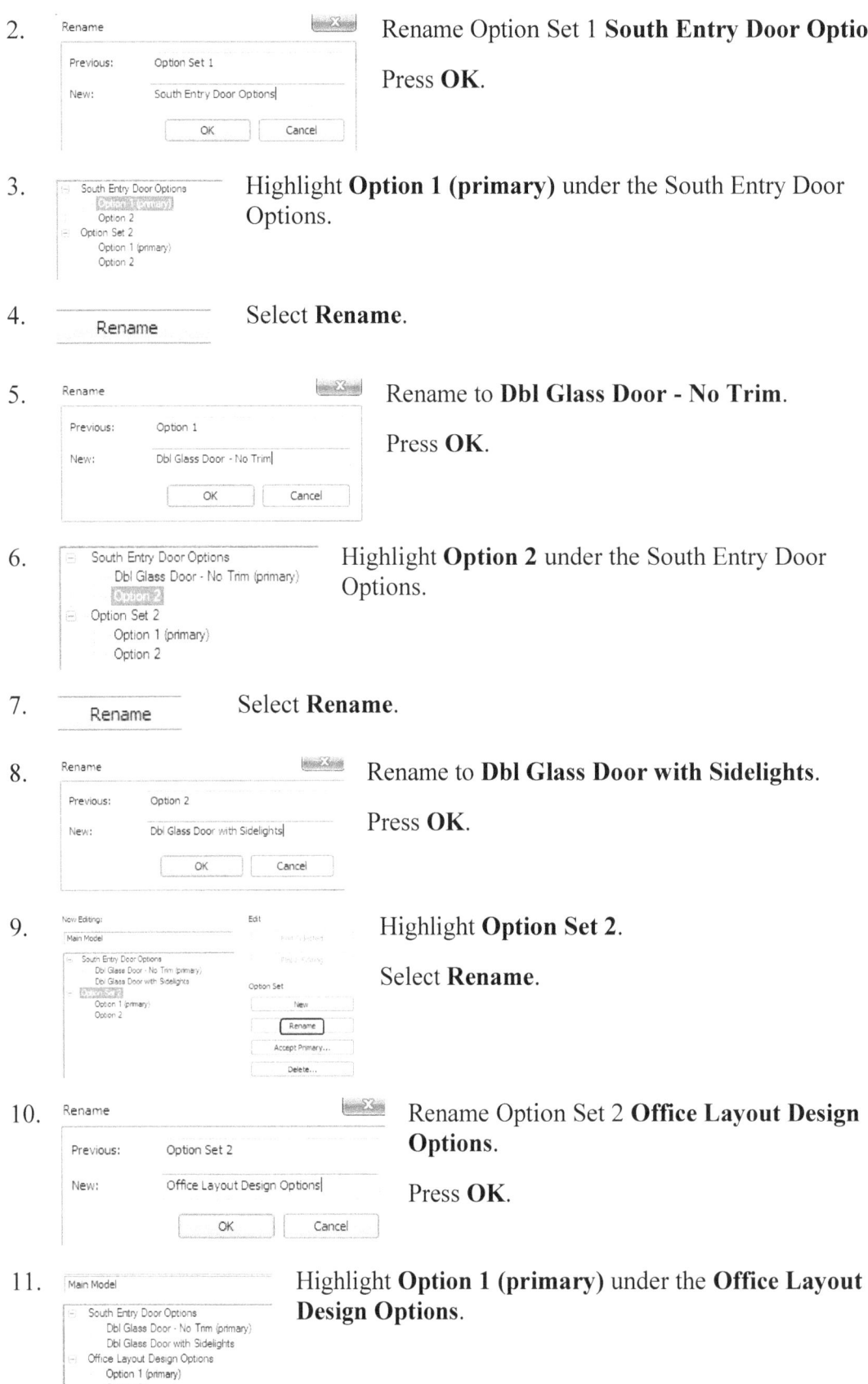

12. Select **Rename**.

13. Rename to **Indented Walls**.

    Press **OK**.

14. Highlight **Option 2**.

    Select **Rename**.

15. Rename Option 2 **Flush Walls**.

    Press **OK**.

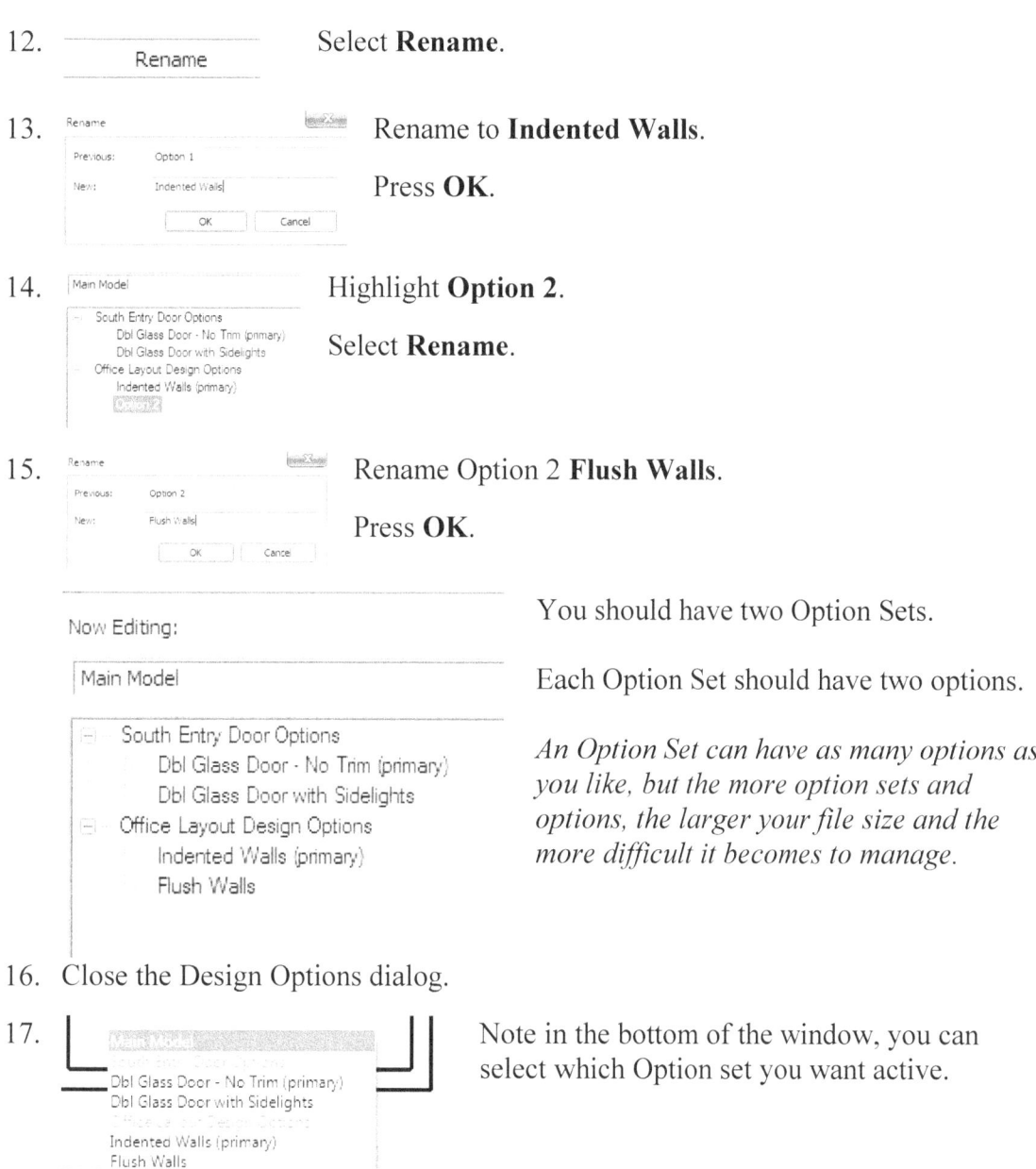

You should have two Option Sets.

Each Option Set should have two options.

An Option Set can have as many options as you like, but the more option sets and options, the larger your file size and the more difficult it becomes to manage.

16. Close the Design Options dialog.

17. Note in the bottom of the window, you can select which Option set you want active.

18. Using **Duplicate View→Duplicate**, create four copies of the Level 1 view.

    Rename the duplicate views:
    Level 1 - Office Layout Indented Walls
    Level 1 - Office Layout Flush Walls
    Level 1 - South Entry Dbl Glass Door – No Trim
    Level 1 - South Entry Dbl Glass Door with Sidelights

    To rename, highlight the level name and press F2.

19. Using **Duplicate View→Duplicate**, create two copies of the South Elevation view.

    Rename the duplicate views:
    South Entry Dbl Glass Door – No Trim
    South Entry Dbl Glass Door with Sidelights

20. Activate **Level 1 - South Entry Dbl Glass Door – No Trim**.

21. In the Properties pane:
    Select Edit for **Visibility/Graphics Overrrides**.

22. Select the **Design Options** tab.

23. Set South Entry Door Options to **Dbl Glass Door - No Trim (primary)**.

    Press **OK**.

24. Set the Design Option to **Dbl Glass Door - No Trim (primary)**.

25. Uncheck **Active Only**.

26. Select the south horizontal wall.

27. Activate the **Manage** ribbon.

    Under Design Options, select **Add to Set**.

    *The selected wall is added to the **Dbl Glass Door - No Trim (primary)** set.*

    We need to add the wall to the set so we can place a

1-18

door. Remember doors are wall-hosted.

28. 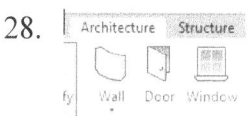 Activate the **Architecture** ribbon.

    Select the **Door** tool from the Build panel.

29. 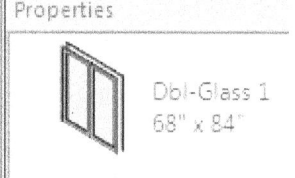 Set the Door type to **Dbl-Glass 1: 68″ x 84″**.

30. 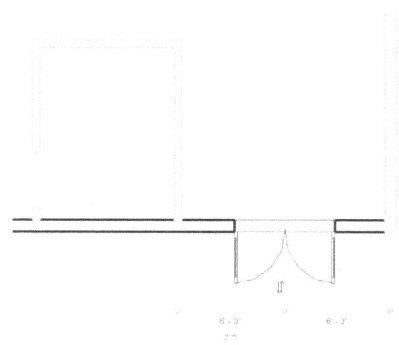 Place the door as shown.

31. Activate **Level 1 - South Entry Dbl Glass Door with Sidelights**.

32. In the Properties pane:
    Select **Edit** Visibilities/Graphics Overrides.

33. Activate the **Design Options** tab.

     Set **Dbl Glass Door with Sidelights** on South Entry Door Options.

    Press **OK**.

34. Set the Design Option to **Dbl Glass Door with Sidelights**.

35. Activate the **Architecture** ribbon.

    Select the **Door** tool from the Build panel.

36. 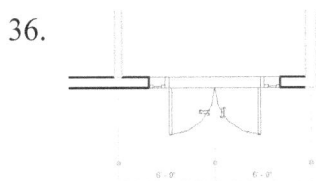 Place a **Double-Raised Panel with Sidelights: 68″ x 80″** door as shown.

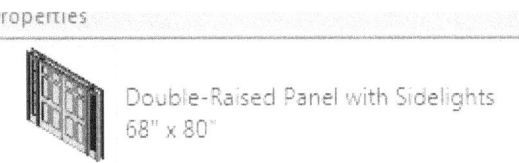

37. Activate the **South Entry Dbl Glass Door – No Trim** elevation.

38. In the Properties pane:
    Select **Edit** Visibilities/Graphics Overrides.

39. Activate the **Design Options** tab.

    Set **Dbl Glass Door - No Trim** on South Entry Door Options.

    Press **OK**.

40. Activate the **South Entry Dbl Glass Door with Sidelights** elevation.

41. In the Properties pane:
    Select **Edit** Visibilities/Graphics Overrides.

42. Activate the **Design Options** tab.

    Set **Dbl Glass Door with Sidelights** on South Entry Door Options.

    Press **OK**.

43. Activate the Sheet named **South Entry Door Options**.

44. 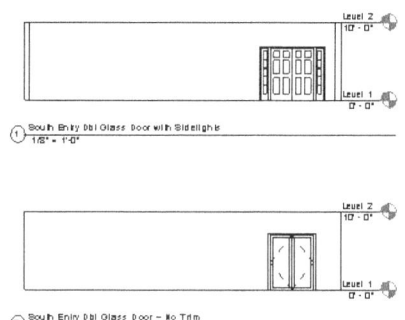 Drag and drop the two South Entry Option elevation views on the sheet.

45. Switch to 3D view.

46. Use **Duplicate View→Duplicate** to create two new 3D views.

    Rename the views:
    3D - South Entry Dbl Glass Door - No Trim
    3D - South Entry Dbl Glass Door with Sidelights

47. Activate **3D - South Entry Dbl Glass Door - No Trim**.

48. In the Properties pane:
    Select **Edit** Visibilities/Graphics Overrides.

49. Activate the **Design Options** tab.

    Set **Dbl Glass Door - No Trim** on South Entry Door Options.
    Press **OK**.

50. Activate **3D - South Entry Dbl Glass Door with Sidelights**.

51. In the Properties pane:
    Select **Edit** Visibilities/Graphics Overrides.

52. Activate the **Design Options** tab.

    Set **Dbl Glass Door with Sidelights** on South Entry Door Options.
    Press **OK**.

53. Activate the Sheet named **South Entry Door Options**.

54. Drag and drop the 3D views onto the sheet.

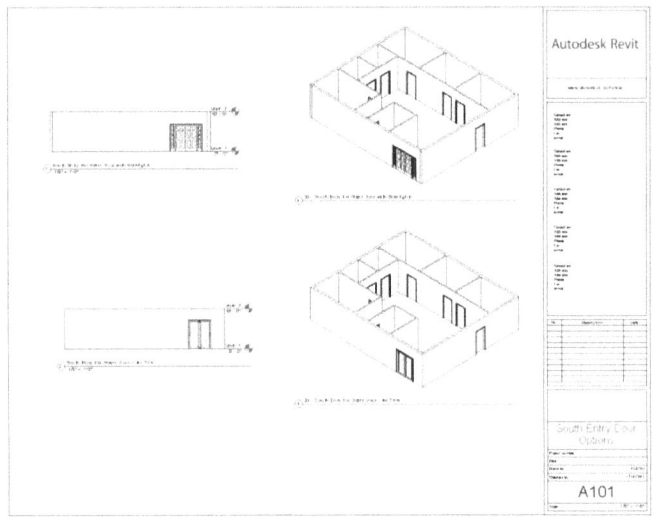

55. Activate **Level 1 - Office Layout Indented Walls**.

56. In the Properties pane: Select Edit for **Visibility/Graphics Overrrides**.

57. Select the **Design Options** tab.

58. Set Office Layout Design Options to **Indented Walls (primary)**.

    Press **OK**.

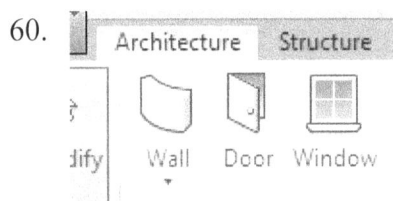

59. Set the Design Option to **Indented Walls (primary)**.

60. Select the Wall tool from the Architecture ribbon.

    Select the Basic Wall: Interior - 5" Partition (2-hr).

61. Place the two walls indicated.

    The vertical wall is placed at the midpoint of the small horizontal wall.

    The horizontal wall is aligned with the wall indicated by the red line.

62. Activate the **Architecture** ribbon.

    Select the **Door** tool from the Build panel.

63. Place a **Sgl Flush 36″ x 80″** door as shown.

64. Activate **Level 1 - Office Layout Flush Walls**.

65. In the Properties pane:
    Select Edit for **Visibility/Graphics Overrrides**.

66. Select the **Design Options** tab.

67. Set Office Layout Design Options to **Flush Walls**.

    Press **OK**.

68. Set the Design Option to **Flush Walls**.

69. Uncheck **Active Only**.

70. Activate the **Architecture** ribbon.

    Select the **Wall** tool from the Build panel.

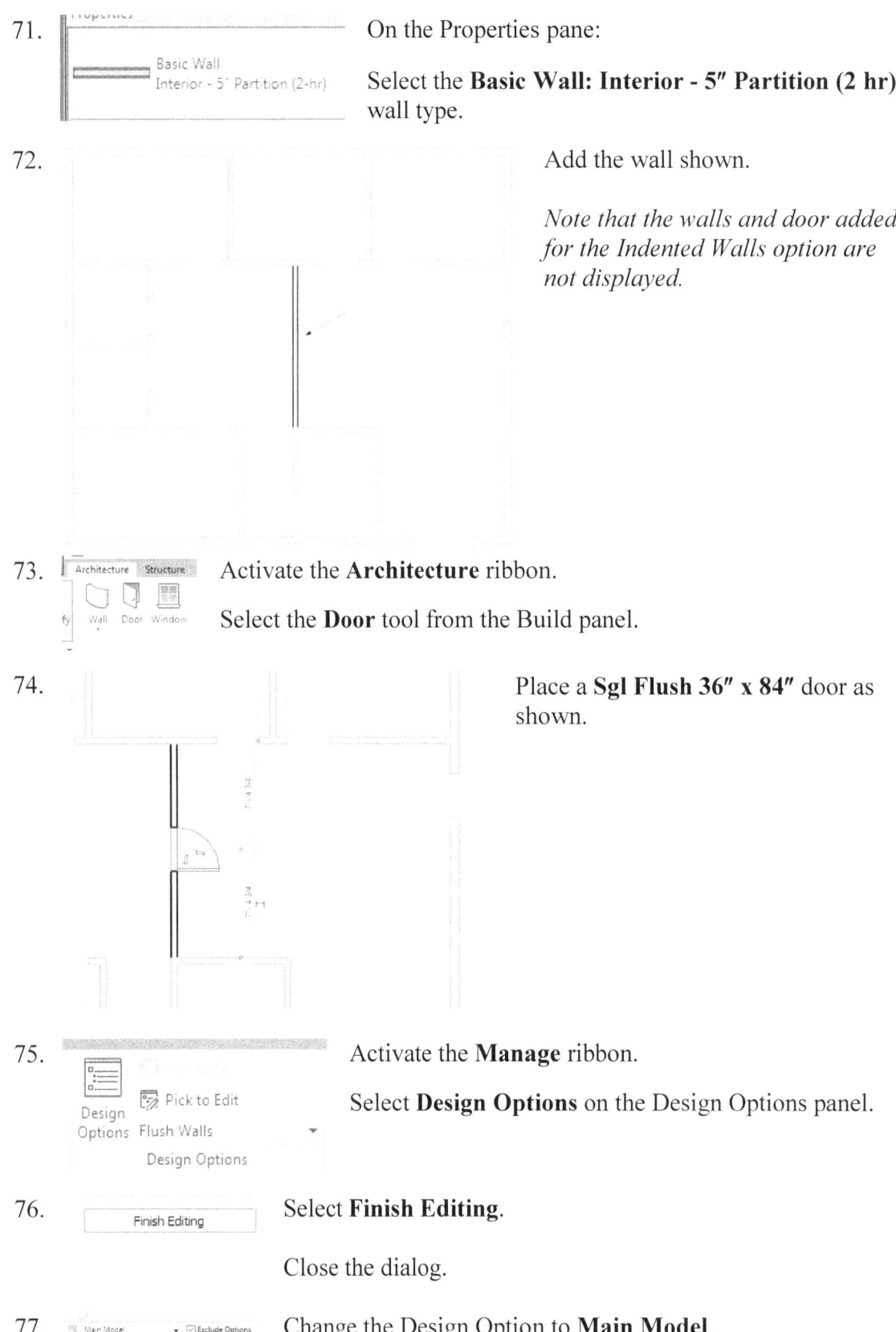

71. On the Properties pane:

    Select the **Basic Wall: Interior - 5″ Partition (2 hr)** wall type.

72. Add the wall shown.

    Note that the walls and door added for the Indented Walls option are not displayed.

73. Activate the **Architecture** ribbon.

    Select the **Door** tool from the Build panel.

74. Place a **Sgl Flush 36″ x 84″** door as shown.

75. Activate the **Manage** ribbon.

    Select **Design Options** on the Design Options panel.

76. Select **Finish Editing**.

    Close the dialog.

77. Change the Design Option to **Main Model**.

78. Note that if you hover your mouse over the element, it will display which Option set it belongs to.

    This only works if Active Only or Exclude Options is disabled.

79. Activate the Sheet named **Office Layout Options**.

80. Drag and drop the two Office Layout options onto the sheet.

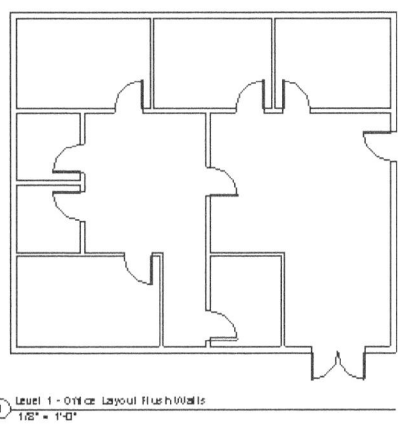
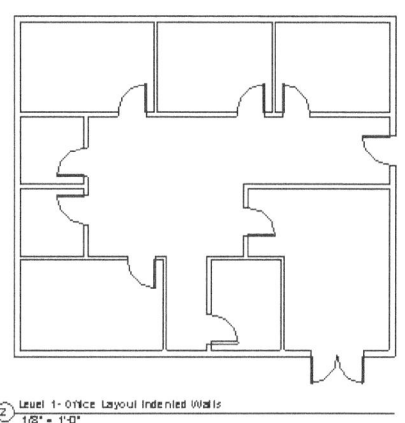

81. Activate the Manage ribbon.

    Select **Design Options**.

82. *Let's assume that the client decided they prefer the flush walls option.*

    Highlight the **Flush Walls** option.

83. Select **Make Primary**.

Note that (primary) is now next to Flush Walls.

If you see an error message, you can click to ignore it.

84. Highlight the **Office Layout Design Options**.

Select **Accept Primary**.

85. Press **Yes**.

86. The view userd with the design option can also be deleted.

Press **Delete**.

87. Now only the design options for the doors remain.

Close the Design Options dialog.

88. Close without saving.

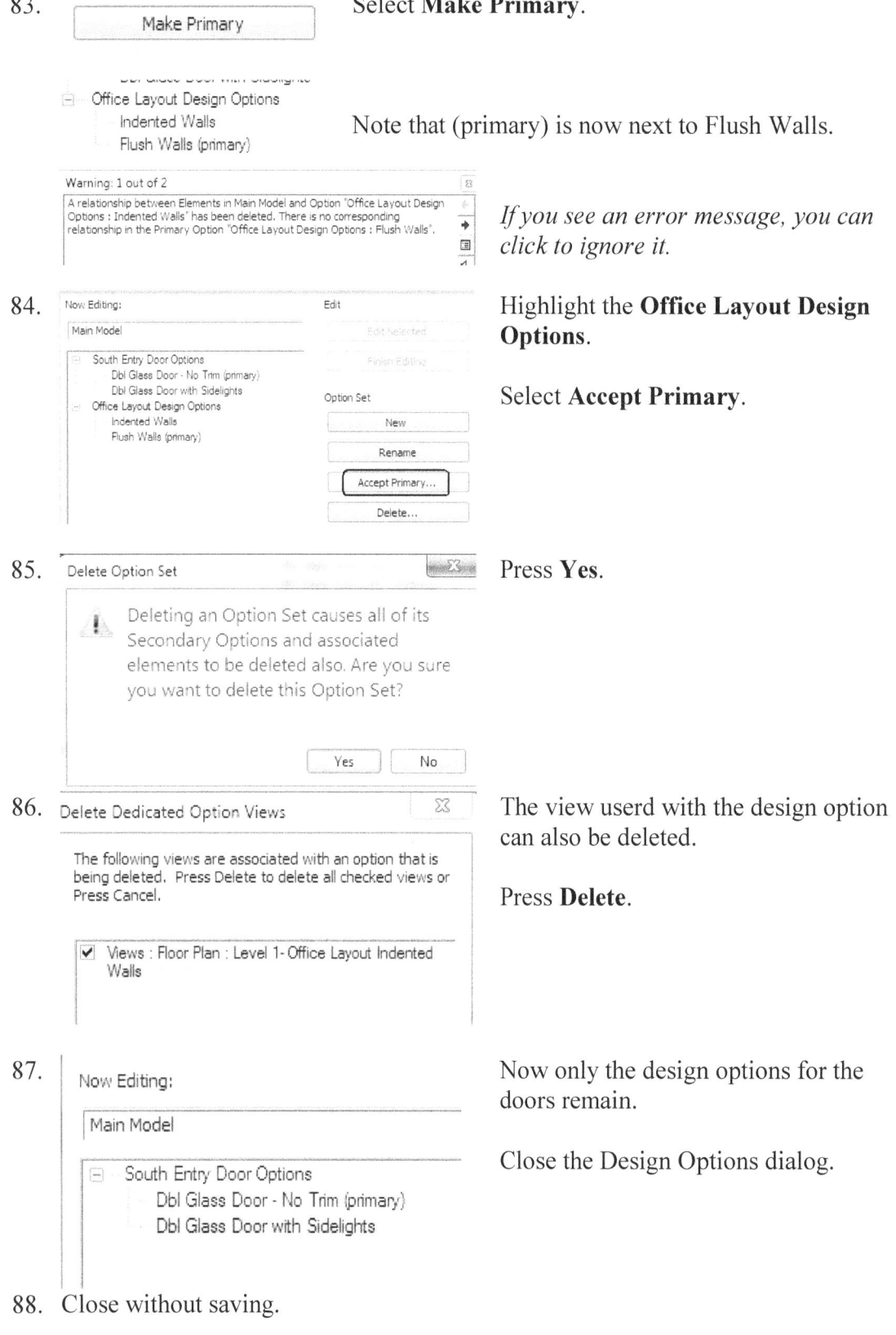

Building Information Modeling and Revit Basics

Exercise 1-5
Phases

Drawing Name: c_phasing.rvt
Estimated Time: 75 minutes

This exercise reinforces the following skills:

- Properties
- Filter
- Phases
- Rename View
- Copy View
- Graphic Settings for Phases

1. Select the **Open** tool.

2. File name: c_Phasing.rvt   Locate the *c_phasing.rvt* file.

   Select **Open**.

3. Activate **Level 1** under Floor Plans.

   Views (all)
   Floor Plans
   Level 1
   Level 2
   Site

4. 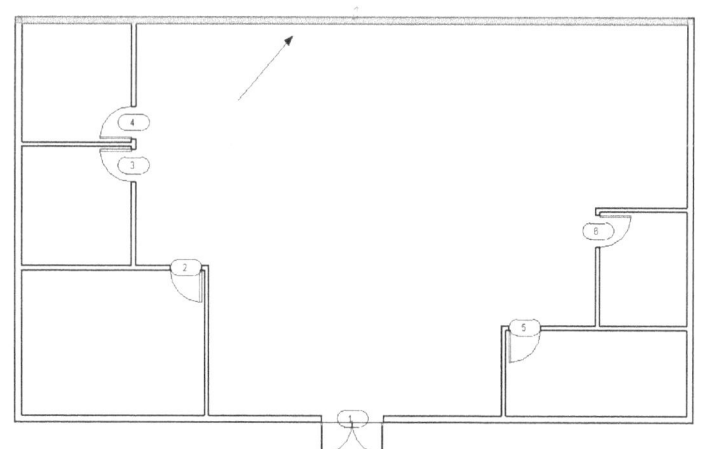 Select the wall indicated.

   It should highlight.

5. Phasing
   Phase Created   New Construction
   Phase Demolished None

   Scroll down to the Phasing category in the Properties panel on the upper left.

6. This wall was created in the New Construction Phase. Note that it is not set to be demolished.

7. Right click and press **Cancel** to deselect the wall.

1-27

8. Go to the **Manage** ribbon.

   Select **Phasing→Phases**.

9. Rename Existing to **As-Built**.

10. Rename New Construction to **2000 Remodel**.

11. Highlight the **2000 Remodel**. Select **After**.

12. 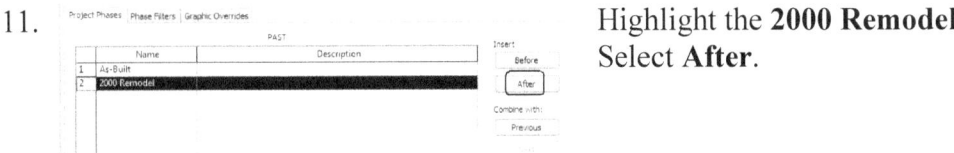 Name the new phase **2010 Remodel**.

    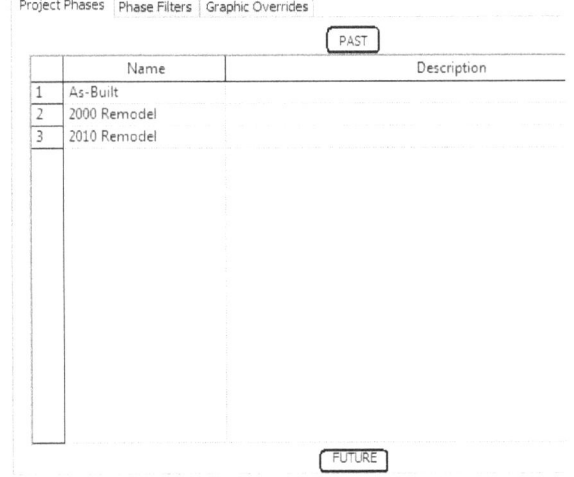 Note that the top indicates the past and the bottom indicates the future to help orient the phases.

13. Select the **Graphic Overrides** tab.

Building Information Modeling and Revit Basics

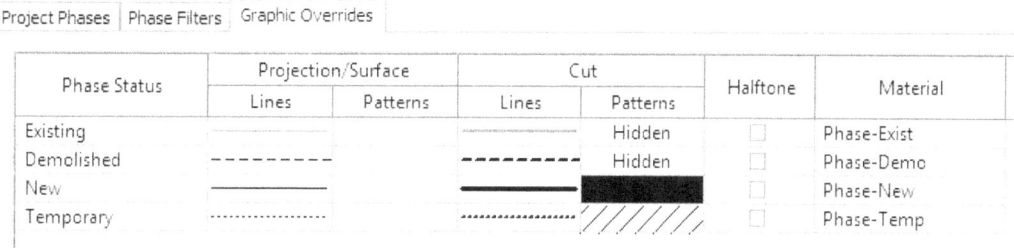

14. Note that in the Lines column for the Existing Phase, the line color is set to gray.

15. 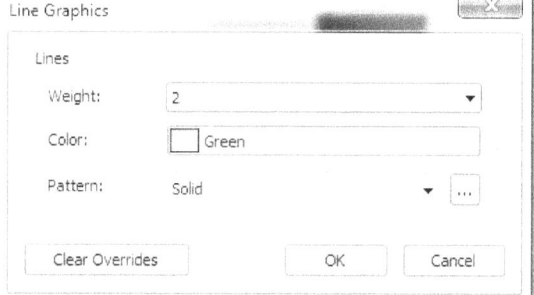 Click in the **Lines** column and the Line Graphics dialog will display.

Projection/Surface is what is displayed in the floor plan views.
Cut is the display for elevation or section views.
Override indicates you have changed the display from the default settings.

16. Set the Color to **Green** for the Existing phase by selecting the color button.

Set the Color to **Blue** for the Demolished phase.

Set the Color to **Magenta** for the New phase.

Change the colors for both Projection/Surface and Cut.

17. Select the **Phase Filters** tab.

18. 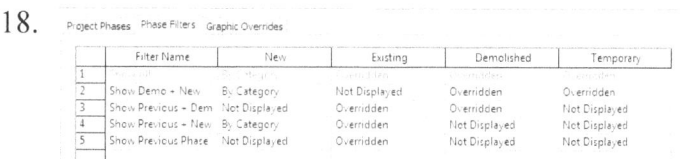 *Note that there are already phase filters pre-defined that will control what is displayed in a view.*

19. Pres the **New** button on the bottom of the dialog.

1-29

20. Change the name for the new phase filter to **Show Existing**

    Show Previous + Demo will display existing plus demo elements, but not new.

    Show Previous + New will display existing plus new elements, but not demolished elements.

21. In the New column, select **Overridden**.

    In the Existing column, select **Overridden**.

    This means that the default display settings will use the new color assigned.

    In the Demolished column, select **Not Displayed**.

22. Use Overridden to display the colors you assigned to the different phases.
    *Verify that in the Show Previous + Demo phase New elements are not displayed.
    Verify that in the Show Previous + New phase Demolished elements are not displayed.*
    Press **Apply** and **OK** to close the Phases dialog.

23. Window around the entire floor plan.

    Select the **Filter** button.

24.  Uncheck **Door Tags**.

    Tags and annotations are not affected by phases.

    Press **OK.**

1-30

Building Information Modeling and Revit Basics

25. Set the Phase Created to **As-Built**.

26. Note that the view changes to display in Green.

This is because we set the color Green to denote existing elements.

27. Next we create three Level 1 floor plan views for each phase.

Highlight **Level 1** under Floor Plan.

Right click and select **Rename**.

28. Rename the view **Level 1- As Built**.

Press **OK**.

29. Press **No**.

30. Highlight **Level 1- As Built** under Floor Plan.

Right click and select **Duplicate View→Duplicate**.

1-31

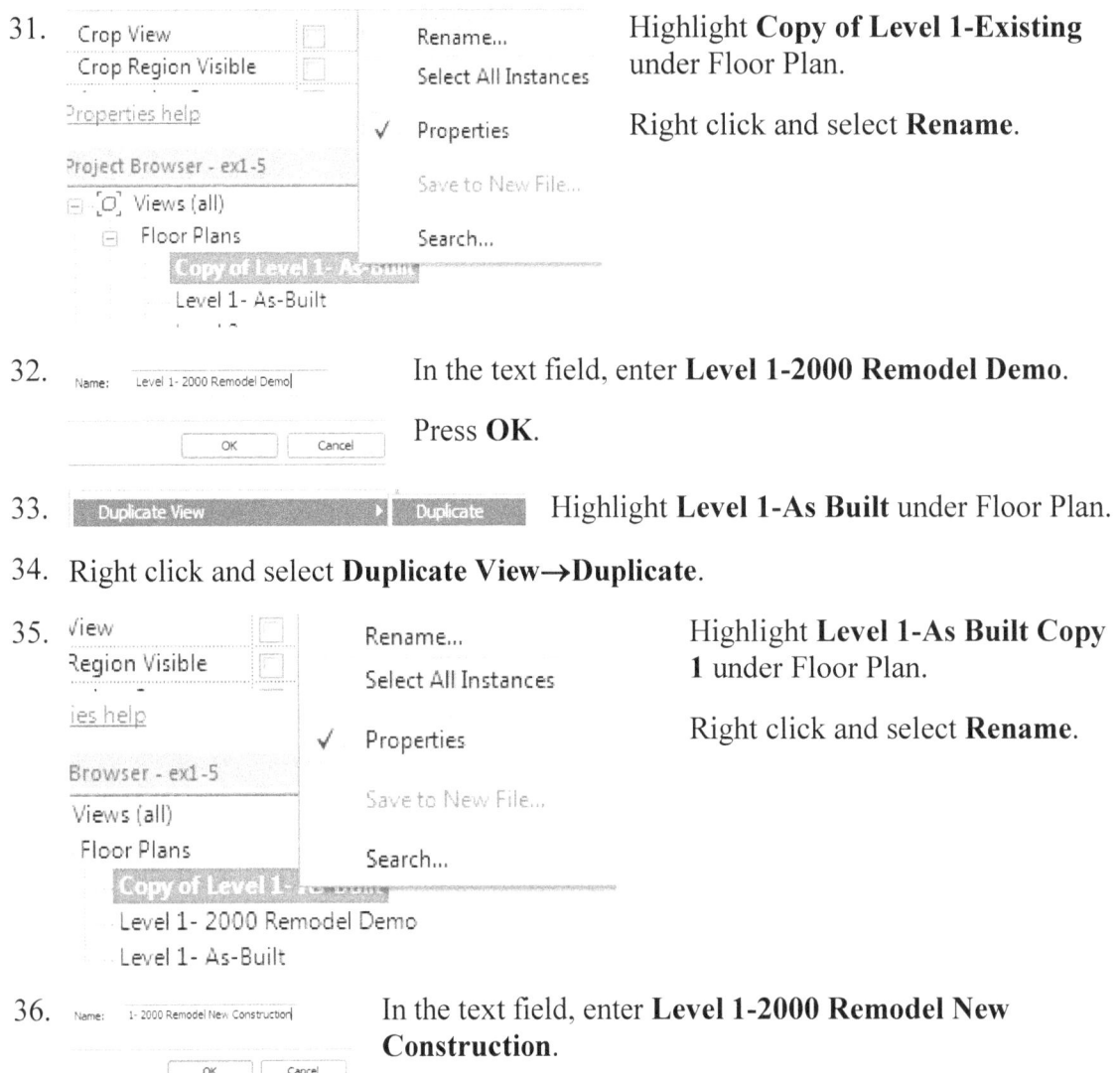

31. Highlight **Copy of Level 1-Existing** under Floor Plan.

    Right click and select **Rename**.

32. In the text field, enter **Level 1-2000 Remodel Demo**.

    Press **OK**.

33. Highlight **Level 1-As Built** under Floor Plan.

34. Right click and select **Duplicate View→Duplicate**.

35. Highlight **Level 1-As Built Copy 1** under Floor Plan.

    Right click and select **Rename**.

36. In the text field, enter **Level 1-2000 Remodel New Construction**.

    Press **OK**.

37. You should have three floor plan views listed:
    As Built
    2000 Demo
    2000 New Construction.

38. Activate the **Level 1-2000 Remodel Demo** view.

39. In the Properties dialog:

    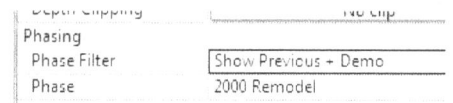

    Set the Phase Filter to **Show Previous + Demo**.

    The previous phase to demo is As-Built. This means the view will display elements created in the existing and demolished phase.

    Set the Phase to **2000 Remodel**.

40. Activate the **Level 1-As Built** view.

    The display does not show the graphic overrides. By default, Revit only allows you to assign graphic overrides to phases AFTER the initial phase. Because the As-Built view is the first phase in the process, no graphic overrides are allowed. The only work-around is to create an initial phase with no graphic overrides and go from there.

41. In the Properties dialog:

    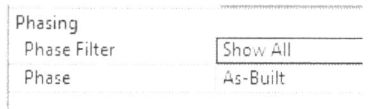

    Set the Phase Filter to **Show All**.
    Set the Phase to **As-Built**.

42. Activate the **Level 1-2000 Remodel New Construction** view.

43. In the Properties dialog:

    Set the Phase Filter to **Show Previous + New**.
    This will display elements created in the Existing Phase and the New Phase, but not the Demo phase.
    Set the Phase to **2000 Remodel**.

44. Activate the **Level 1 - 2000 Remodel Demo** view.

45. 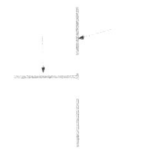 Hold down the Ctrl button.

    Select the two walls indicated.

46. In the Properties pane:

    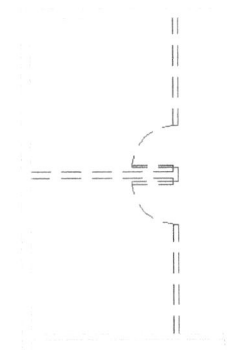

    Scroll down to the bottom.
    In the Phase Demolished drop-down list, select **2000 Remodel**.

47. Press **OK**.

48. The demolished walls change appearance based on the graphic overrides.
    Release the selected walls using right click→Cancel or by pressing ESCAPE.

49.  Activate the **Modify** ribbon.

    Use the **Demolish** tool on the Geometry panel to demolish the walls indicated.

50. 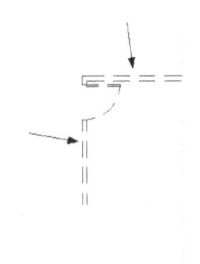 Note that the doors will automatically be demolished along with the walls. If there were windows placed, these would also be demolished. That is because those elements are considered *wall-hosted*.

    Right click and select Cancel to exit the Demolish mode.

51. This is how the Level 1- 2000 Remodel Demo view should appear.

    If it doesn't, check the walls to verify that they are set to Phase Created: As Built, Phase Demolished: 2000 Remodel.

Phasing	
Phase Created	As-Built
Phase Demolished	2000 Remodel

52. 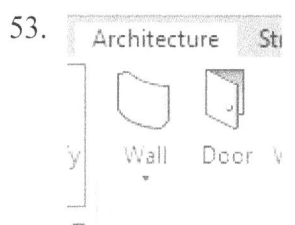 Activate the **Level 1 - 2000 New Construction** view.

53. 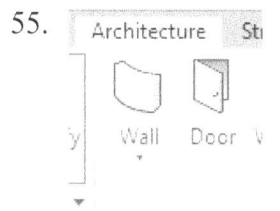 Select the **Wall** tool from the Architecture ribbon.

54. 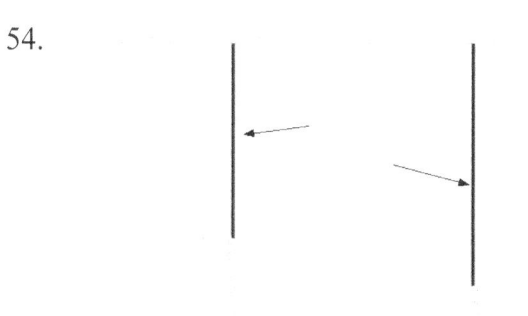 Place two walls as shown. Select the end points of the existing walls and simply draw up.

    Right click and select **Cancel** to exit the Draw Wall mode.

55. Select the **Door** tool under the Build panel on the Architecture ribbon.

56. 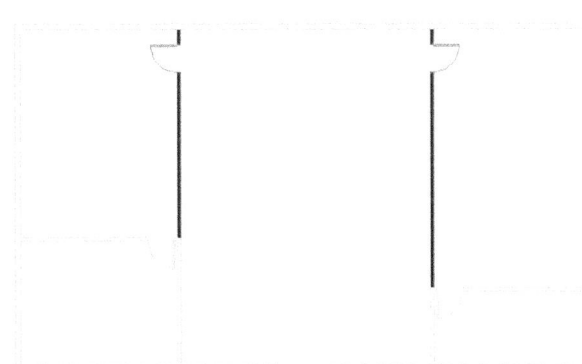 Place two doors as shown. Set the doors 3′ 6″ from the top horizontal wall. Flip the orientation of the doors if needed.

    You can press the space bar to orient the doors before you left click to place.

    Note that the new doors and walls are a different color than the existing walls.

57. 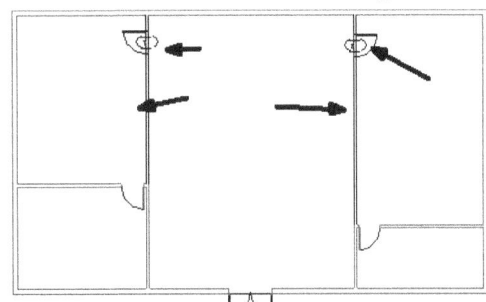 Select the doors and windows you just placed.

    You can select by holding down the CONTROL key or by windowing around the area.

    Note: If Door Tags are selected, you will not be able to access Phases in the Properties dialog.

58. Look in the Properties panel and scroll down to Phasing.

    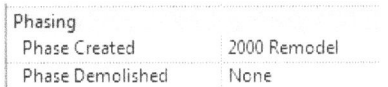 Note that the elements are already set to **2000 Remodel** in the Phase Created field.

59. Switch between the three views to see how they display differently.

60. Highlight **Sheets** in the Project Browser. Right click and select **New Sheet**.

61.  Press **OK** to accept the default title block.

1-36

Building Information Modeling and Revit Basics

62. 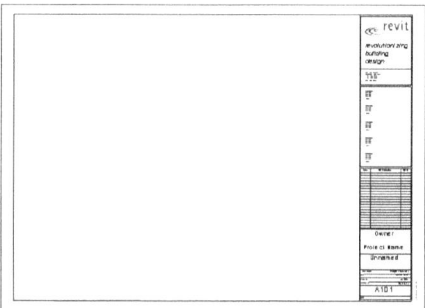 A view opens with the new sheet.

63. Highlight the Level 1 - As Built Floor plan.

    Hold down the left mouse button and drag the view onto the sheet. Release the left mouse button to click to place.

64.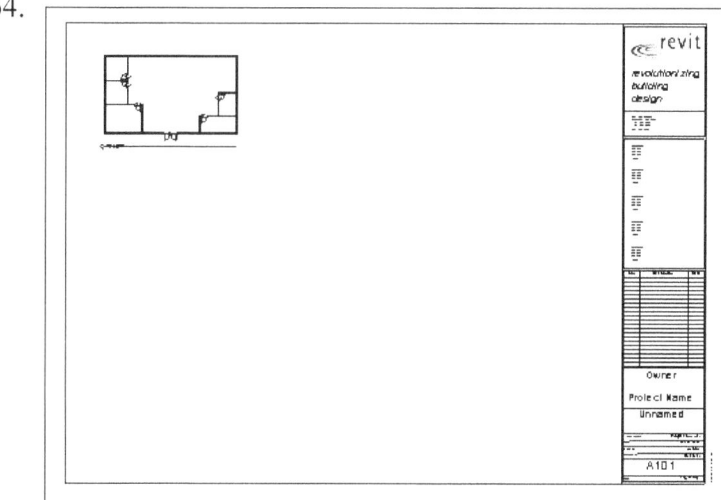

    A preview will appear on your cursor.

    Left click to place the view on the sheet.

65. Highlight the Level 1 - 2000 Remodel Demo Floor plan.

    Hold down the left mouse button and drag the view onto the sheet. Release the left mouse button to click to place.

1-37

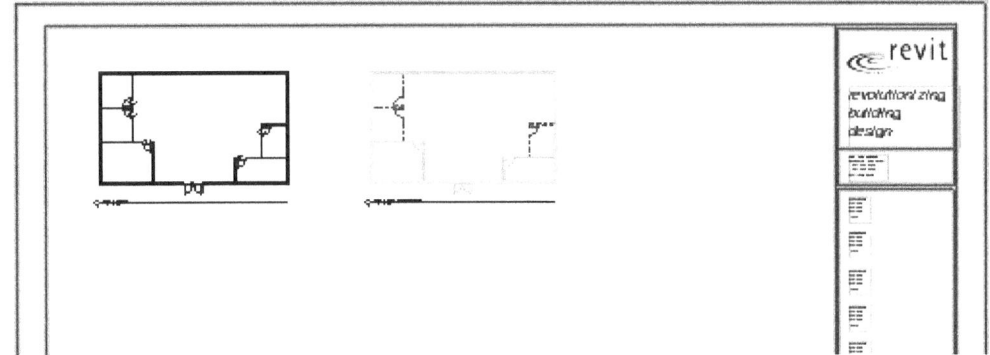

The two views appear on the sheet.

66. Highlight the Level 1 - 2000 New Construction plan.

   Hold down the left mouse button and drag the view onto the sheet. Release the left mouse button to click to place.

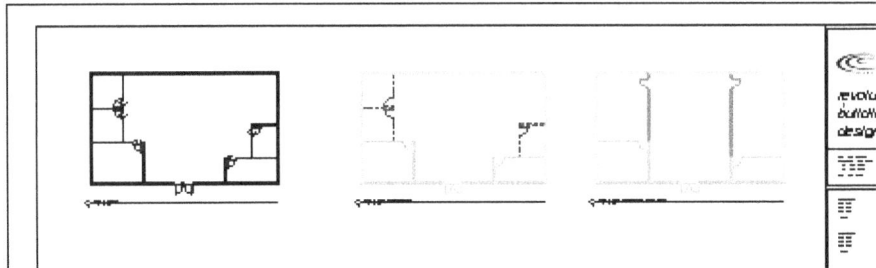

67. Zoom in to inspect the views.

68. Save as *ex1-5.rvt*.

Challenge Exercise:

Create two more views called Level 1 2010 Remodel Demo and Level 1 2010 Remodel New Construction.

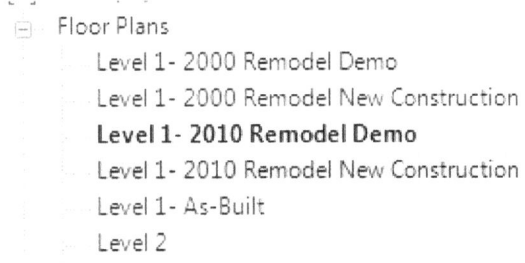

Set the Phases and phase filters to the new views.

The 2010 Remodel Demo view should be set to:

Phasing	
Phase Filter	Show Previous + Demo
Phase	2010 Remodel

The 2010 Remodel New Construction view should be set to:

Phasing	
Phase Filter	Show Previous + New
Phase	2010 Remodel

On the 2010 Remodel Demo view: Demo all the interior doors.

For the 2010 remodel new construction, add the walls and doors as shown.

Note you will need to fill in the walls where the doors used to be.

Add the 2010 views to your sheet.

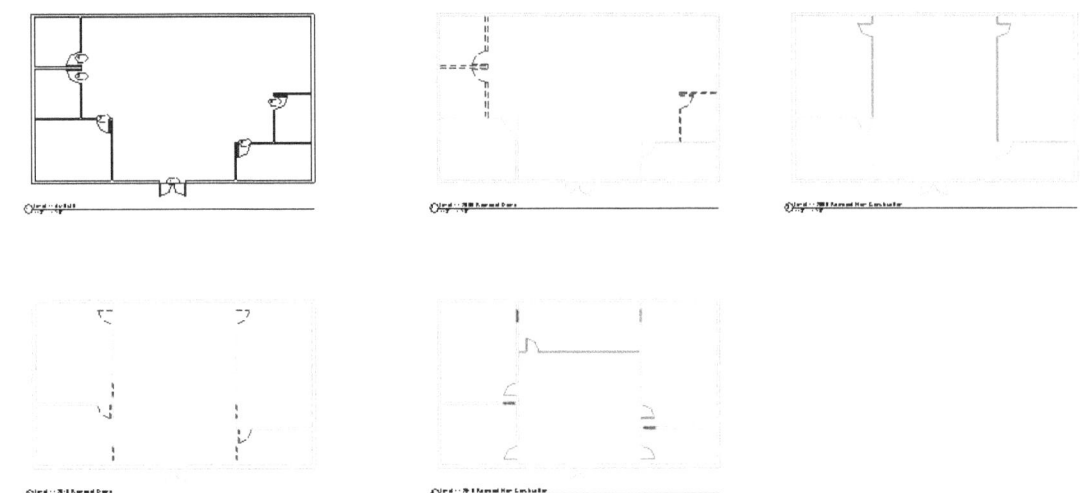

Answer this question:

When should you use phasing as opposed to design options?

Practice Exam

1. Select the answer which is NOT an example of bidirectional associativity:
    A. Flip a section line and all views update
    B. Draw a wall in plan view and it appears in all other views
    C. Change an element type in a schedule and the change is displayed in the floor plan view as well
    D. Flip a door orientation so the door swing is on the exterior of the building.

2. Select the answer which is NOT an example of a parametric relationship:
    A. A floor is attached to enclosing walls. When a wall moves, the floor updates so it remains connected to the walls.
    B. A series of windows are placed along a wall using an EQ dimension. The length of the wall is modified and the windows remain equally spaced.
    C. A door is placed in a wall. The wall is moved and the door remains constrained in the wall.
    D. A shared parameter file is loaded to the server

3. Which tab does NOT appear on Revit's ribbon?
    A. Architecture
    B. Basics
    C. Insert
    D. View

4. Which item does NOT appear in the Project Browser?
    A. Families
    B. Groups
    C. Callouts
    D. Notes

5. Which is the most recently saved backup file?
    A. office.0001
    B. office.0002
    C. office.0003
    D. office.0004

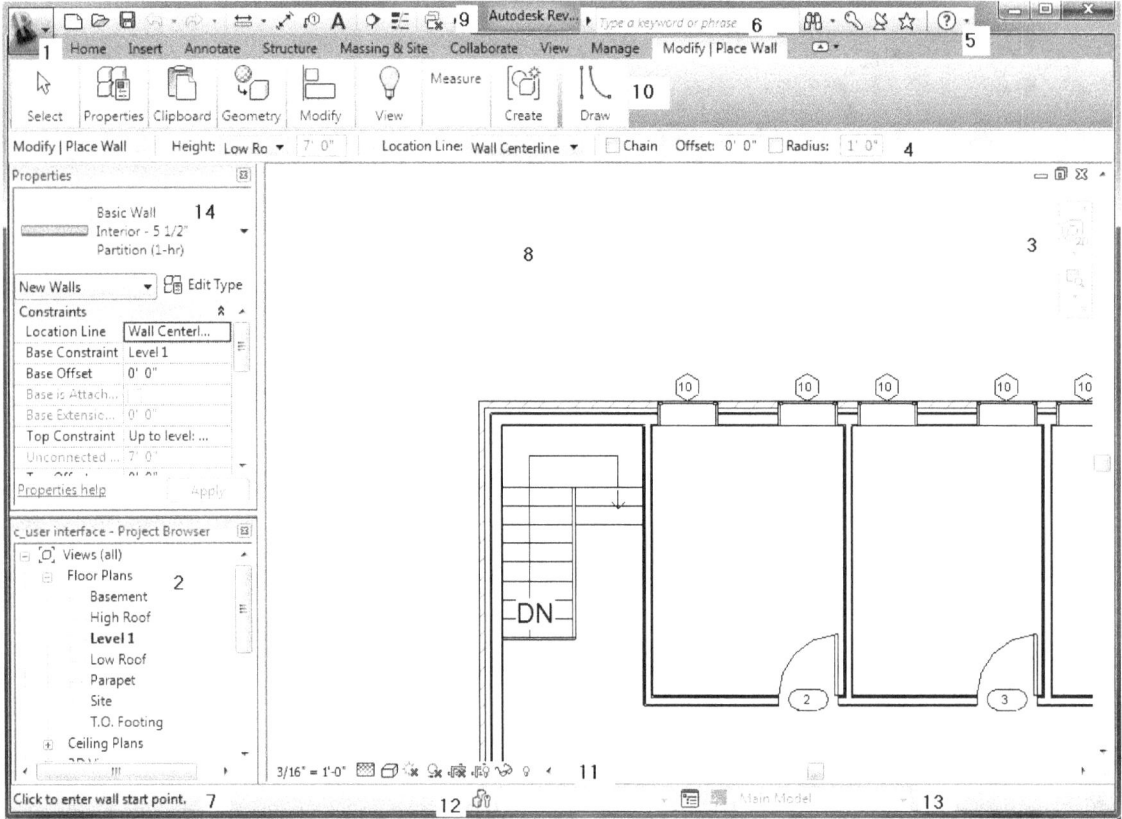

6. Match the numbers with their names.

View Control Bar	InfoCenter
Project Browser	Status Bar
Navigation Bar	Properties Pane
Options Bar	Application Menu
Design Options	Drawing Area
Help	Quick Access Toolbar
Ribbon	Worksets

7. When using design options, the active option is the _____
   A. preferred design option in the design option set
   B. part of the building that is not defined using design options
   C. design options currently being edited
   D. collection of all design options

Answers:
1) D; 2) D; 3) B; 4) D; 5) D; 6) 1- Application Menu, 2- Project Browser, 3- Navigation Bar, 4- Options Bar, 5- Help, 6- InfoCenter, 7- Status Bar, 8- Drawing Area, 9- Quick Access Toolbar, 10- Ribbon, 11- View Control Bar, 12- Worksets, 13- Design Options; 7) C

Lesson Two

The Basics of Building a Model

This lesson addresses the following User and Professional level exam questions:

- Wall Properties
- Compound Walls
- Stacked Walls
- Doors and Windows
- In-Place Mass

In the Professional exam, most of the wall problems follow these steps:

- Place a wall of a specific element type. (Be able to select wall type.)
- Place a wall by setting the location line. (Understand how to use the location line setting.)
- Place a wall using different Option Settings. (Understand how to use the Options Settings when placing a wall.)
- After placing the wall, place a dimension to determine if the wall was placed correctly.
- After placing the wall, inspect the element properties to determine if the wall was placed correctly.

In the User exam, the user will need to be familiar with the different parameters in walls and compound walls. The user should also know which options are applied to walls and when those options are available.

Command Exercise
Exercise 2-1 – Wall Options

Drawing Name: **i_firestation_basic_plan.rvt**
Estimated Time to Completion: 10 Minutes

Scope

Exploring the different wall options

Solution

1. Floor Plans
   Ground Floor
   Lower Roof
   Main Floor
   Main Roof
   Site
   T.O. Footing
   T.O. Parapet

   Activate the **Ground Floor** floor plan.

2.

   Zoom into the area where the green polygon is.

3.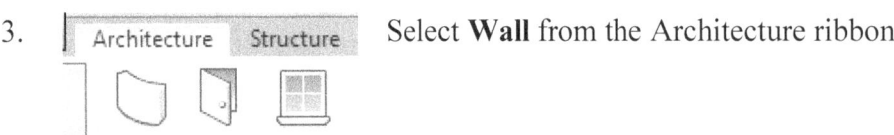

   Select **Wall** from the Architecture ribbon.

4. 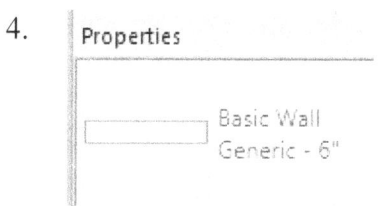 Set the Wall Type to **Generic – 6″** in the Properties pane.

5. Set the Location Line to **Core Face:Exterior**.

6. Select the **Rectangle** tool on the Draw panel.

7. 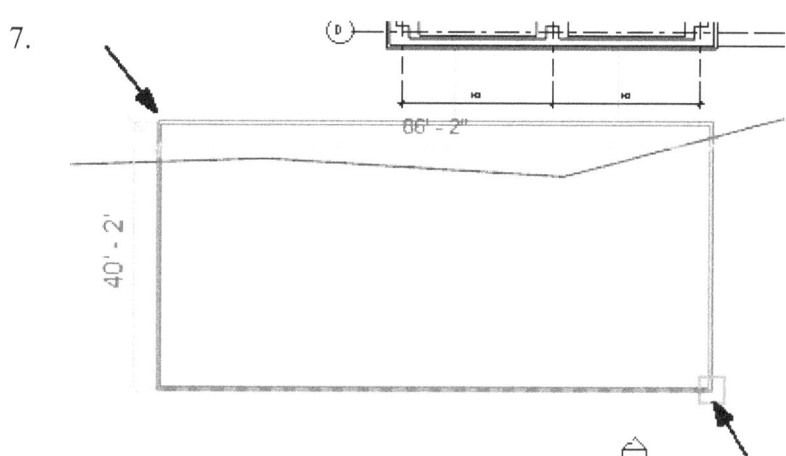 Select the two points indicated to place the rectangle.

8. 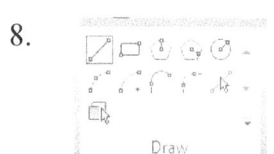 Select the **Line** tool from the Draw panel.

9.  Start the line at the midpoint of the lower horizontal wall.

10. 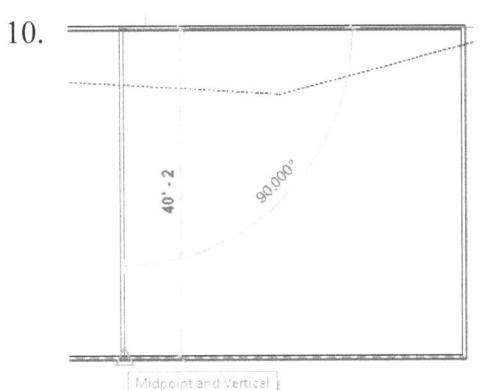 Bring the line end up to the midpoint of the upper horizontal wall.

    Left click to finish placing the wall.

    Exit the wall tool.

11. 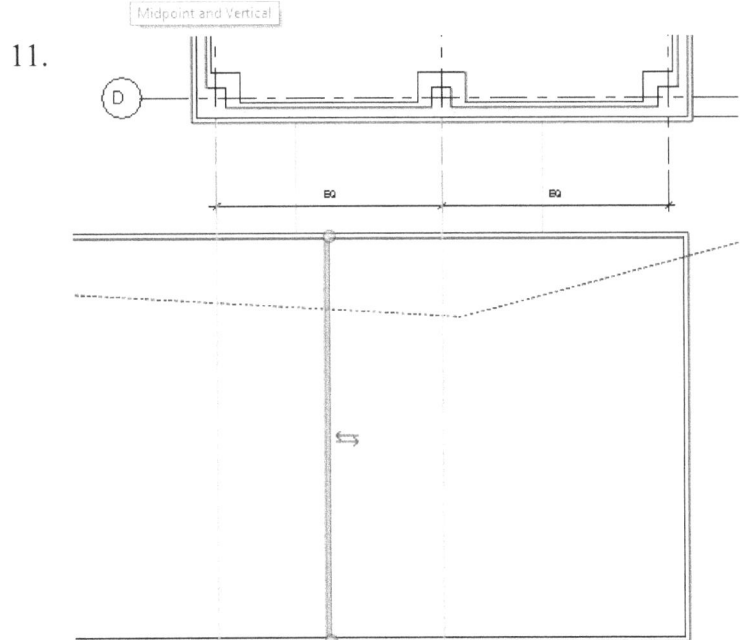 Select the vertical wall. Two temporary (listening) dimensions will appear.

    Change the right dimension to **12′ [3600 mm]**.

12. 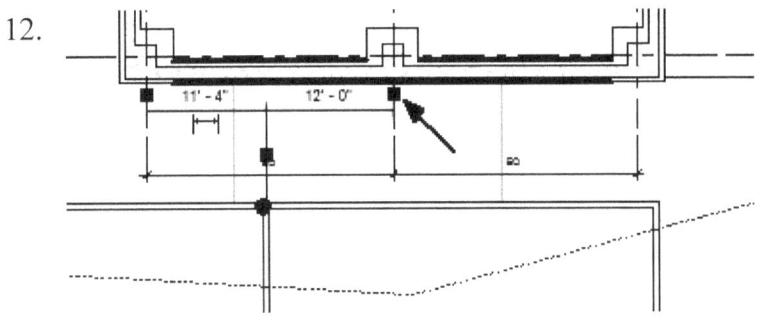 Select the witness grip point indicated.

13. 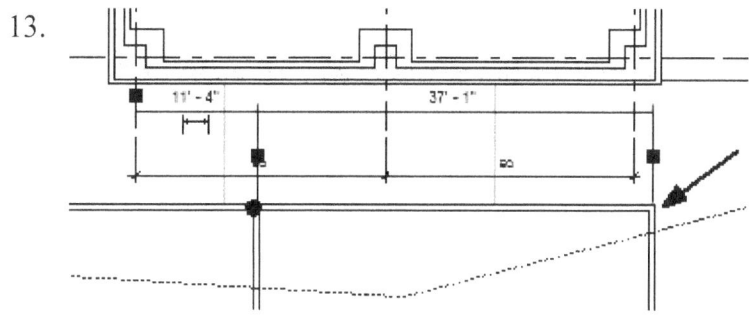 Move the witness line to the right vertical wall.

Note that the dimension updates.

The dimension should display 37' 1". This is the answer you would enter in the exam. If you did not get that answer, check your location line setting.

14. Close the file without saving.

Command Exercise

Exercise 2-2 – Placing a Wall Sweep

Drawing Name: **walls.rvt**
Estimated Time to Completion: 20 Minutes

Scope

Placing a wall sweep.

Solution

1. Activate **Level 1** Floor Plan.

2. Select the **Wall** tool from the Architecture ribbon on the Build panel.

3. Set the wall type to **Exterior - Brick on Mtl. Stud** using the Type Selector on the Properties pane.

4. 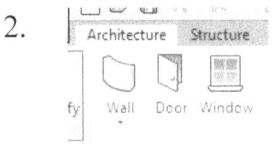 Set the Location Line to **Finish Face: Exterior**.

5.  Select the **Pick Line** mode from the Draw panel.
   Select the four green lines.

   Note that when you pick the lines, the side of the line you use determines which side of the line is used for the exterior side of the wall.

The Basics of Building a Model

6.  The lines should be aligned to the exterior side of the walls.

   Set the Detail Level to **Medium**.

7. 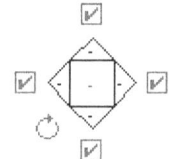 Activate the **View** ribbon.

   Select the **Elevation** tool on the Create panel.

8. Place an elevation in the center of the room.
   Right click and select **Cancel** to exit the command.

   Place a check mark on each box to create an elevation for each interior wall.

9. In the Project Browser, you will see that four elevation views have been created.

10. 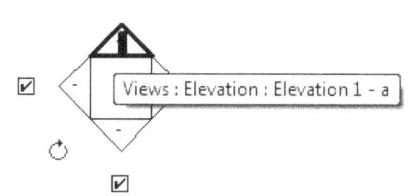 If you hover your mouse over a triangle, a tooltip will appear with the name of the linked view.

11. Rename the elevation views to East Interior, North Interior, South Interior and West Interior.

    *Pressing **F2** is a shortcut for Rename.*

2-7

12. 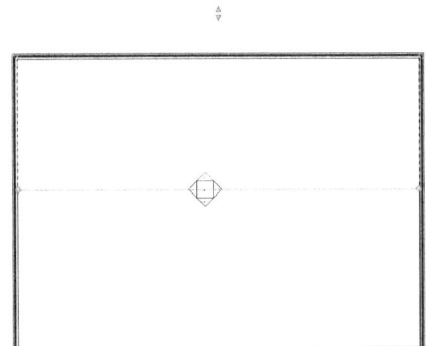 If you pick on the triangle part of the elevation, you will see the view depth (Far Clip Offset) of that elevation view.

13. Activate the **South Interior** View.

14. Use the grips to extend the elevation view beyond the walls.

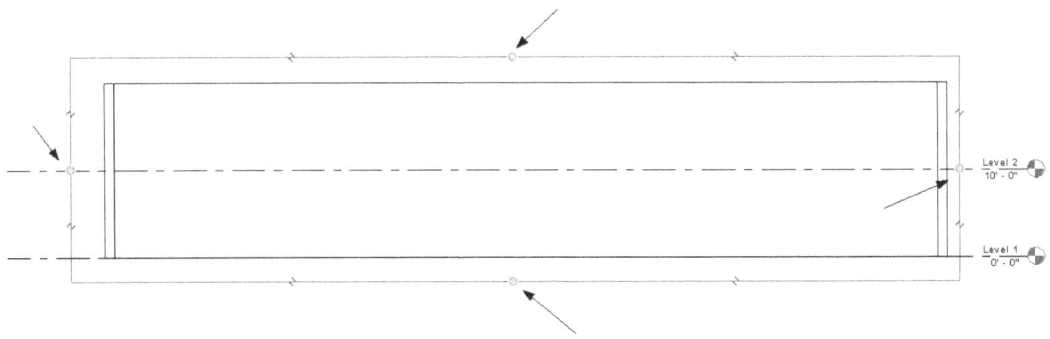

15. 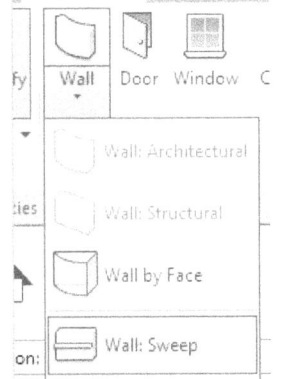 Activate the **Architecture** ribbon.

    Select the **Wall Sweep** tool.

    The Wall Sweep tool is only available in elevation, 3D or section views.

16. Place the sweep so it is toward the top of the wall.

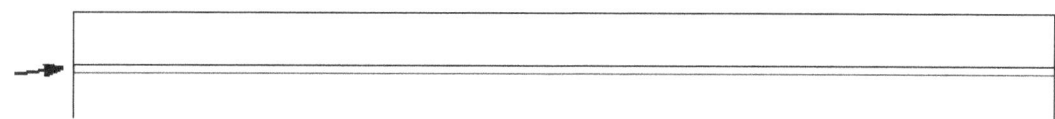

17. 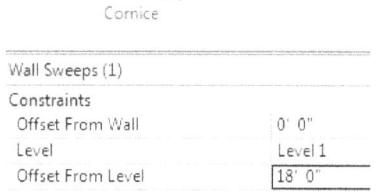 Select the wall sweep.

    In the Properties pane, adjust the Offset from Level to **18′ 0″**.

18. 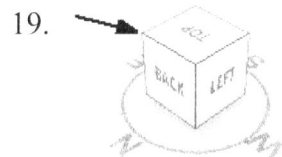 Switch to a 3D view.

19. Select the top corners of the view cube to orient the view so you can see the wall sweep.

20. Select the wall sweep that was placed.

    It will highlight when selected.

21. 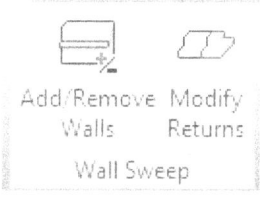 Select **Add/Remove Walls** from the ribbon.

    Select the other walls.
    Orbit around to inspect.

22. Save as *ex2-2.rvt*.

Command Exercise

Exercise 2-3 – Create a Wall Sweep Style

Drawing Name: **ex2-2.rvt**
Estimated Time to Completion: 15 Minutes

Scope

Creating a wall sweep style.
Loading a Profile.

Solution:

1. Activate the **South Interior** View.

2. Activate the **Insert** ribbon.

   Select **Load Family**.

3. Load the following profiles:

   Base-3.rfa
   Crown 1.rfa

   These can be downloaded from this link:
   http://1drv.ms/1ByXaSF

   You can load more than one file at a time by holding down the CTL key.

5. Press **Open**.

The Basics of Building a Model

6. Activate the **Architecture** ribbon.

   Select the **Wall Sweep** tool.

   The Wall Sweep tool is only available in elevation, 3D or section views.

7. Select **Edit Type** from the Properties pane.

8. Select **Duplicate**.

9. Enter **Base Moulding** in the Name field.

   Press **OK**.

10. Set the Profile to **Base 3 : 3 1/2" x 9/16"**.

    Press **OK** to exit the dialog.

11. 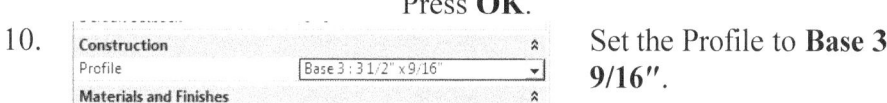 Place the baseboard on the bottom of the wall.

12. Save as *ex2-3.rvt*.

2-11

Command Exercise
Exercise 2-4 – Create a Custom Profile

Drawing Name: **ex2-3.rvt**
Estimated Time to Completion: 20 Minutes

Scope

Creating a custom profile.
Using the custom profile in a wall sweep.

Solution

1. 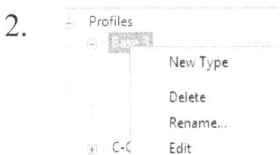 Browse under Profiles in the Project Browser.

   Locate the **Base 3** profile family.

2. Right click on the **Base 3** family.
   Select **Edit**.

3. Save the file as *Base 4.rfa*.

   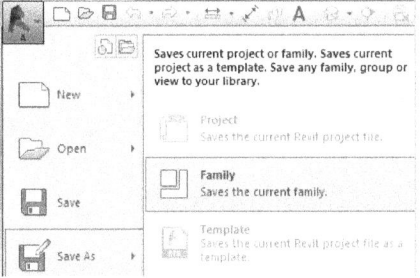

4. 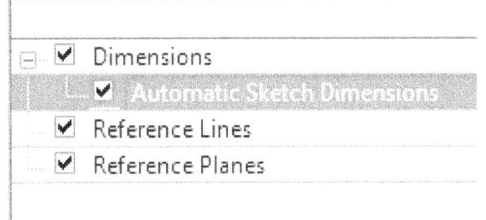 Bring up the Visibility/Graphics dialog.
   You can do this by typing VV.

   Select the Annotation Categories tab.

   Enable all the Annotation Categories.

   Press **OK**.

5. 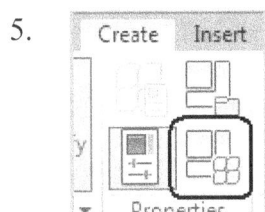 Activate the Create ribbon.

   Select the **Family Types** tool on the Properties pane.

2-12

The Basics of Building a Model

6.   Note that several sizes are available for this profile.

Use the Apply button to see how the profile changes depending on the size selected.

Set the size to **5 1/2" x 9/16"**.

Press **OK** to close the Types dialog.

7.   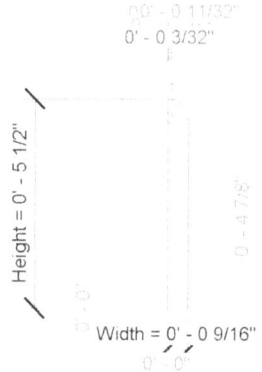 Modify the profile.

I eliminated the offset on the left and simplified the top. You can delete any unnecessary dimensions, but be sure to keep the Height and Width Dimensions.

Verify that the profile still flexes properly using the different types.

8.   Save the new profile.

9.   Activate the Modify ribbon.

Select **Load into Project**.

10. Close the family file.

11. Activate the **South Interior** View.

12. Select the base wall sweep.

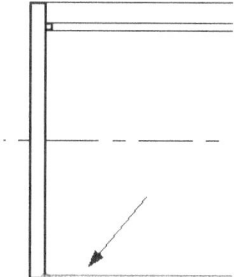

2-13

13. 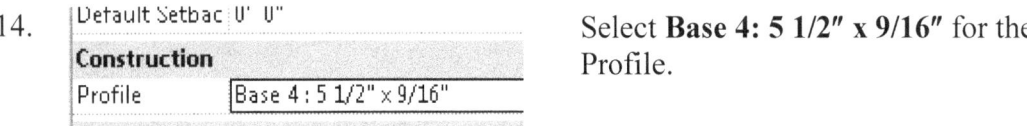 Select **Edit Type** from the Properties pane.

14. Select **Base 4: 5 1/2″ x 9/16″** for the Profile.

    This is the new profile you just created and loaded into the project.

    Press **OK**.

15. Save as *ex2-4.rvt*.

Command Exercise
Exercise 2-5 – Stacked Walls

Drawing Name: **i_stacked_walls.rvt**
Estimated Time to Completion: 60 Minutes

Scope

Defining a stacked wall structure

Revit defines a stacked wall as a wall that has 2 or more horizontal layers, each consisting of different materials and surfaces.

Solution

1. Open *i_stacked_walls.rvt*.

2. Activate **Level 1**.

3. Select the left vertical wall.

4. Select **Edit Type** on the Properties pane.

5. Select **Duplicate**.

6. Type **Exterior - Concrete Foundation**. Press **OK**.

7. Select **Edit Structure**.

8. Switch the view to **Section: Modify Type**.

2-15

9.

	Function	Material	Thickness
		EXTERIOR SIDE	
1	Finish 1 [4]	Masonry - Brick	0' 6"
2	Structure [1]	Concrete	0' 6"
3	Core Boundary	Layers Above Wrap	0' 0"
4	Substrate [2]	Wood - Sheathing - plywood	0' 2"
5	Thermal/Air Layer [3]	Misc. Air Layers - Air Space	0' 1"
6	Structure [1]	Wood - Stud Layer	0' 6"
7	Substrate [2]	Wood - Sheathing - plywood	0' 2"
8	Core Boundary	Layers Below Wrap	0' 0"
9	Finish 2 [5]	Gypsum Wall Board	0' 0 3/4"
		INTERIOR SIDE	

Add Layers as follows:
Layer 1: Finish 1 [4] Masonry - Brick 6"
Layer 2: Structure [1] Concrete 6"
Layer 3: Core Boundary
Layer 4: Substrate [2] Wood - Sheathing 2"
Layer 5: Thermal Air Lay - Misc Air Layers - Air Space 1"
Layer 6: Structure [1] Wood - Stud Layer 6"
Layer 7: Substrate [2] Wood - Sheathing 2"
Layer 8: Core Boundary
Layer 9: Finish 2 [5] Gypsum Wall Board 3/4"

10. Select **Split Region**.

11.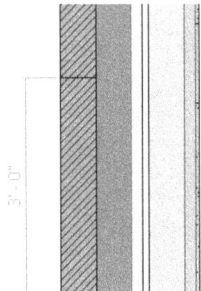

Cut the Layer 1: brick layer 3'-0" from the base.

Toggle the 2D button to see the hatch patterns on the layers.

12.

	Function	Material
1	Finish 1 [4]	Masonry - Brick
2	**Structure [1]**	**Concrete**

Highlight the **Layer 2: Concrete** Layer.

13. Pick on the **Assign Layers** button.

14. 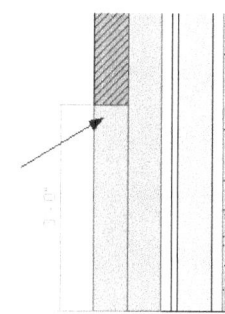 Select the lower region of the brick layer that was just split.

    Left pick just below the cut line.

    The upper region will now be brick and the lower region will be concrete.

    It may take some practice before you are able to do this.

15. 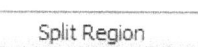 Select **Split Region**.

16. 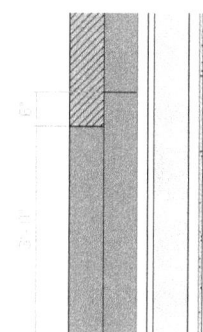 Cut the Layer 2: concrete 3'-6" from the base.

17. 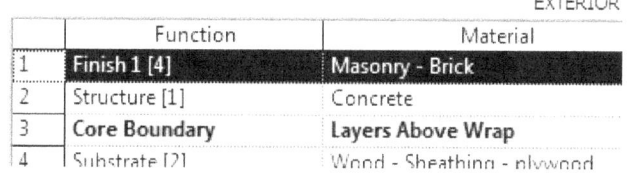 Highlight the **Layer 1: Masonry Brick** Layer.

18.  Pick on the **Assign Layers** button.

19. 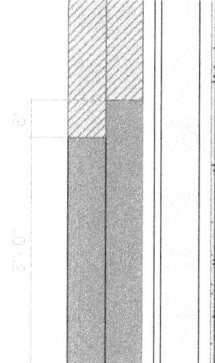 Highlight the Masonry brick layer.

    Select the upper region of layer 2.

    Left pick slightly above the cut line.

    The upper region will now be brick and the lower region will be concrete.

    It may take some practice before you are able to do this.

20. 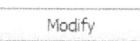 Select **Modify**.

21. 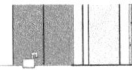 Select the base of the concrete Layer 1 component.

22. 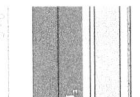 A small lock will appear.

    Click on the lock to unlock it.

23. Select the base of Layer 2: Concrete.

    Click on the lock to unlock it.

24. Press **OK** to close the dialogs.

25. 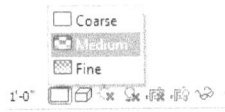 Select the wall with the Exterior - Concrete Foundation wall type.

    In the Properties pane:

    Set the Base Extension Distance to **-3' 0"**.

    Left Click in the display window to release the selection.

26.  Set the display to Medium or Fine to see the wall layers.

27. Activate the View ribbon.

    Select the Section Tool from the Create panel.

28. Place a small section on the wall you just defined.

    Activate the section view.

29. Set the display to Medium or Fine to see the wall layers.

2-18

30. 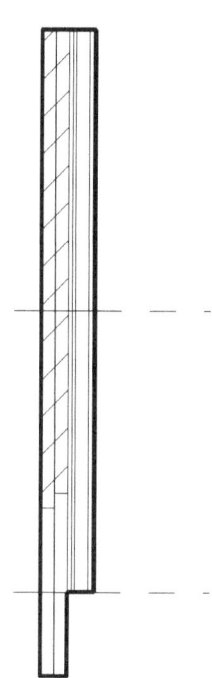 Note the concrete section is below the base level.

Select the wall.

31. 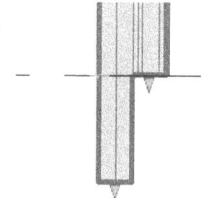 You can use the grips to adjust the base depth of the concrete section.

This type of wall is called a compound wall, because you have split wall layers and modified the layers.

32. Activate **Level 1**.

33. In the Project Browser, locate the two Stacked Wall types.

These are the stacked walls available in the current project.

34. 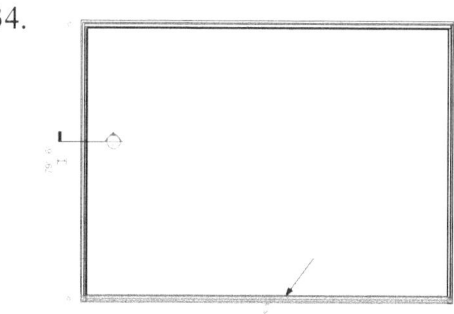 Select the south wall.

35. Switch the wall to **Stacked Wall 1** using the Type Selector on the Properties panel.

Select **Edit Type**.

36. Select **Duplicate**.

37. Rename **Exterior - Brick with Concrete Foundation**.

Press **OK**.

38. Select **Edit** Structure.

39.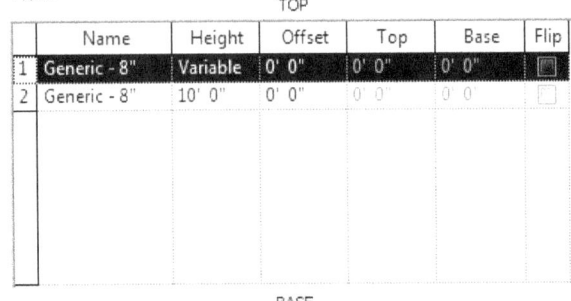

Note the stacked wall uses different layers going from Top to Base instead of Exterior to Interior.

Each layer is a wall type instead of a component material.

40.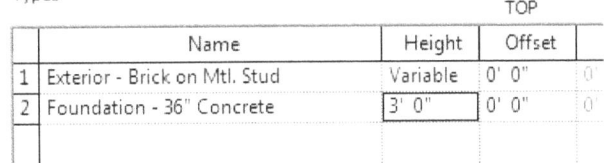

Change Layer 1 to **Exterior Brick on Mtl. Stud**.
Change Layer 2 to **Foundation - 36" Concrete**.

Set the Height of Layer 2 to **3'-0"**.

41. Select **Insert**.

Position the new layer between the existing layers.

42. Set the new layer to:

Foundation- 12" Concrete.
Set the Height to **3' 6"**.
Set the Offset to **-7/8"**.

2-20

The Basics of Building a Model

43. You can zoom into the preview window to check the offset value.

44. The height of the Top Layer is set to Variable so the user can set the wall height.

    Press **OK** twice to exit the dialog.

45. Switch to a 3D view so you can inspect the two wall types.

    Note that when you hover the mouse over the first wall you defined it displays as a Basic Wall.

    The other wall displays as a Stacked Wall.

46.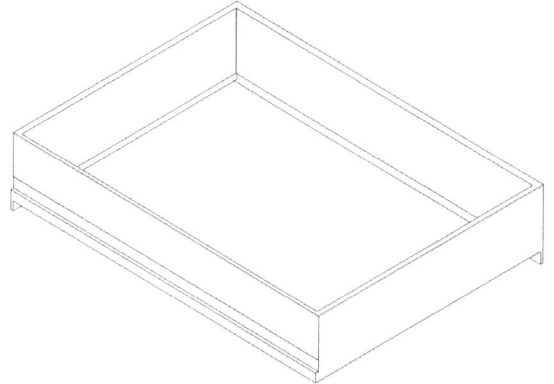
    Activate the Level 1 view.

    Select the North wall.

47.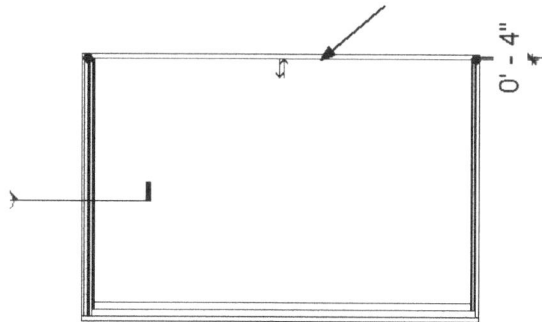
    Use the Type Selector drop-down to set the Type to **Exterior - Siding with Wood Stud**..

2-21

48. 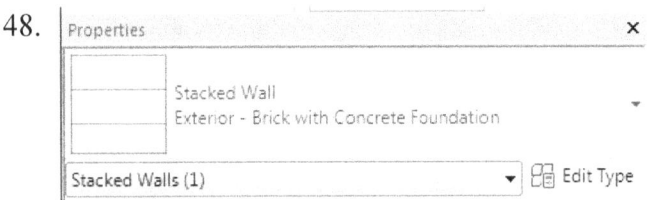   Select the South Wall (the stacked wall).

    Select **Edit Type**.

49. Select **Duplicate**.

50. Rename **Exterior - Siding with Concrete Foundation**.

    Press **OK**.

51. Select **Edit** Structure.

52. Set Layer 1 to the new wall type: **Exterior - Siding with Wood Stud**.

    Adjust the Offset for Layer 2: Foundation -12" Concrete to **1"**.

    Press **OK** to close all dialogs.

53. 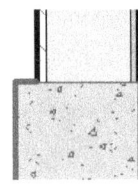 You can zoom into the preview window to check the offset.

    Press OK twice to close the dialogs.

54. 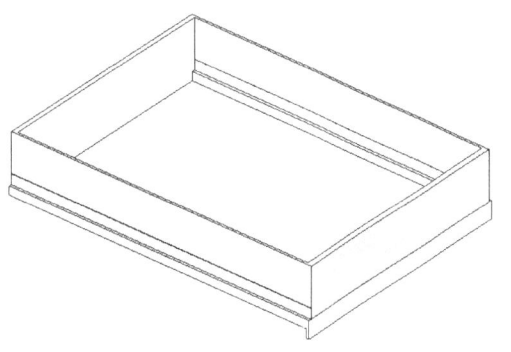 Switch to a 3D view.

    How many stacked walls are there?

    How many basic walls?

    What is the difference between a compound wall and a stacked wall?

    Is a compound wall a basic wall or a stacked wall?

55. Close without saving.

Command Exercise

Exercise 2-6 – Dividing a Wall into Parts

Drawing Name: **wall_parts.rvt**
Estimated Time to Completion: 45 Minutes

Scope

Use of parts to apply materials to a wall

Solution

1. Activate the **South** elevation view.

2. Select the wall so it highlights.

3. 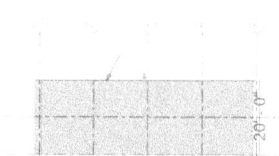 Select the **Create Parts** tool on the Create panel.

4. 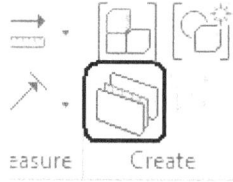 Select **Divide Parts** on the Part panel.

5. Enable **Pick a plane**.

   Press **OK**.

6.  Pick the front of the wall for the work plane.

7. Select the **Add** tool from the Divided Parts panel.

8. Select **Intersecting References** from the References panel.

9. Set Filter to **All**.

   This allows you to see Grids and Levels.

10. 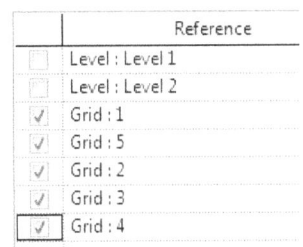 Place a check next to the **Grids**.

    Press **OK**.

11. 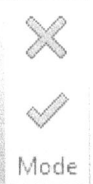 Select the **Green Check** on the Mode panel to finish dividing the parts.

12. Switch to a 3D view.

13. 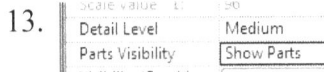 In the Properties pane:

    Set the Detail Level to **Medium**.
    Set the Parts Visibility to **Show Parts**.

The Basics of Building a Model

14. 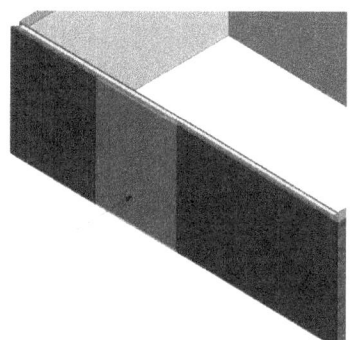 Select the second panel/part.

15. In the Properties panel:

    Uncheck **Material by Original**.
    Left click in the Material column to assign a material.

16.  Select the **Concrete - Precast Concrete** material.

    Press **OK**.

    Left click in the window to release the selection.

17.  The wall changes to show the new material.

18. Select the fourth panel/part.

19. In the Properties panel:

    Uncheck **Material by Original**.
    Left click in the Material column to assign a material.

2-25

20. 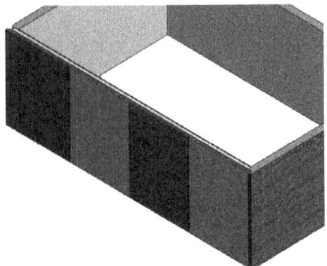 Select the **Concrete - Precast Concrete** material.

    Press **OK**.

    Left click in the window to release the selection.

21.  The wall changes to show the new material.

22. Hold down the CTRL key and select the two concrete panels.

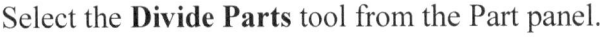

23. Select the **Divide Parts** tool from the Part panel.

24. Select **Intersecting References** from the References panel.

25. 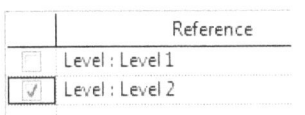 Place a check next to the **Level 2**.

    Press **OK**.

2-26

26. Select the **Green Check** on the Mode panel to finish dividing the parts.

27. The concrete panels are now divided into two sections.

    Shown in wireframe so you can see the divisions easily.

28. 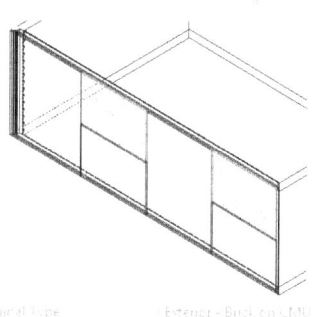 Hold down the CTRL key and select the two top sections of the concrete panels.

29. In the Properties panel:

    Left click in the Material column to assign a material.

30.  Select the **Masonry-Stone** material.

    Press **OK**.

    Left click in the window to release the selection.

31. 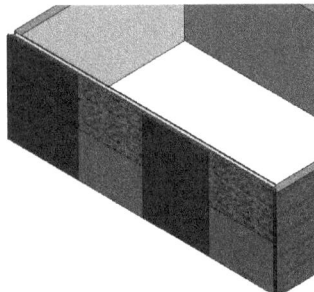 Set the display to **Realistic**.

32. 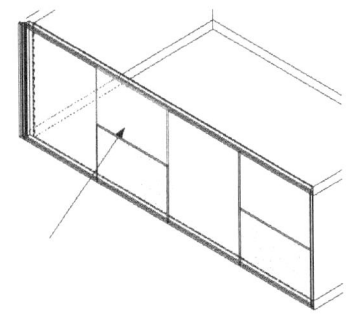 The new materials are assigned.

33. 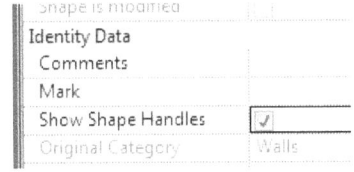 Select the top part indicated.

    Note that no grips are available when a part is selected.

    Shown in wireframe so you can see the divisions easily. Use the TAB key to cycle through selection choices.

34. 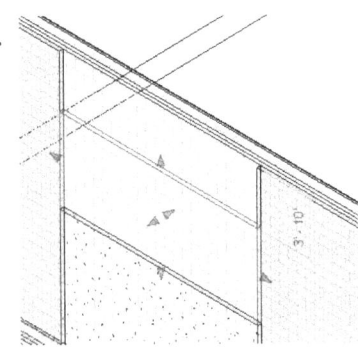 Enable **Show Shape Handles** on the Properties pane.

35. Use the top shape handle to lower the material 3' -10" from the top of the wall.

    Left click in the window to release the selection.

36. 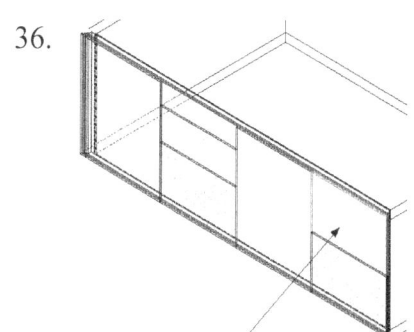 Select the top part indicated.

    Note that no grips are available when a part is selected.

    Shown in wireframe so you can see the divisions easily.

37. 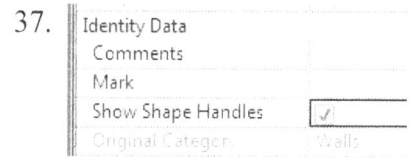 Enable **Show Shape Handles** on the Properties pane.

38. 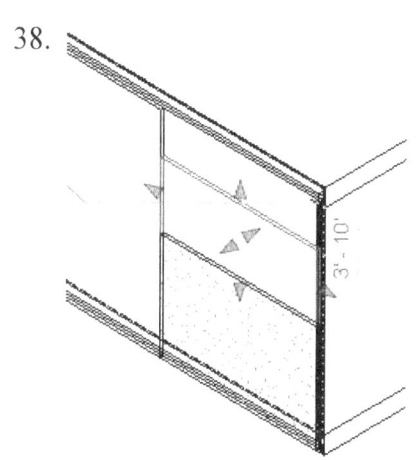 Use the top shape handle to lower the material 3'-10" from the top of the wall.

39.  Switch to a **Realistic** display.

    Orbit the model to inspect your new wall.

40. Select the bottom part of the second panel.

41.  Click in the Material column.

42. Set the Material to **Masonry -Brick**.

    Press **OK**.

43. 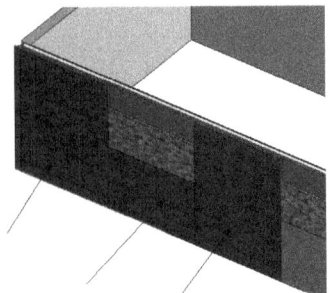 Hold down the Control key and select the first panel, the bottom part of the second panel and the third panel.

44. Select **Merge Parts**.

    Left click to release the selection

45.  Inspect the wall.

46. Save as *ex2-6.rvt*.

Command Exercise

Exercise 2-7 – Creating an In-Place Mass

Drawing Name: **in_place_mass.rvt**
Estimated Time to Completion: 60 Minutes

Scope

Use of in-place masses to create a conceptual model

Solution

1. Activate the **Site** plan view.

2. Activate the **Massing & Site** ribbon.

   Select the **In-Place Mass** tool from the Conceptual Mass panel.

3. Revit has enabled the Show Mass mode, so the newly created mass will be visible.

   To temporarily show or hide masses, select the Massing & Site ribbon tab and then click the Show Mass button on the Massing panel.

   Masses will not print or export unless you make the Mass category permanently visible in the View Visibility/Graphics dialog.

   Revit displays a message indicating that visibility of masses has been turned on.

   Press **Close**.

4. Name your first mass **Building 1**.

   Press **OK**.

5.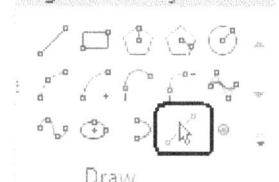

   Select the **Pick Line** tool from the Draw panel.

2-31

6. Pick the lines for the building in the upper right quadrant.

7. When you select the sketch, it should form a closed boundary.

   Make sure there are no overlapping lines.

8. Switch to a 3D view.

9. 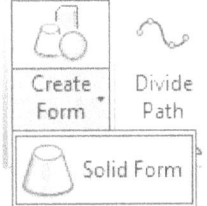 Select the sketch.

   Select **Form→Create Form→Solid Form**.

10. Switch to the **East** elevation.

11. There are two dimensions. The bottom dimension displays the overall height of the mass. The top dimension displays the distance from the top of the mass to the level above it.

12. Click on the bottom dimension controlling the overall height of the mass.
    Change it to **70′ 0″**.

    Press **ENTER**.
    Left click in the display window to release the selection.

13. Select **Finish Mass** from the In-Place Editor panel.

14. Activate the **Site** plan view.

15. Activate the **Massing & Site** ribbon.

16. Select the **In-Place Mass** tool from the Conceptual Mass panel.

17. Name your mass **Building 2**.
    Press **OK**.

18. Select the **Pick Line** tool from the Draw panel.

2-33

19. 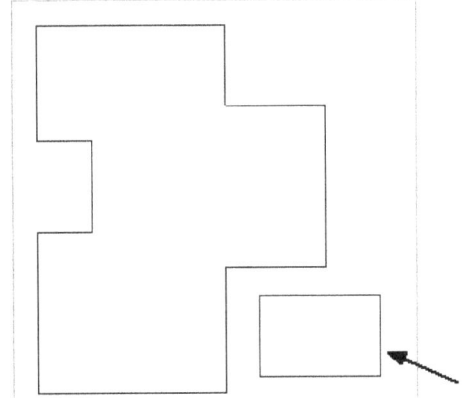 Pick the lines for the small building in the upper right quadrant.

    You can also use the rectangle tool.

20. 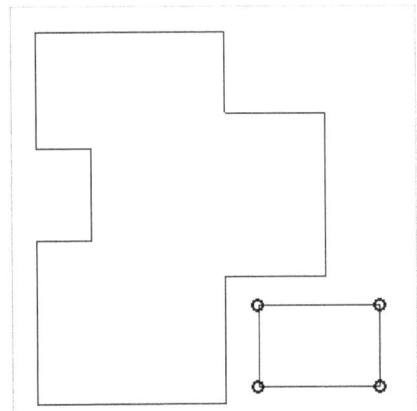 When you select the sketch, it should form a closed boundary.

    Make sure there are no overlapping lines.

21. 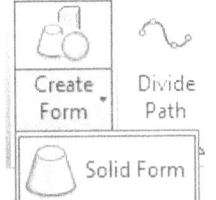 Switch to a **3D** view.

22. 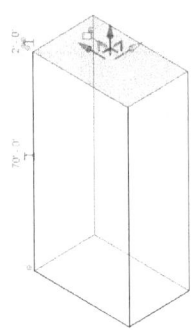 Select the sketch.

    Select **Form→Create Form→Solid Form**.

23. Select the Blue Z-axis.

    Drag the building up until the dimension displays **70'-0"**.

    Left click in the display window to release the selection.

The Basics of Building a Model

24. Select **Finish Mass** from the In-Place Editor panel.

25. Activate the **Site** plan view.

26. Activate the **Massing & Site** ribbon.

27. Select the **In-Place Mass** tool from the Conceptual Mass panel.

28. Name your mass **Building 3**.

    Press **OK**.

29. Select the **Pick Line** tool from the Draw panel.

30. Pick the lines for the building in the upper left quadrant.

    Select both the inner and outer boundaries.

2-35

31.  When you select the sketch, it should form a closed boundary.

    Make sure there are no overlapping lines.

32. 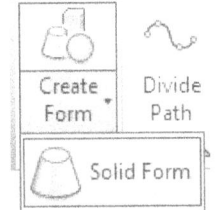 Switch to a **3D** view.

33. 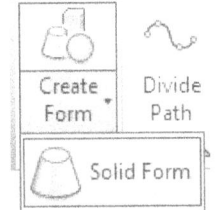 Select the outer sketch.

    Select **Form→Create Form→Solid Form**.

34. 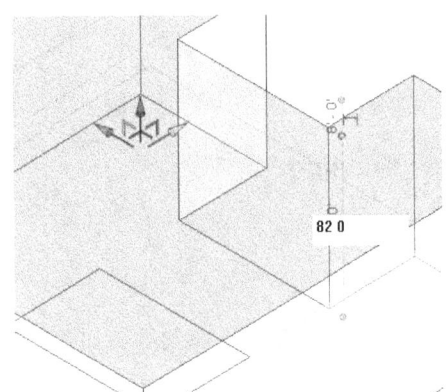 Select the Blue Z-axis.

    Drag the building up until the dimension displays **82′-0″**.

    Left click in the display window to release the selection.

35. 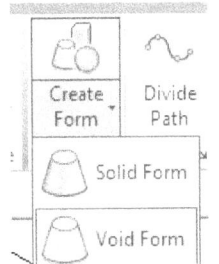 Select the inner sketch.

    Select **Form→Create Form→Void Form**.

The Basics of Building a Model

36.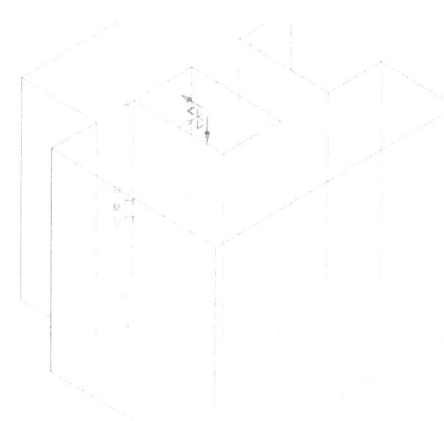

Select the Blue Z-axis.

Drag the void down until the dimension displays **82'-0"**.

Left click in the display window to release the selection.

37.

Select **Finish Mass** from the In-Place Editor panel.

38. Activate the **Site** plan view.

39. Activate the **Massing & Site** ribbon.

40.

Select the **In-Place Mass** tool from the Conceptual Mass panel.

41.

Name your mass **Building 4**.

Press **OK**.

42.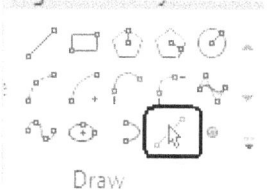

Select the **Pick Line** tool from the Draw panel.

2-37

43. 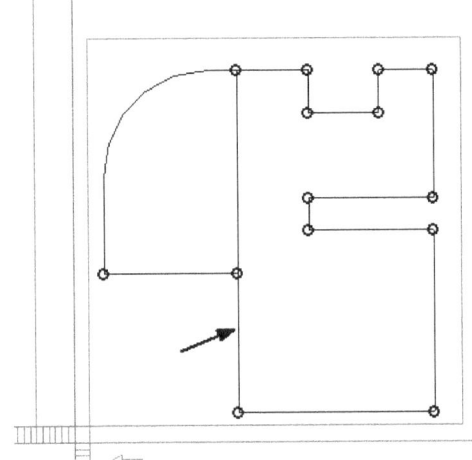 Pick the lines for the building in the lower left quadrant.

    You will need to use the TRIM tool from the Modify panel to trim the lower left side of the sketch.

44.  When you select the sketch, it should form a closed boundary.

    Make sure there are no overlapping lines. Arrows indicate the sketch components.
    I have x'd out the line which needs to be trimmed out.

45. Switch to a **3D** view.

46. Select the sketch.

47. 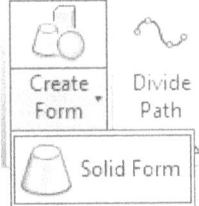 Select **Form→Create Form→Solid Form**.

The Basics of Building a Model

48. Select the Blue Z-axis.

    Drag the building up until the dimension displays **94′-0″**.

    Left click in the display window to release the selection.

49. Select **Finish Mass** from the In-Place Editor panel.

50. Activate the **Site** plan view.

51. Activate the **Massing & Site** ribbon.

52. Select the **In-Place Mass** tool from the Conceptual Mass panel.

53. Name the mass **Towers**.

    Press **OK**.

54. Select the **Rectangle** tool from the Draw panel.

2-39

55. 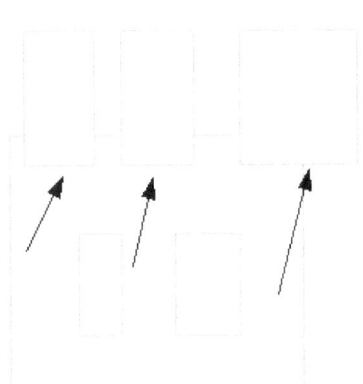 Trace over the top three rectangles in the fourth quadrant.

56. Switch to a **3D** view.

57. Select the left rectangle.

    You can only use one closed polygon at a time for a solid form.

58. 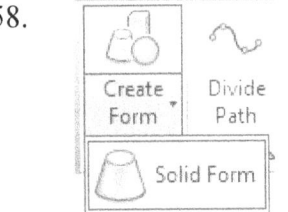 Select **Form→Create Form→Solid Form**.

59. 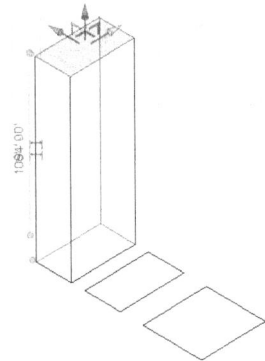 Select the Blue Z-axis.

    Drag the building up until the dimension displays **106′-0″**.

    Left click in the display window to release the selection.

60.  Select the middle rectangle.

The Basics of Building a Model

61. 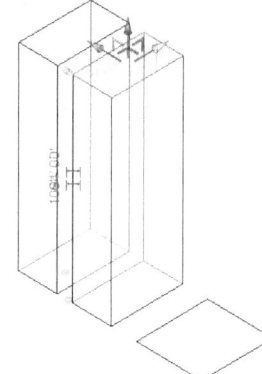 Select **Form→Create Form→Solid Form**.

62. 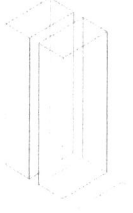 Select the Blue Z-axis.

    Drag the building up until the dimension displays **106'-0"**.

    Left click in the display window to release the selection.

63. 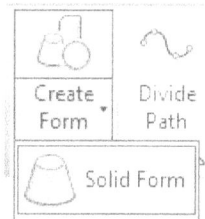 Select the right rectangle.

64. 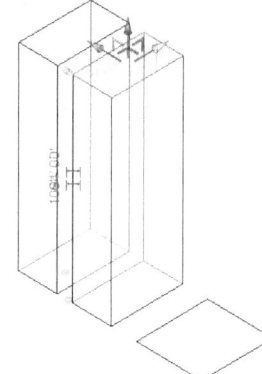 Select **Form→Create Form→Solid Form**.

65. 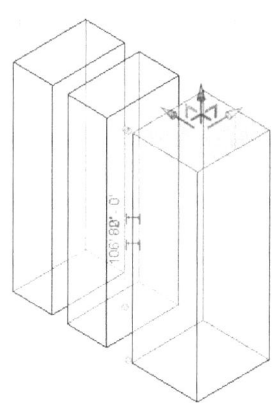 Select the Blue Z-axis.

    Drag the building up until the dimension displays **106'-0"**.

    Left click in the display window to release the selection.

66.		Select **Finish Mass** from the In-Place Editor panel.
67.		Activate the **Site** plan view.
68.		Activate the **Massing & Site** ribbon.
69.		Select the **In-Place Mass** tool from the Conceptual Mass panel.
70.		Name the mass **Building 5**. Press **OK**.
71.		Select the **Pick Line** tool from the Draw panel.
72.		Pick the lines for the sketch in the lower right quadrant. Use **Draw Line** to complete the sketch. *Disable Chain on the Options bar to make placing the lines easier.* Use the RECTANGLE tool to create two sketches for the internal rectangles. *These will be voids.*

73. 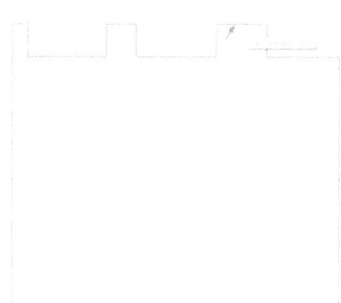 When you select one of the lines of the sketch, you should see the entire sketch highlight.

    If you don't see a continuous loop, there are either missing lines or overlapping/duplicate lines.

    Check the sketch for overlapping lines by deleting a line, then click UNDO if it is not a duplicate.

74. 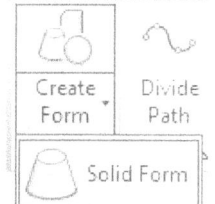 Switch to a **3D** view.

75. Select the outside boundary sketch.

    Select **Form→Create Form→Solid Form**.

76. 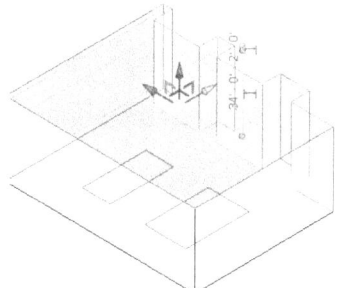 Select the Blue Z-axis.

    Drag the building up until the dimension displays **34′-0″**.

    Left click in the display window to release the selection.

77. 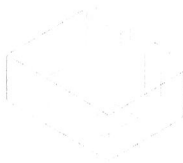 Select the left rectangle.

78. 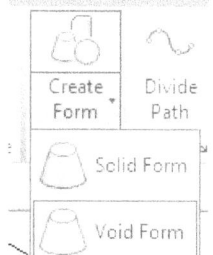 Select **Form→Create Form→Void Form**.

2-43

79. 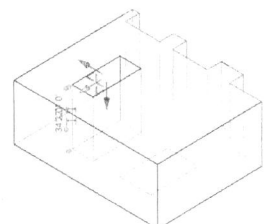 Select the Blue Z-axis.

Drag the building up until the dimension displays **34′-0″**.

Left click in the display window to release the selection.

80. 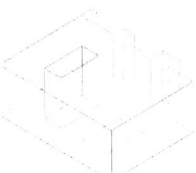 Select the right rectangle.

81. 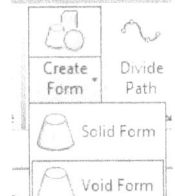 Select the sketch.

Select **Form→Create Form→Void Form**.

82. 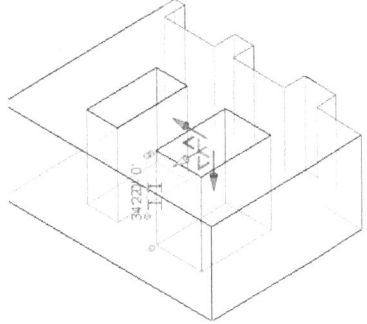 Select the Blue Z-axis.

Drag the building up until the dimension displays **34′-0″**.

Left click in the display window to release the selection.

83.  Select **Finish Mass** from the In-Place Editor panel.

84. Close without saving.

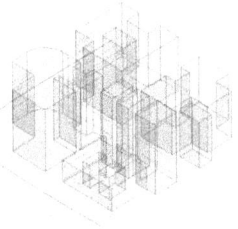

Command Exercise

Exercise 2-8 – Editing an In-Place Mass

Drawing Name: **editing_masses.rvt**
Estimated Time to Completion: 30 Minutes

Scope

Editing in-place masses to develop a conceptual model

Solution

1. 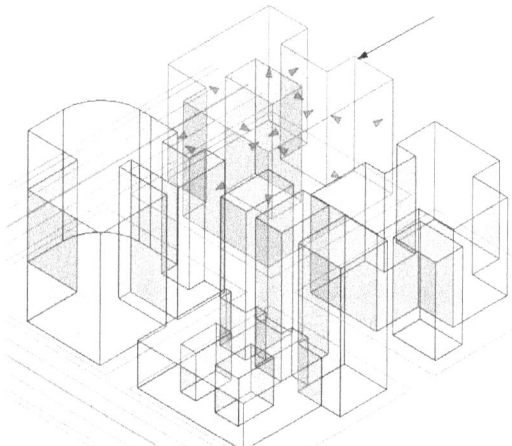 Select **Building 3** in the NW quadrant.

   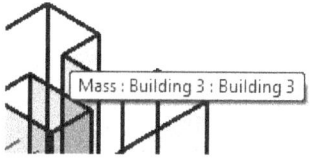

   If you hover your mouse over a mass, it will display the mass name assigned.

2.  Select **Edit In-Place** from the Model panel.

3. 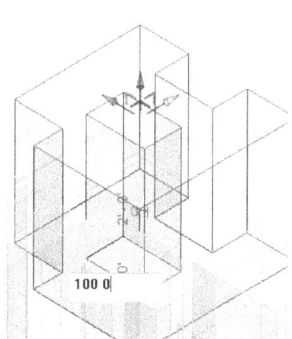 Use the TAB key to cycle through the selections until you have selected the top face.

   Change the height of the mass to **100'-0"**.

4.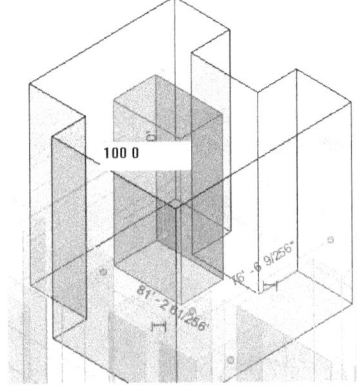

Use the TAB key to cycle through the selections until you have selected the void.

Change the height of the void to **100′-0″**.

5.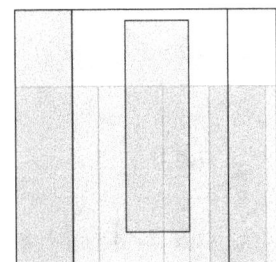

If you switch to a South elevation, you can check the void to see if it really is aligned on the top and bottom.

6.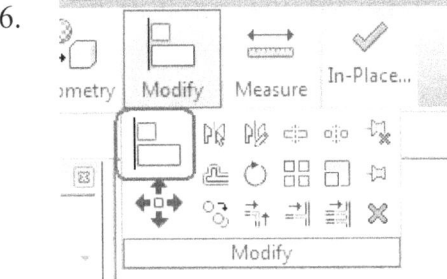

You can also use the ALIGN tool on the Modify panel to align the top of the void with the top of the solid form.

Select ALIGN. Use the TAB key to select the top of the solid form. Then, use the the TAB key to select the top of the void.

Repeat for the bottom.

7.

Select **Finish Mass** on the In-Place Editor panel when you are done editing the mass.

8.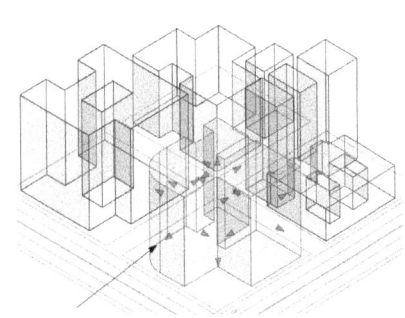

Select Building 4.

9.

10.  Select **Edit In-Place** from the Model panel.

11.  Rotate the view using the ViewCube.

12.  Select **Set Work Plane** from the Work Plane panel on the Create ribbon.

13. 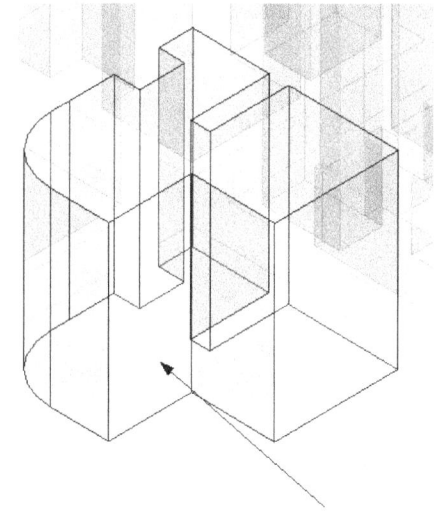 Select the face indicated.

14.  Select **Show Work Plane** from the Work Plane panel.

   The color of the selected face will change.

15. 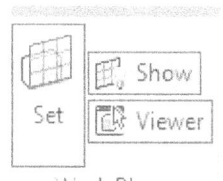 Select **Viewer** from the Work Plane panel.

16.		A window will open with a normal (perpendicular) view to the active work plane.
17.	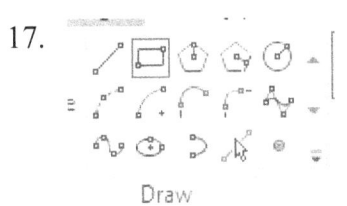	Select the **Rectangle** tool from the Draw panel.
18.	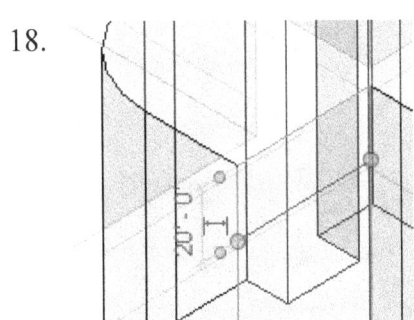	Draw a rectangle that is 20' high using the Viewer window. *To adjust the dimension, select the bottom line and the temporary dimension will appear.* *You can also use the ALIGN tool to set the sides of the rectangle collinear to the mass.*
19.	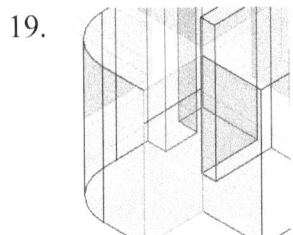	Check the placement of the rectangle in the 3D view. Close the Viewer window.
20.	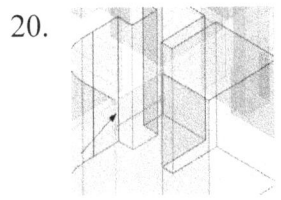	Select the sketch.
21.	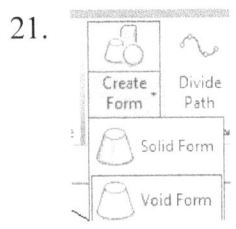	Select **Form→Create Form→Void Form**.

2-48

22. 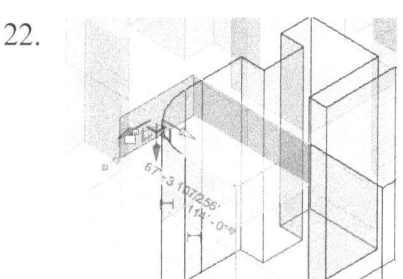 Use the Green Axis to drag the void form through the existing mass.

23.  Select **Finish Mass** from the In-Place Editor panel.

24. 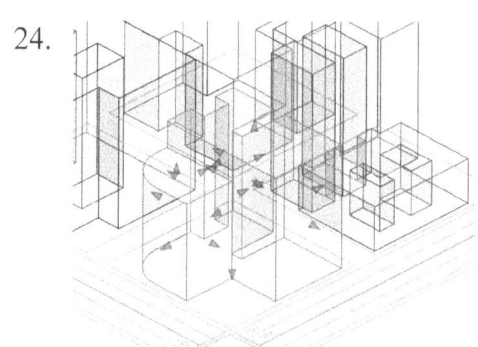 Select **Building 4**. (This is the building you just modified.)

25.  Select **Mass Floors** from the Model panel.

26. 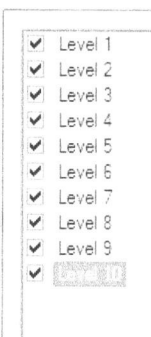 Enable all the Levels.

    Press **OK**.

27. 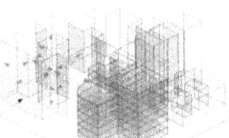 Floors are placed at each level.

28.  Select Building 3.

29. 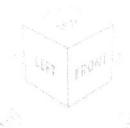 Select **Edit In-Place** from the Model panel.

30. 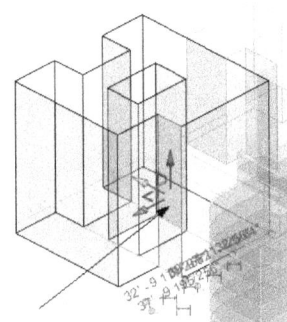 Orient your view cube as shown.

31.  Select the face indicated.

    You should see the shape handle axis tool.

    You can use the TAB key to cycle through the selection until the face is highlighted then left click.

32. Adjust the dimension using the red axis.

    It may be difficult to see the distance. You can also select the temporary dimension and set it to 60' 0". The arrows indicate the edge which is being adjusted.

The Basics of Building a Model

33. To verify that the face was moved properly, switch to the Site floor plan view.

    Then use the MEASURE tool to verify the dimension.

34. Select **Finish Mass** from the In-Place Editor panel.

35. Save the file as *ex2-8.rvt*.

Command Exercise

Exercise 2-9 – Mass Properties

Drawing Name: **new**
Estimated Time to Completion: 15 Minutes

Scope

Modifying a conceptual mass

Solution

1. Start a new project using the Architectural template.

   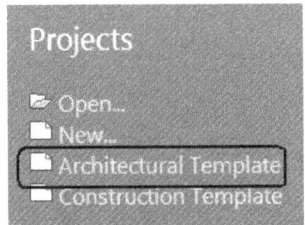

2. Activate the Massing & Site ribbon.

   Enable **Show Mass Form and Floors** from the Conceptual Mass panel.

3. Activate the Architecture ribbon.

   Select the **Component→Place a Component** tool on the Build panel.

4.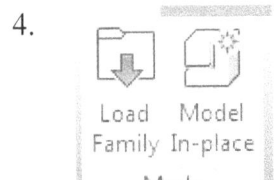

   Select **Load Family** from the Mode panel.

The Basics of Building a Model

5. Browse to the *Mass* folder.

6. Select the *Rectangle-Blended* mass.

   Press **Open**.

7. On the ribbon:

   Select **Place on Work Plane** on the Placement panel.

8. Place in the window.

   Right click and select Cancel twice to exit the command.

9. Switch to a 3D view.

10. Activate the Architecture ribbon.

    Select the **Wall→Wall by Face** tool on the Build panel.

11. Set the wall type to **Generic - 12"** using the Type Selector.

2-53

12. 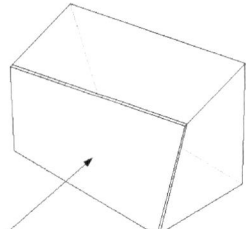 Select the face indicated.

    Left click in the window to exit the command.

13. 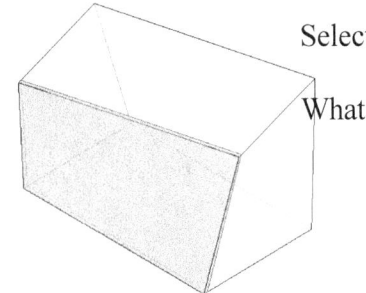 Select the wall you just placed.

    What is the area of the wall?

14. You should see the Volume and Area in the Properties pane.

15. Close without saving.

Things to remember about Masses:

- The default material for a mass is 5 percent transparent.
- Masses will not print unless the category is enabled in Visibility/Graphics Overrides.
- Masses are created from a single closed profile.
- Masses are a nested entity. In order to modify the profile, you have to open the mass up for editing and then open the desired form component up for editing.
- Masses can be comprised of multiple forms, a combination of voids and solids.

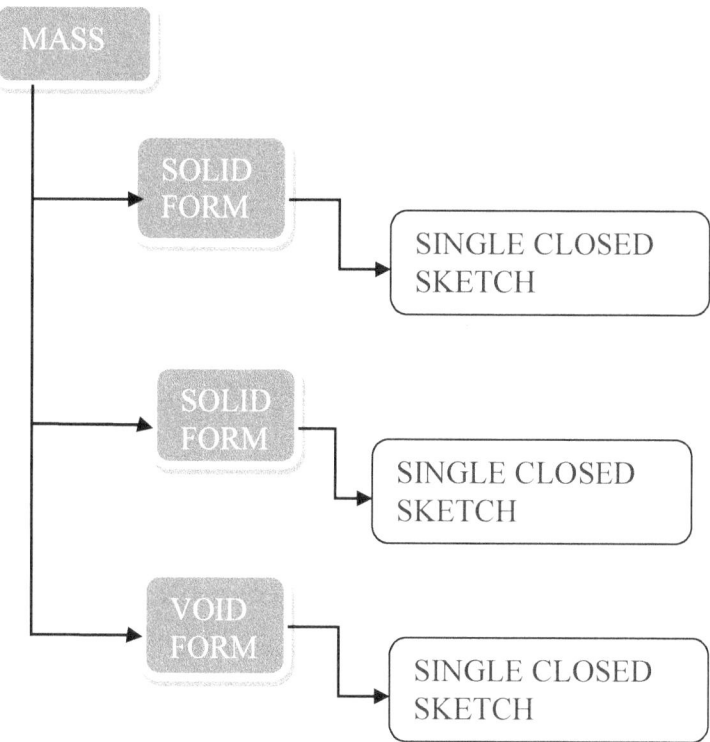

Practice Exam

1. Which of the following can NOT be defined prior to placing a wall?

    A. Unconnected Height
    B. Base Constraint
    C. Location Line
    D. Profile
    E. Top Offset

2. Identify the stacked wall.

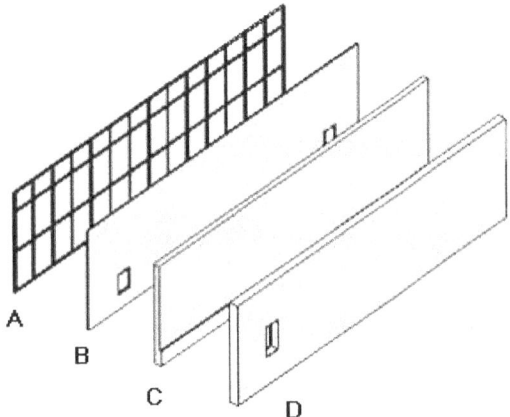

3. Walls are system families. Which name is NOT a wall family?

    A. BASIC
    B. STACKED
    C. CURTAIN
    D. COMPLICATED

4. Select the TWO which are wall type properties:

    A. COARSE FILL PATTERN
    B. LOCATION LINE
    C. TOP CONSTRAINT
    D. FUNCTION
    E. BASE CONSTRAINT

5. Select ONE item that is used when defining a compound wall:

    A. MATERIAL
    B. SWEEPS
    C. GRIDS
    D. LAYERS
    E. FILL PATTERN

6. A Mass Face:

    A. Can be used to generate a building element, such as a wall or roof.
    B. Can be moved.
    C. Can be assigned a material.
    D. Can be deleted without deleting the mass.

7. Use this key to cycle through selections:
   A. TAB
   B. CTL
   C. SHIFT
   D. ALT

8. When working with a mass, you can use levels to define mass _____
   A. Roofs
   B. Floors
   C. Ceilings
   D. Walls

9. To create a mass that is unique in a project, use the _____ Mass tool.
   A. In-Place
   B. Component
   C. System
   D. By Face

10. Selecting a work plane:
    A. Automatically changes the relative coordinate system
    B. Changes the project location
    C. Changes the level
    D. Determines the depth of an extrusion

11. The construction of a stacked wall is defined by different wall _____.
    A. Types
    B. Layers
    C. Regions
    D. Instances

12. To change the structure of a basic wall you must modify it's:
    A. Type Parameters
    B. Instance Parameters
    C. Structural Usage
    D. Function

13. Select THREE element types than can be created using mass faces:
    A. Doors
    B. Walls
    C. Levels
    D. Floors
    E. Roofs

14. Select TWO methods used to create a conceptual design using mass families;

    A. Go to the Applications Menu and select New→Conceptual Mass.
    B. Go to the Massing & Site ribbon and select Place Mass.
    C. Go to the Applications Menu and select New→Family
    D. Go to the Massing & Site ribbon and select In-Place Mass
    E. Go to the Applications Menu and select New→Project

15. In order to place a wall or floor on a mass face, the face must be:

    A. Horizontal
    B. Vertical
    C. Either Horizontal or Vertical
    D. Curved or spherical
    E. None of the above

16. To divide a floor or wall into parts, you can use the following (select all that apply):

    A. Lines
    B. Levels
    C. Grids
    D. Circles
    E. Arcs

17. To display parts in a view:

    A. Go to the Massing & Site ribbon and select Show Mass.
    B. On the View Properties pane: set Parts Visibility to Show Parts
    C. Go to the Visibility/Graphics dialog and enable Parts.
    D. Go to Temporary Hide/Isolate and Reset

18. To assign a different material to a part, select the part and:

    A. On the Properties pane: Enable Material by Original
    B. Right click and select Assign Material
    C. On the Modify ribbon, select Paint from the Geometry Panel.
    D. On the Properties pane: Uncheck Material by Original, then assign a material in the material field.

Answers:
1) D; 2) C; 3) D; 4) A & D; 5) D; 6) A & B; 7) A; 8) B; 9) A; 10) A; 11) A; 12) A; 13) B, D, & E; 14) A & D; 15) E; 16) A, B, C, & D; 17) B; 18) D

Lesson Three

Component Families

This lesson addresses the following User and Professional exam questions:

- Load Component Families
- Adding Components to a Project
- Creating families

Users should be able to understand the difference between a hosted and non-hosted component. A hosted component is a component that must be placed or constrained to another element. For example, a door or window is hosted by a wall. You should be able to identify what components can be hosted by which elements. Walls are non-hosted. Whether or not a component is hosted is defined by the template used for creating the component. A wall, floor, ceiling or face can be a host.

Some components are level-based, such as furniture, site components, plumbing fixtures, casework, roofs and walls. When you insert a level-based component, it is constrained to that level and can only be moved within that infinite plane.

Components must be loaded into a project before they can be placed. Users can pre-load components into a template, so that they are available in every project.

Users should be familiar with how to use Element and Type Properties of components in order to locate and modify information.

There are three kinds of families in Revit Architecture:

- system families
- loadable families
- in-place families

System families are walls, ceilings, stairs, floors, etc. These are families which can only be created by using an existing family, duplicating, and redefining. These families are loaded into a project using a project template.

Loadable families are external files. These include doors, windows, furniture, and plants.

In-place families are components which are created inside of a project and are unique to that project.

The Unofficial Revit 2015 Certification Exam Guide

Command Exercise
Exercise 3-1 – Level-Based Component

Drawing Name: **i_components.rvt**
Estimated Time to Completion: 10 Minutes

Scope

Moving a component from one level to the next.

Solution

1. Activate the **Main Floor** floor plan.

   Floor Plans
   - Ground Floor
   - Lower Roof
   - **Main Floor**
   - Main Roof
   - Site
   - T. O. Footing
   - T. O. Parapet

2. Select the table element indicated.

   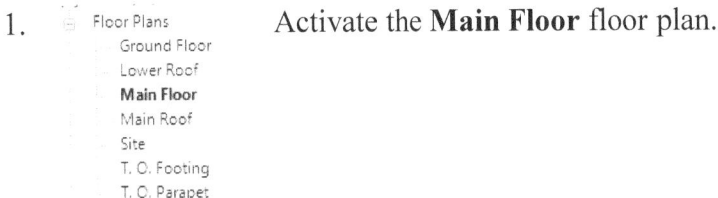

3. On the Properties pane:

   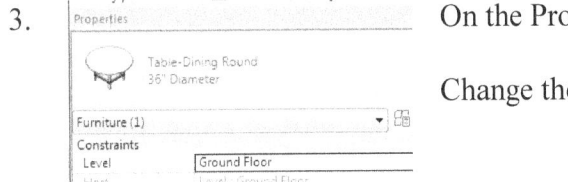

   Change the Level parameter to **Ground Floor**.

4. Activate **Ground Floor**.

   Views (all)
   Floor Plans
   - **Ground Floor**
   - Lower Roof
   - Main Floor
   - Main Roof
   - Site
   - T. O. Footing
   - T. O. Parapet

3-2

5. 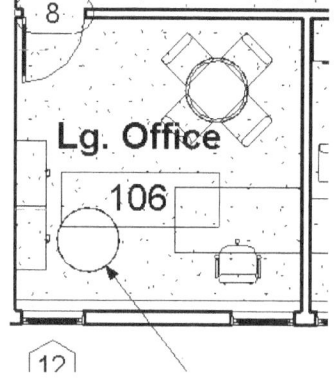   The table is in the same location on the Ground Floor.

6. Close the file without saving.

Command Exercise

Exercise 3-2 – Creating a Family

Drawing Name: **park bench.dwg**
Estimated Time to Completion: 80 Minutes

Scope

Create a Revit family from an AutoCAD symbol.

Solution

1. Close any open files.

2. 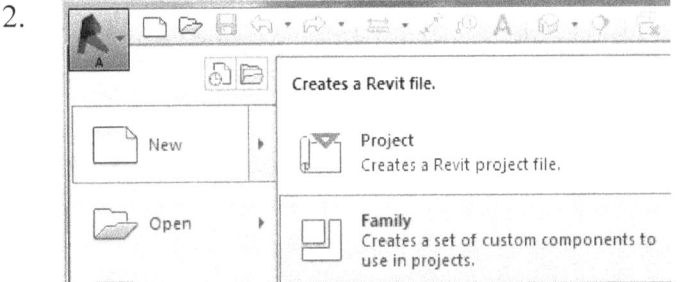 Go to **New→Family**.

3. Select the *Furniture.rft* from the Imperial templates.

   Press **Open**.

4. Activate the **View** ribbon.

   Select **Tile** under the Window panel.

5. 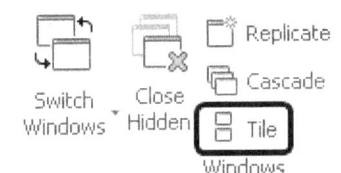 Double click the title for the **Elevation:Right** window to enlarge it.

6. 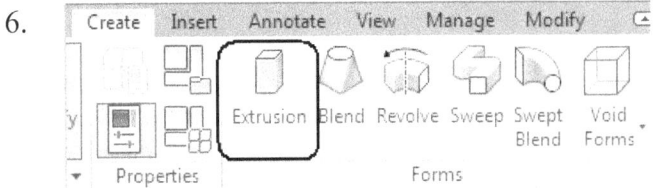 Activate the **Create** ribbon.

   Select **Forms→Extrusion**.

Component Families

7.  Activate the **Insert** ribbon.

   Select the **Import CAD** tool from the Import panel.

8.  Locate the *park bench.dwg* file.
   Set the positioning to Manual-Origin.
   Press Open.

9. 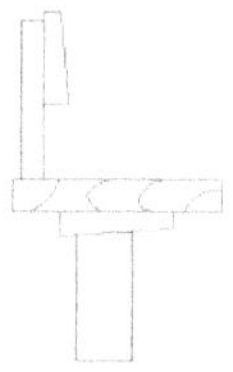 The park bench will appear as shown.

10. 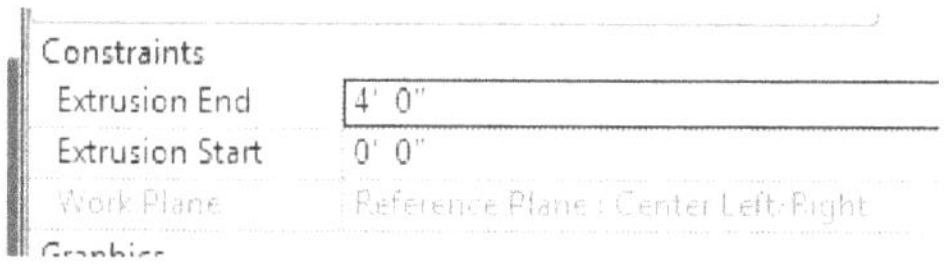 On the Properties pane:

    Set the Extrusion End to **4′-0″**.

11. 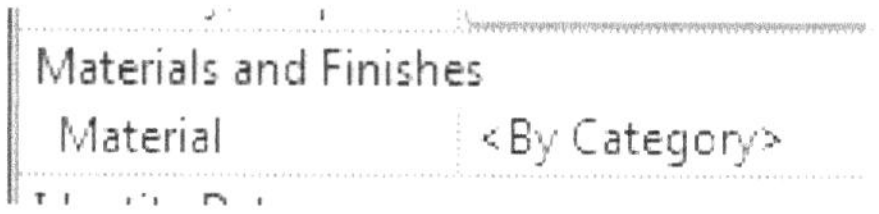 Left click in the **Material** column to assign a material.

12.
13. 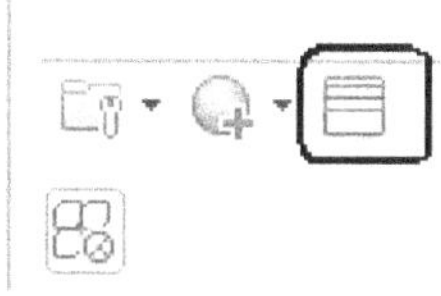 Launch the **Asset Browser**.

14. Type **redwood** in the search field.

3-5

15. Highlight the Default material in the Project Materials list.
Highlight the Redwood material in the Asset Browser.
Click on the far right of the Asset Browser to copy the asset browser properties over to the Default material definition.

Close the Asset Browser.

16. Rename the Default material **Redwood.**

The Redwood material is displayed as a Document Material.

Highlight the material and press **OK**.

17. Redwood is listed in the Properties pane.

18. Clean up the sketch so that there are no self-intersecting lines.

Use the TRIM and SPLIT tools.

19. Select the **Green Check** under Mode to **Finish Extrusion**.

Component Families

20. 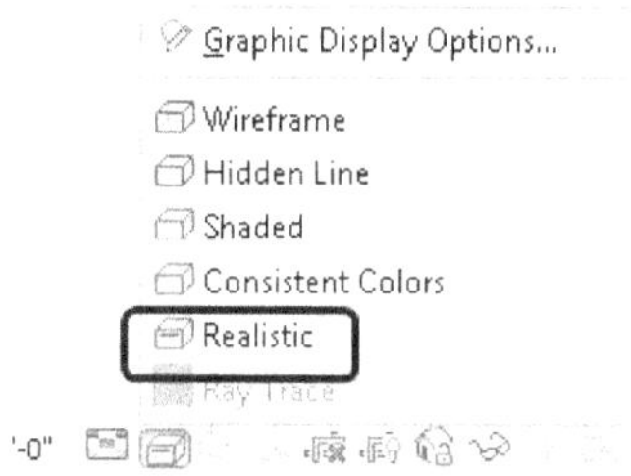 Switch to a 3D view.

Switch to a Realistic display.

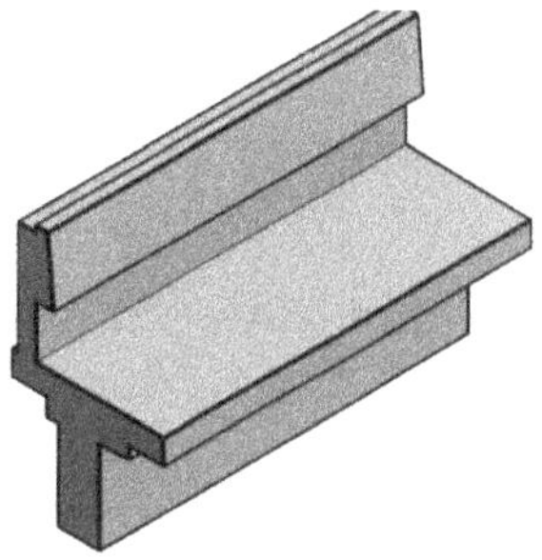

21.  Activate the **Back Elevation** view.

22. 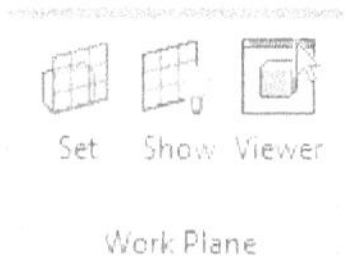 Activate the **Create** ribbon.

Select the **Set** tool under the Work Plane panel.

23. 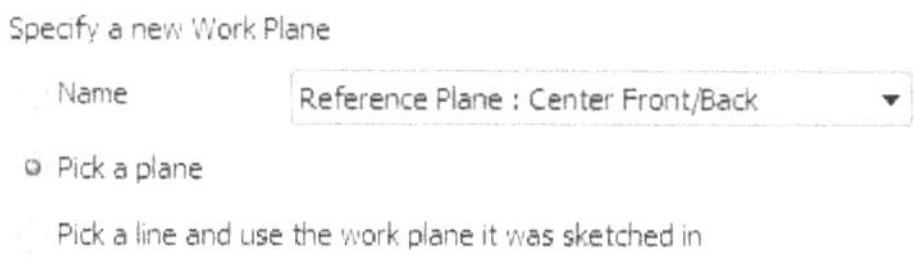 Enable **Pick a plane**.

Press **OK**.

24. 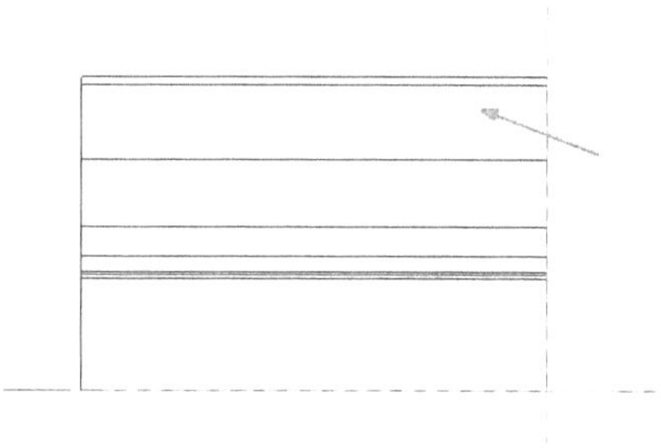 Select the front face of the back support.

25. 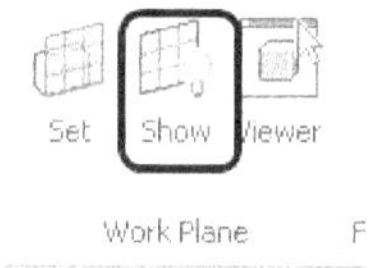 Select the **Show** tool on the Work Plane panel.

26. 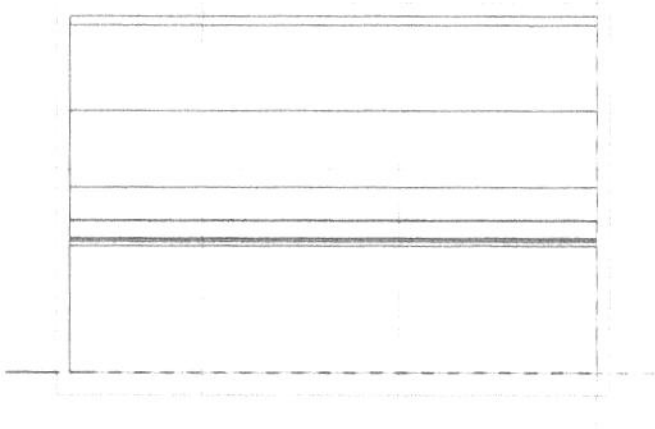 The active work plane will display as a light blue rectangle.

3-7

27. 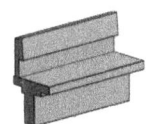 Switch to a 3D view and turn on the active work plane.

    Verify that the active work plane is the front face of the back support.

28. 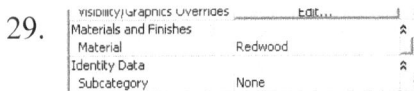 Select **Forms→Extrusion**.

29.  Select **Material** from the Properties Pane.

30. 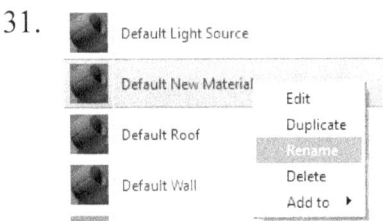 Select **Create New Material** at the bottom of the Material Editor dialog.

31.  Highlight the new material.
    Right click and select **Rename**.

32. 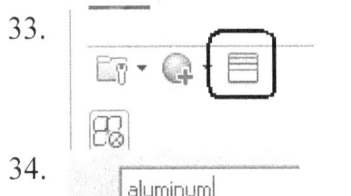 Rename **Aluminum**.

33. Launch the **Asset Browser**.

34. Type **aluminum** in the search field.

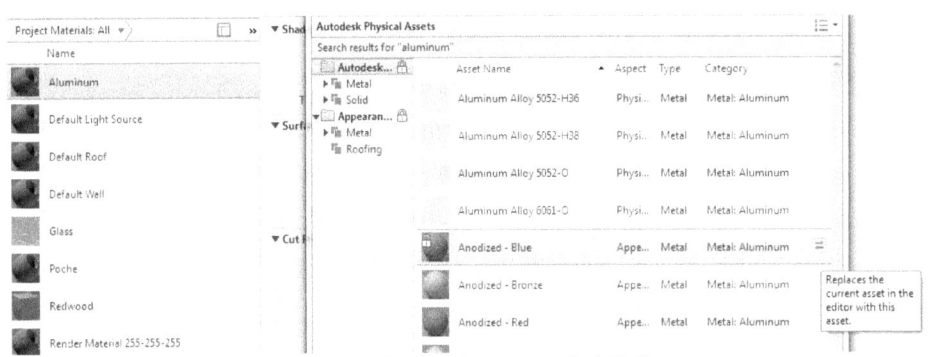

35. Highlight the Aluminum material in the Material Editor.
    Highlight the **Aluminum, Anodized Blue** material and **replace the current asset in the editor with this asset**.

Component Families

36. 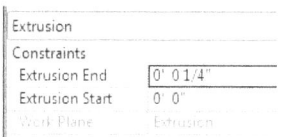 The aluminum material has been copied to the Project Materials list.

Close the Asset Browser.
Highlight the Aluminum material to select.
Press **OK**.

37. 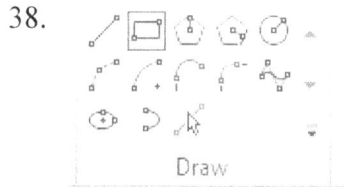 You should see the aluminum material listed in the Material field in the Properties pane.

Set the Extrusion End to **1/4"**.

38. Select the **Rectangle** tool from the Draw panel.

39. Draw a rectangle on the seat face.

Use dimensions to center it on the surface.

Remember temporary dimension values control the values of permanent dimensions.

40. Use the Fillet Arc tool to fillet the corners of the rectangle.

41.  Set the corners to 1" radius on the Options Bar.

3-9

42. 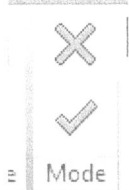 Select the **Green Check** under Mode to **Finish Extrusion**.

43. 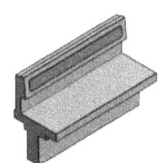 Switch to a 3D view to inspect your model.

44. Activate the **Left Elevation** view.

45. Activate the **Create** ribbon.

    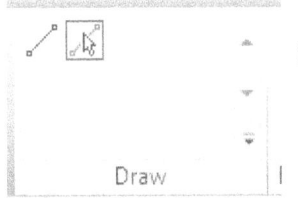 Select the **Reference Plane** tool.

46. Select the **Pick** tool.

47. 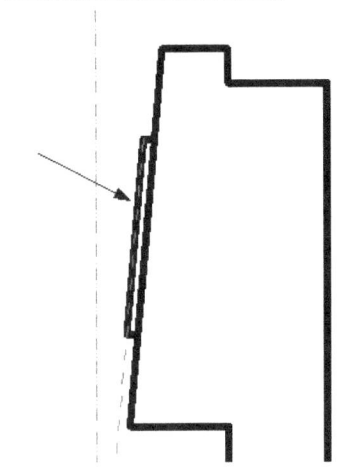 Pick the outside line of the aluminum plate.

Component Families

48. Use the end grips to extend the reference plane.

    Use the TAB key to select the reference plane.

49. Rename the reference plane **Front of Plate**.

50. Select the **Set** tool under the Work Plane panel.

51. Select the Front of Plate from the list of Names.

    Press **OK**.

52. Select **Elevation: Back**.

    Press **Open View**.

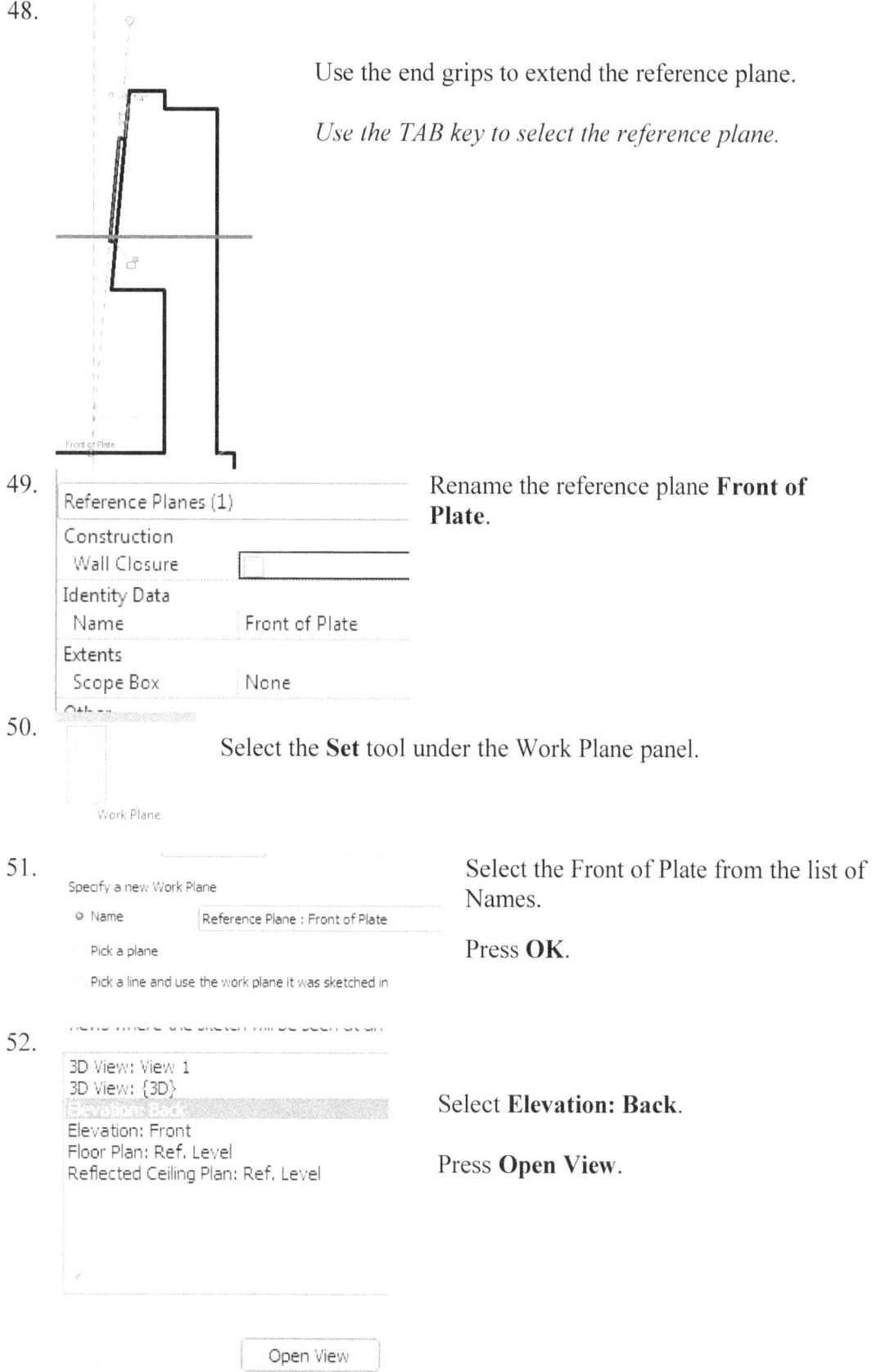

3-11

53. Select **Model Text** from the Model panel on the Create ribbon.

54. Type **In Memory of Our Fallen Soldiers** in the Edit Text dialog.

    Press **OK**.

55. Click to place in the view.

    Right click and select **Cancel** to exit the command.

    The space bar is not available to rotate the text in the Model Text command.

56. Select the Model text you just created.

    Select **Edit Type** in the Properties Pane.

57. Select **Duplicate**.

58. Name the new type **Model Text 2**.

    Press **OK**.

59. Change the Text Size to **2"**.

    Press **OK**.

60. Use the Rotate and Move tools from the Modify panel to position the text.

61. Set the Depth to **1/8"**.

62. Select **Material** from the Properties Pane.

63.  Select **Create New Material** at the bottom of the Material Editor dialog.

64.  Highlight the new material.
Right click and select **Rename**.

65.  Rename **Brass**.

66. Launch the **Asset Browser**.

67. Type **brass** in the search field.

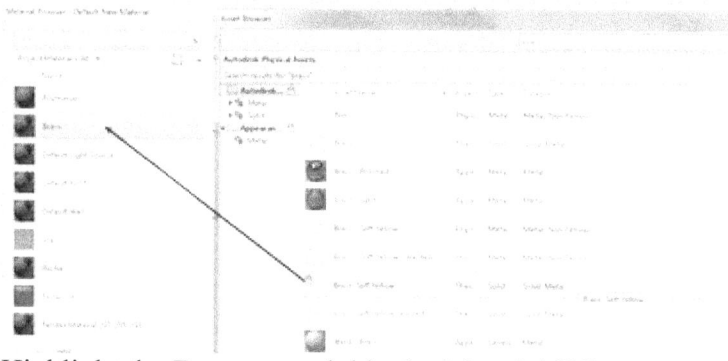

68. Highlight the **Brass** material in the Material Editor.
Highlight the **Brass, Soft Yellow** material and **replace the current asset in the editor with this asset**.

69. The yellow material has been copied to the Project Materials list.
Close the Asset Browser.
Highlight the brass material to select.
Press **OK**.

70. The material is listed in the Properties Pane.

71.  Switch to a 3D View.

72. 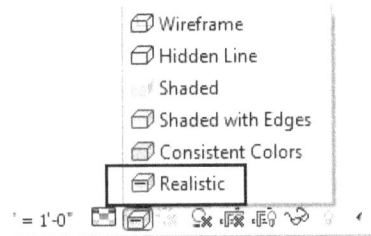 Turn off the visibility of the work plane by clicking on the Show Work Plane button.

73.  Set the Display to **Realistic**.

74. 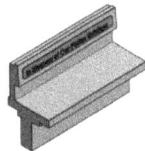 The model so far.

75. Activate the **Create** ribbon.

    Select the **Set** tool under the Work Plane panel.

76. Enable **Pick a plane**.

    Press **OK**.

77.  Select the front surface of the bench leg area.

    The active work plane will shift to the new location.

78.  Activate the **Back Elevation** view.

3-14

Component Families

79. Select **Forms→Void Extrusion** from the Architecture ribbon.

80. Use the Rectangle tool from the Draw panel to place a rectangle centered on the face.

81. Use the Fillet Arc tool to fillet the corners of the rectangle.

82. Set the corners to 1″ radius.

83. Set the depth of the void extrusion **8″**.

    Verify the Work Plane is set to Extrusion.

    Press **Apply**.

84. Select the **Green Check** under Mode to finish the void.

3-15

85. 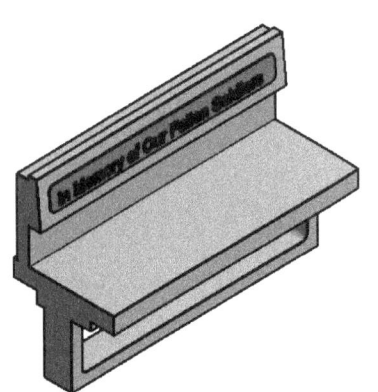 Switch to a wire frame view to see the void.

Use the grips to adjust the void position so it goes entirely through the park bench.

86. On the Properties pane:

Left click in the OmniClass Number field:

87. Assign the OmniClass Number:

23.40.10.11.14

Press **OK**.

88. Notice how the OmniClass Title updates.

Component Families

89.   Save as *park bench.rfa*.

This is a loadable family which can be loaded and used in any project.

Creating a Door Family
I assign this exercise as a homework assignment in my classes.

Scope

1. Use the Generic Model-Face Based template for the door panel.

2. Create the materials for the frame/door panel/glazing.

Add the following materials to the family document:

Mahogany
Canvas Paint (R244,G236,B215)
Frosted Glass

3. Create material parameters for the frame/door panel/glazing.
   Link the material parameters with the materials created.

4. Create the door panel - one extrusion for exterior side and one extrusion for interior side.

5. Assign the materials to the extrusions.

6. Open the Door template to start a new door family.

Create three door sizes as family types:

6' 10" x 3' 10"
6' 10" x 3' 2"
7' 2" x 3' 10"

7. Place a reference line to host the door panel.

8. Insert the door panel family.

8. Flex the model.

9. Add symbolic lines for the door swing.

10. Test the model family.

Component Families

Command Exercise

Exercise 3-3 – Creating a Door Panel

Drawing Name: **new**
Estimated Time to Completion: 1 hours 30 minutes

Scope: Create a door panel using the Generic Model template
Create extrusions
Create voids
Define materials
Assign materials to extrusions
Define custom parameters

Solution:

1. In order to use a door swing, the easiest way is to create the door panel as a generic model face based and insert it into a door family.

   Go to **New** →**Family**.

2. Select **Generic Model face based**.

   By selecting face based, we can control the position of the door using a reference line.

3. Activate the **Right** Elevation.

4. Add a horizontal reference plane and two vertical reference planes to the **left** of the existing reference plane.

   The intersection of the pre-existing work planes control the insertion point of the family when it is placed. We want to insert the door panel at the hinge point of the door.

5. Add an EQ dimension between the vertical reference planes.

   Add an overall vertical dimension and an overall horizontal dimension.

3-19

6. 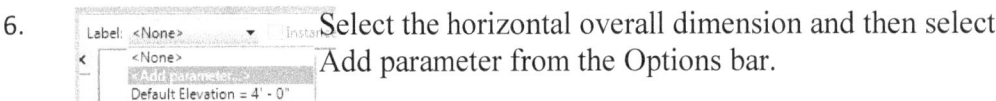 Select the horizontal overall dimension and then select Add parameter from the Options bar.

7. Type **Thickness** for Name.
   Enable **Type**.
   Press **OK**.

8. Select the vertical dimension and then select Add parameter from the Options bar.

9. Type **Height** for Name.
   Enable **Type**.
   Press **OK**.

10.  Name the middle vertical reference plane **Midplane**.

11. Name the far left vertical reference plane **Interior**.

12. 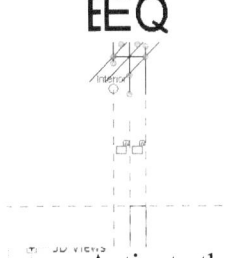 Activate the **Front** Elevation.

13. Add a vertical reference plane to the **left** of the existing reference plane.

Component Families

14. Add a horizontal dimension between the reference planes.

15. 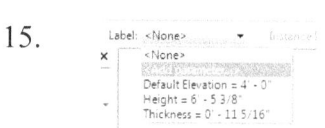 Select the horizontal dimension and select **Add parameter** from the Options bar.

16. Type **Width** for Name.
    Enable **Type**.
    Press **OK**.

17. 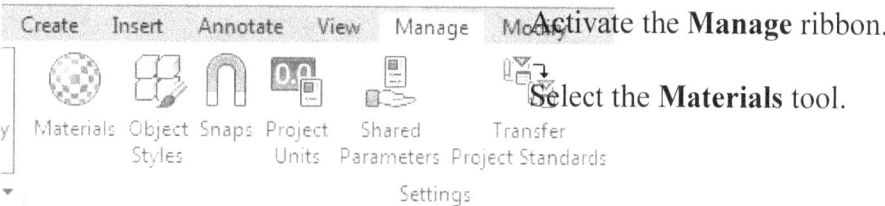 Activate the **Manage** ribbon.
    Select the **Materials** tool.

Door Panel Exterior	Mahogany	These are the materials we want to create.
Door Panel Interior	Canvas Paint (R244,G236,B215)	
Glazing	Frosted	

18. Do a search for Mahogany.

19. 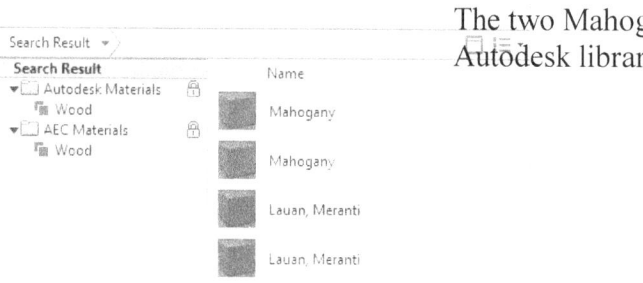 The two Mahogany materials in the Autodesk library will be displayed.

3-21

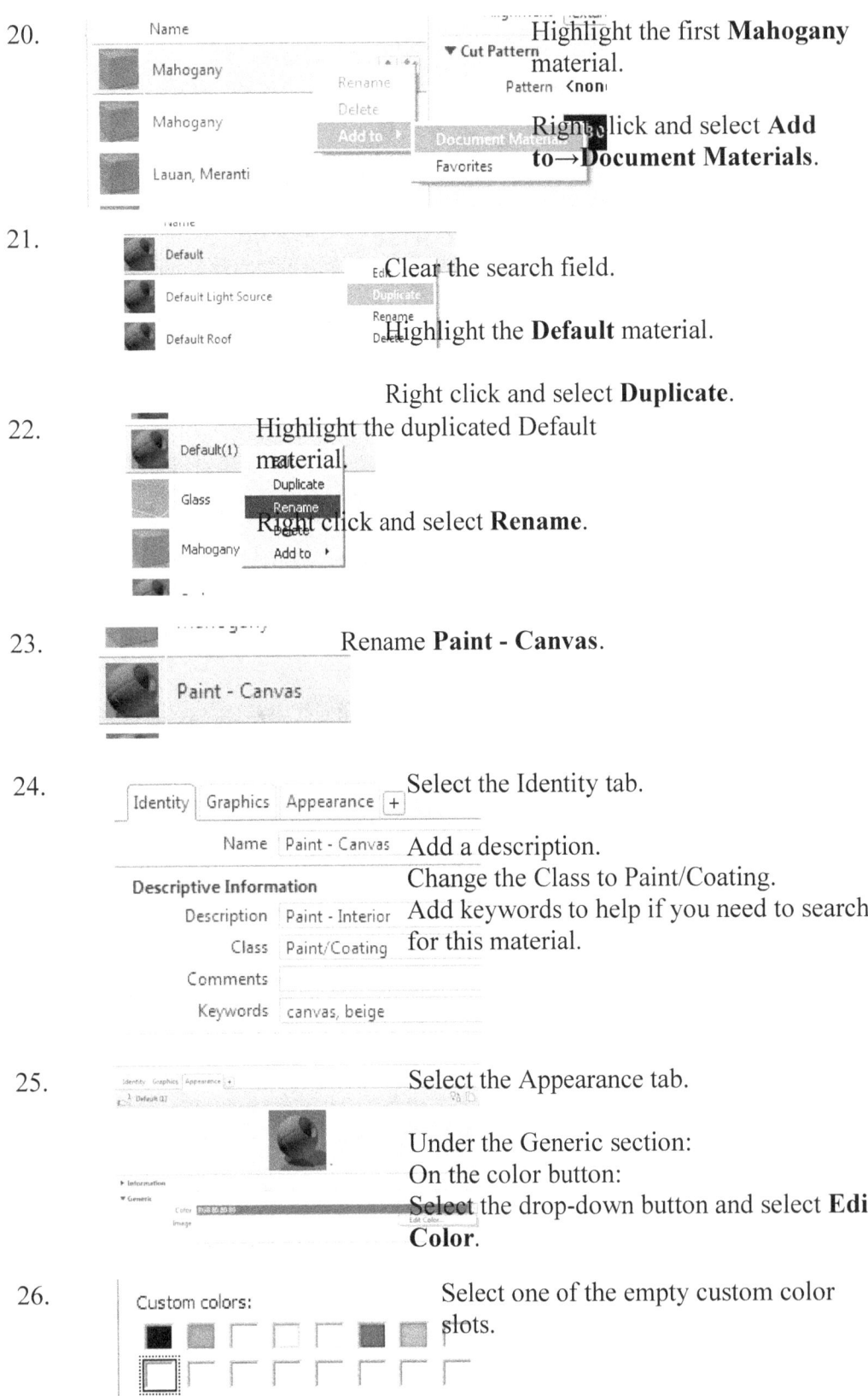

20. Highlight the first **Mahogany** material.

    Right click and select **Add to→Document Materials**.

21. Clear the search field.

    Highlight the **Default** material.

    Right click and select **Duplicate**.

22. Highlight the duplicated Default material.

    Right click and select **Rename**.

23. Rename **Paint - Canvas**.

24. Select the Identity tab.

    Add a description.

    Change the Class to Paint/Coating.

    Add keywords to help if you need to search for this material.

25. Select the Appearance tab.

    Under the Generic section:
    On the color button:
    Select the drop-down button and select **Edit Color**.

26. Select one of the empty custom color slots.

Component Families

27. 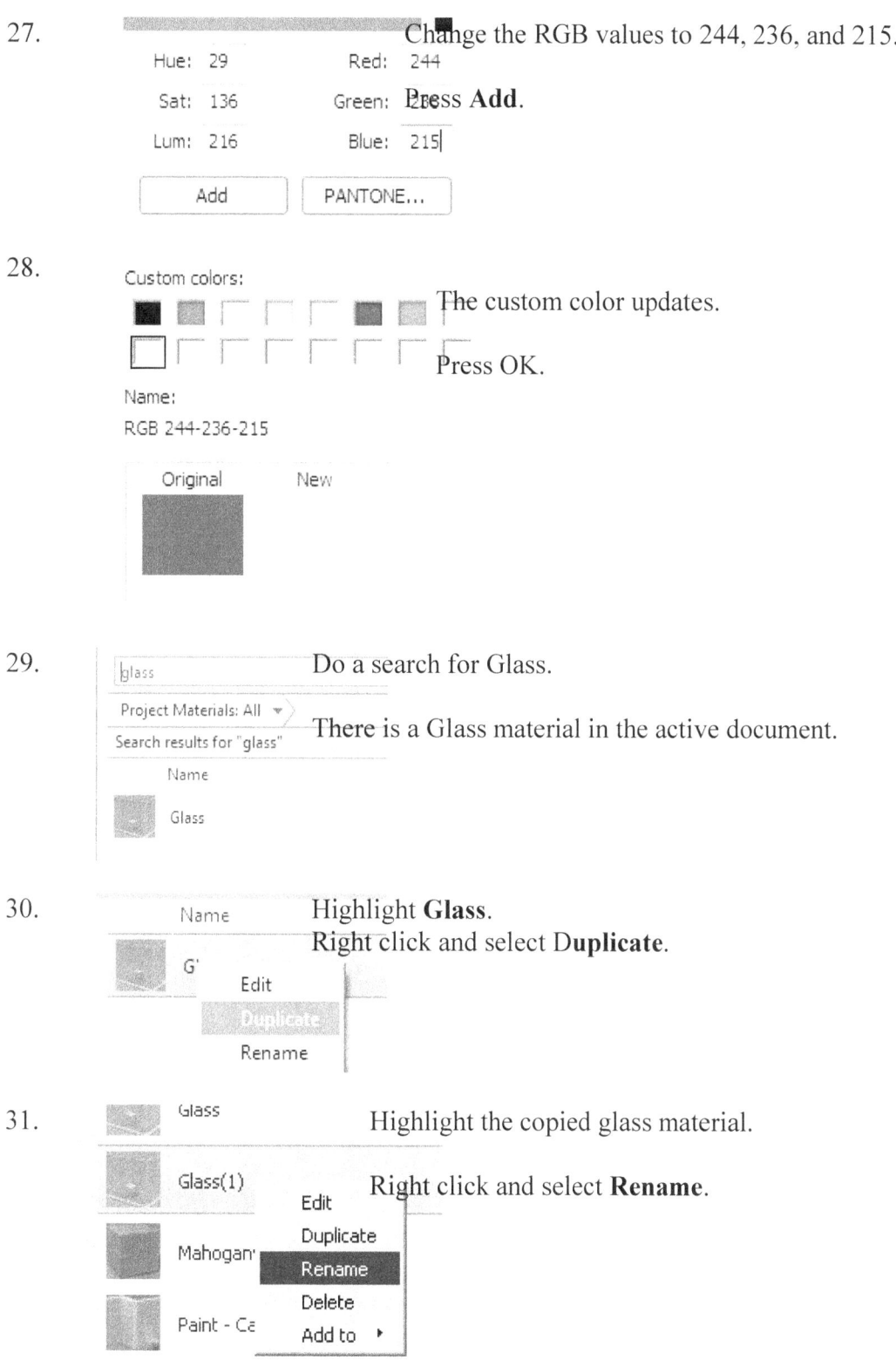 Change the RGB values to 244, 236, and 215.

    Press **Add**.

28. The custom color updates.

    Press OK.

29. Do a search for Glass.

    There is a Glass material in the active document.

30. Highlight **Glass**.

    Right click and select **D**uplicate.

31. Highlight the copied glass material.

    Right click and select **Rename**.

3-23

32. 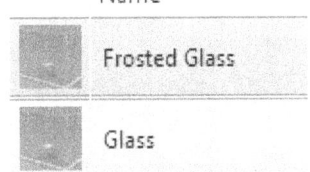 Rename **Frosted Glass**.

33.  Launch the Asset Browser.

34. Perform a search for frosted glass in the Asset Browser.

35. Highlight the Glass Appearance row. Select **Replace this asset in this material**.

36. Close the Asset Browser.

37. Select the Appearance tab.

    In the Information section:
    Change the Name to **Frosted - Blue**.
    In the Solid Glass Section:
    Change the Color to **Blue**.
    In the Relief Pattern section:
    Enable **Relief Pattern**.
    Set the Type to **Rippled**.
    In the Tint Section:
    Enable **Tint**.
    Change the Tint Color to **Cyan**.
    Press **OK**.

Component Families

38. 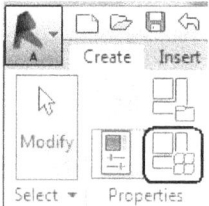 Next we create parameters to control the material settings.

Activate the **Create** ribbon.

Select **Family Types**.

39.
Door Panel Exterior	Mahogany
Door Panel Interior	Canvas Paint (R244,G236,B215)
Glazing	Frosted

We need to create parameters for these materials.

40. Under Parameters:

Select **Add**.

41. For Name, type **Panel Exterior**.
For Type of Parameter, select **Material**.
Group parameter under: **Materials and Finishes**.
Enable **Type**.

Press **OK**.

By default, the Type of Parameter is set to Length. If you go too fast, you will not set the correct parameter type and you will have to delete your parameter and do it over.

42. Under Materials and Finishes:
Locate the **Panel Exterior** parameter.

Click in the column to assign the material.

3-25

43. Select the **Mahogany** material.

    Press **OK**.

44. You should see the Mahogany material listed.

45. Under Parameters:

    Select **Add**.

46. For Name, type **Panel Interior**.
    For Type of Parameter, select **Material**.
    Group parameter under: **Materials and Finishes**.
    Enable **Type**.

    Press **OK**.

47. Under Materials and Finishes:
    Locate the **Panel Interior** parameter.

    Click in the column to assign the material.

48. Select the **Paint -Canvas** material.

    Press **OK**.

49. You should see the Paint-Canvas material listed.

Component Families

50. Under Parameters:

    Select **Add**.

51. For Name, type **Glazing**.
    For Type of Parameter, select **Material**.
    Group parameter under: **Materials and Finishes**.
    Enable **Type**.

    Press **OK**.

52. Under Materials and Finishes:
    Locate the **Glazing** parameter.
    Click in the column to assign the material.

53. Select the **Frosted Glass** material.

    Press **OK**.

54. You should see the Frosted Glass material listed.

    Press **OK**.

55. Rename the Front Elevation **Front-Exterior**.
    Rename the Back Elevation **Back-Interior.**

    I use both names in case a student doesn't rename their elevations, so I can troubleshoot their model.

56. Switch to the **Front-Exterior** Elevation view.

    The exterior door panel will be extruded from the center reference plane to the exterior reference plane.

3-27

57. Activate the Create ribbon.

    Select the **Set Work Plane** tool.

58. Set the Reference Plane is set to Center (Front/Back).

    Press **OK**.

59. Select **Extrusion**.

60. Select the **Rectangle** tool.

61. Draw a rectangle to fit into the opening.

Component Families

62. 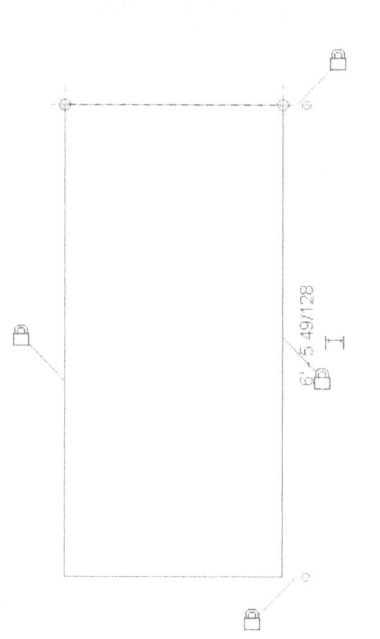 Enable all the locks.

    This will ensure that the door panel size is controlled by the reference planes.

    If you forget to lock the geometry, you can use the ALIGN tool to align the lines with the reference planes and lock each set.

63. Select the small button on the far right of the Material parameter.

64. Select the **Panel Exterior** material.

    Press **OK**.

65. Click the far right button the Extrusion End parameter.

66. Select Add parameter.

67. Enter **Exterior Panel Thickness**.

    Enable **Type**.

    Press **OK**.

3-29

68. 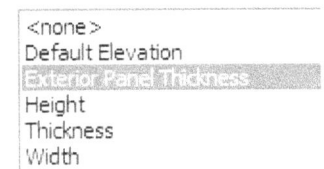 Select the **Exterior Panel Thickness** parameter.

    Press **OK**.

69.  Launch the **Family Types** dialog.

70. Locate the Exterior Panel Thickness parameter.

    In the Formula column:
    Type **Thickness/2**.

    Set the Thickness to 2".

    This will set the Exterior Panel Thickness to **1"**.

    Press **OK**.

71. In the Properties pane:

    Note that the Extrusion End is set to 1".

    It is grayed out because the value is controlled by a parameter.

72. 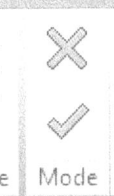 Green Check to finish the extrusion.

73.  Switch to View 1.
    Set the Visual Style to **Realistic**.
    Verify that the material displays correctly.

74. 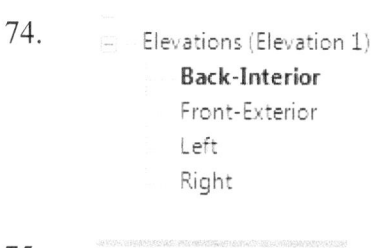 Switch to an **Interior** Elevation view.

75. Activate the Create ribbon.

    Select the **Set Work Plane** tool.

76. 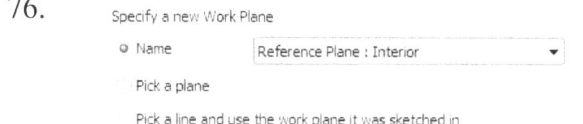 Set the Reference Plane to **Interior**.

    Press **OK**.

77. Select **Extrusion**.

78.  Select the **Rectangle** tool.

79. 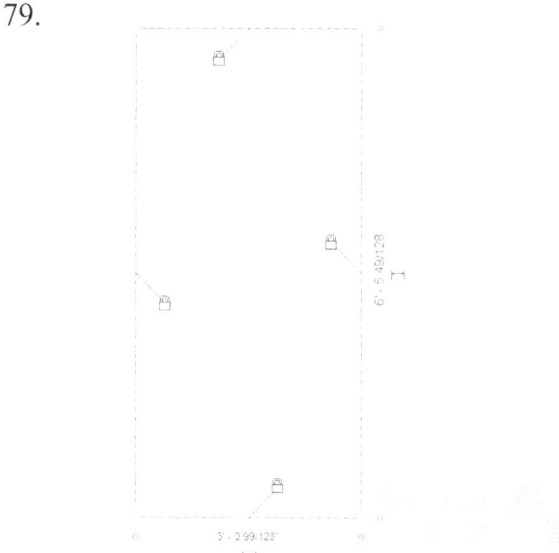 Draw a rectangle to fit into the opening.

    Enable all the locks to ensure that the extrusion is controlled by the reference planes.

3-31

80. 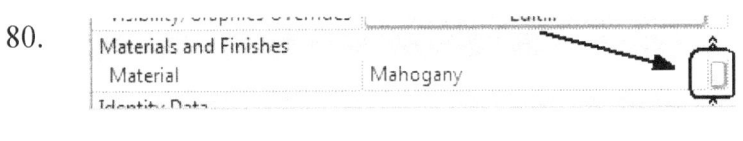 Select the small button on the far right of the Material parameter.

81. Select the **Door Panel Interior** parameter.

    This links the interior door panel to the material assigned to the Door Panel Exterior.

    Press **OK**.

82. 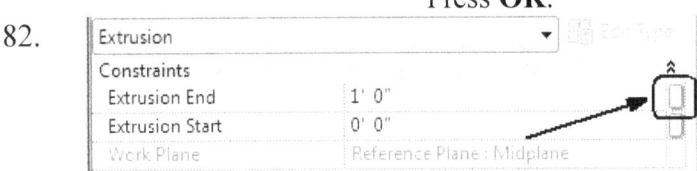 Click the far right button the Extrusion End parameter.

83. Select **Add parameter**.

84. Enter **Interior Panel Thickness**.

    Enable **Type**.

    Press **OK**.

85. Highlight Interior Panel Thickness.

    Press OK.

86. 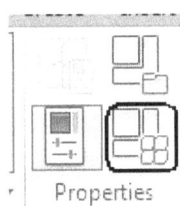 Launch the **Family Types** dialog.

87. Locate the Interior Panel Thickness parameter.

In the Formula column: Type **Thickness/2**.

This will set the Interior Panel Thickness to **1"**.

Press **OK**.

88. **Green Check** to finish the extrusion.

89. Switch to the **Ref. Level**.

90. You should see the two extrusions.
If you don't, they are probably on top of each other. Use the shape grips to move the panels to the correct positions.

91. Switch to **View 1**.

Verify that you see the two different materials.

92. Switch to an exterior elevation view.

3-33

93. 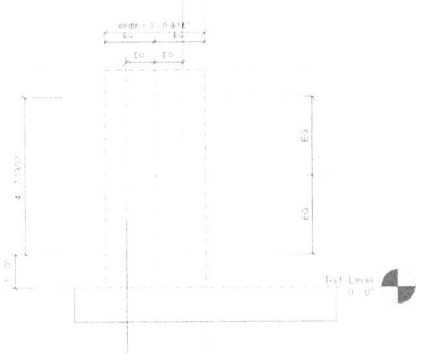 Add reference planes to define locations for the glazing.

    Remember to only dimension between reference planes. Do not select the extrusions!

    Use EQ dimensions to center the middle horizontal reference plane and middle vertical reference planes.

    The lower horizontal reference plane is 1'-0" above the Ref Level to define a kick plate area.

94. 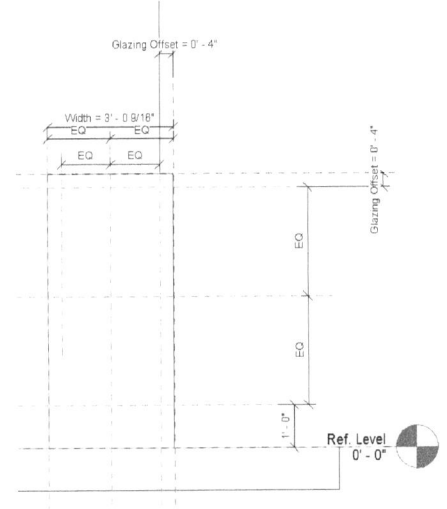 Add dimensions to offset the glazing 4" from the outside and the center.

    Remember that the temporary dimensions set the values for the permanent dimensions. To change the values of the permanent dimensions, select the reference plane and then modify the temporary dimension.

    Do not lock the dimensions or they won't flex properly.

    I created a parameter called Glazing Offset and used it to apply the 4" dimension.

95. 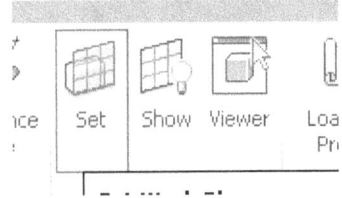 Activate the **Create** ribbon.

    Select the **Set Work Plane** tool.

96. 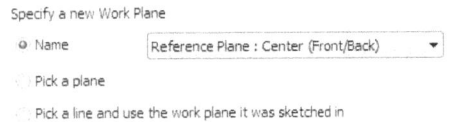 Set the Reference Plane to Center (Front/Back).

    Press **OK**.

Component Families

97. 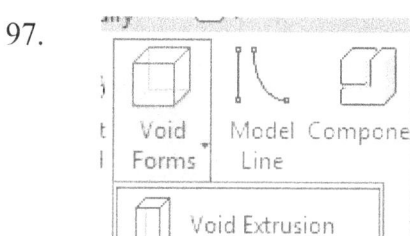 Select **Void Forms→Void Extrusion**.

    First we create an opening for the glass, then we place the glass.

98.

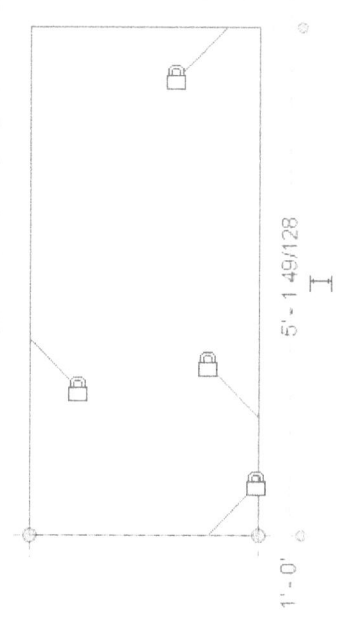

Draw a rectangle in the glazing outline.

Lock to the reference planes.

99. Set the Extrusion End to Thickness.
    Set the Extrusion Start to 0".

    Select the Green Check to finish the void.

3-35

100. Switch to View 1.
The void should go through the panel.

101. Switch to an **Interior** Elevation.

102. Activate the **Create** ribbon.

Select the **Set Work Plane** tool.

103. Set the Reference Plane to **MidPlane**.

Press **OK**.

104. Select the **Extrusion** tool.

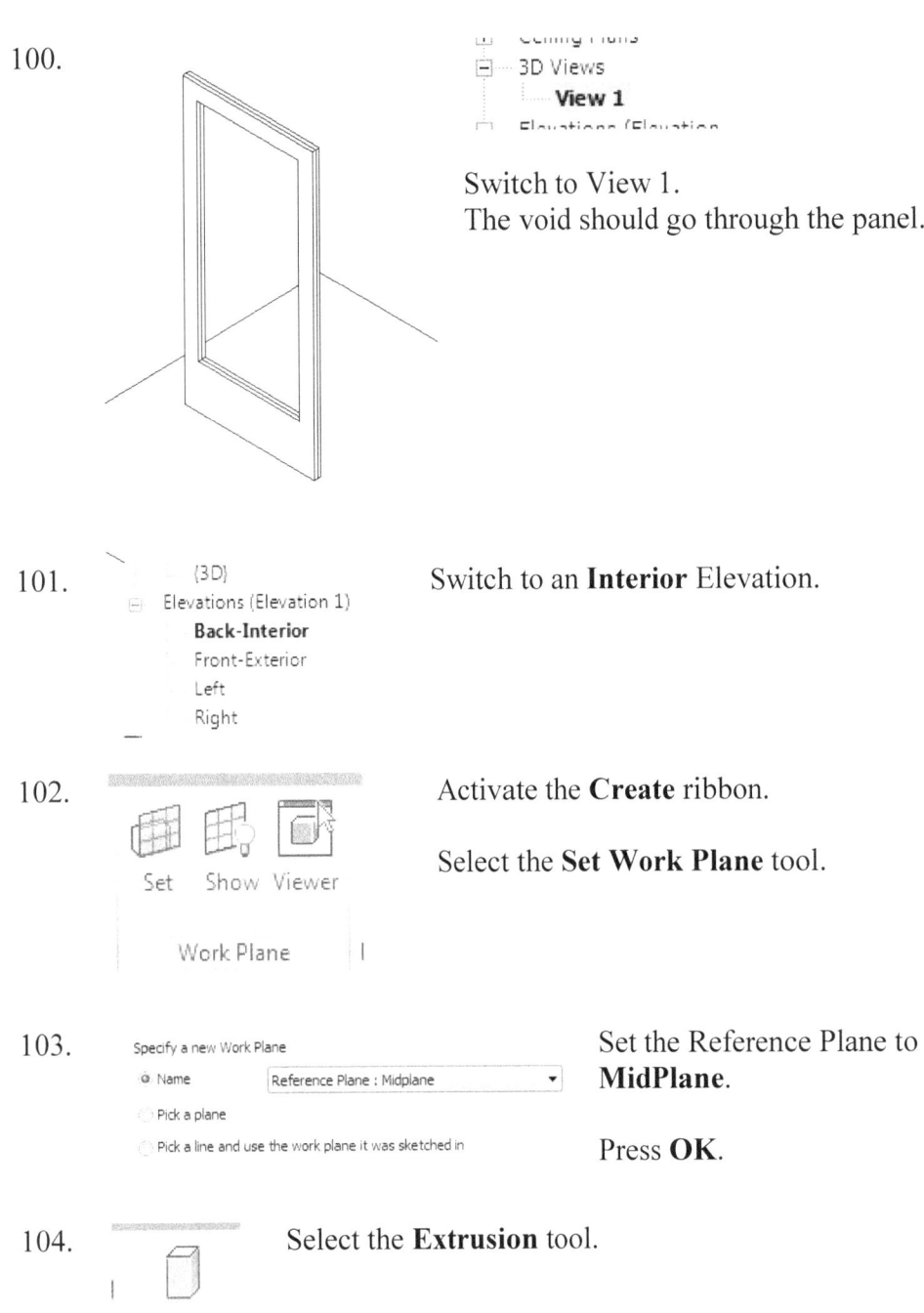

105. 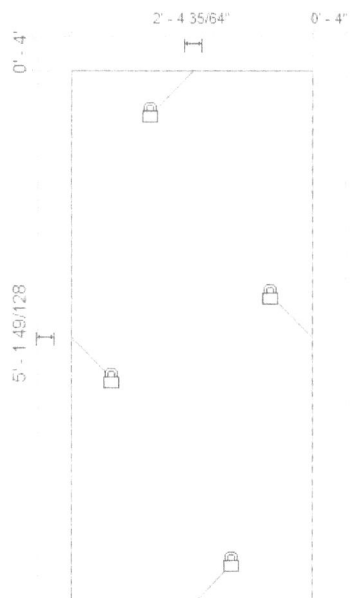 Draw a rectangle in the upper left area where a void is located.

Lock the sides to be constrained to the reference planes.

106. Set the Extrusion End to the **Interior Door Panel Thickness**.
Set the Extrusion Start to 0".

107. Set the Material to the **Glazing** parameter.

108. 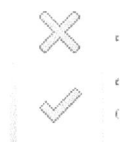 Select Green Check to finish the extrusion.

109. 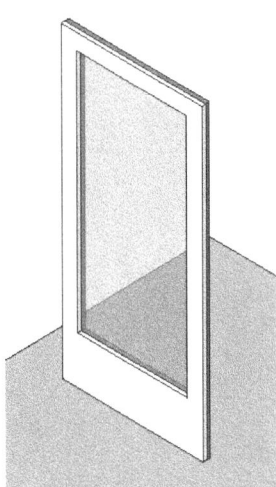 Switch to View 1 to inspect your door.

Save as **Door Panel Mahogany**.rfa.

Next we create the door family and insert the door panel.

Component Families

Command Exercise

Exercise 3-4 – Create a Nested Door Family

Drawing Name: **new**
Estimated Time to Completion: 1 hours 15 minutes

Scope: Create a door family
Insert a generic model into a door family
Apply constraints to control the generic model's position and size
Link the generic model parameters to the door family parameters

Solution:

1.  In order to use a door swing, the easiest way is to create the door panel as a generic model face based and insert it into a door family.

    When you insert a family into another family, it is considered a *nested family*.

    Go to **New** →**Family**.

2. Select the **Door** Family template.

3. Select the **Interior** Elevation view.

4. 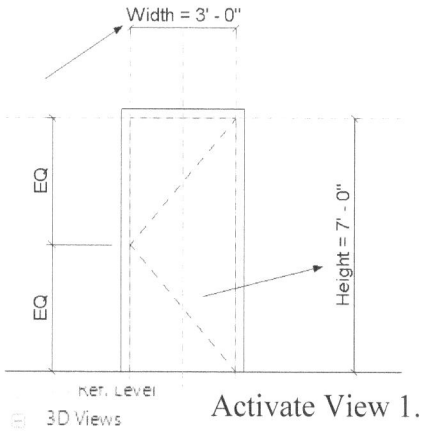 The template already has some parameters set up.

5. Activate View 1.

3-39

6. 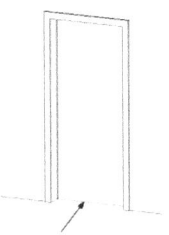 Select the rectangle.

7. 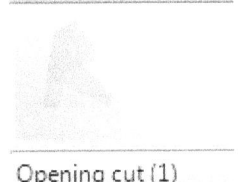 If you look at the Properties pane, you see this is the opening for the door, not the door panel.

8. 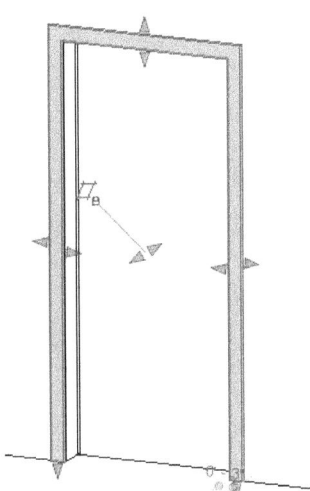 Select the extrusion surrounding the door.

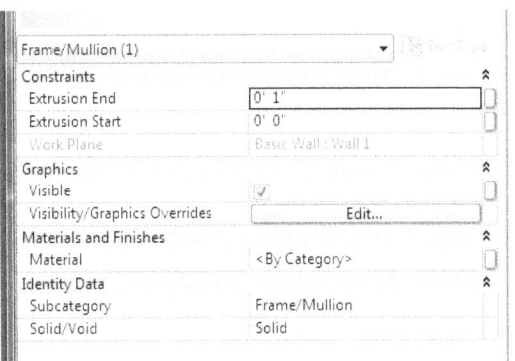

In the Properties pane, it is designated as Frame/Mullion.

9.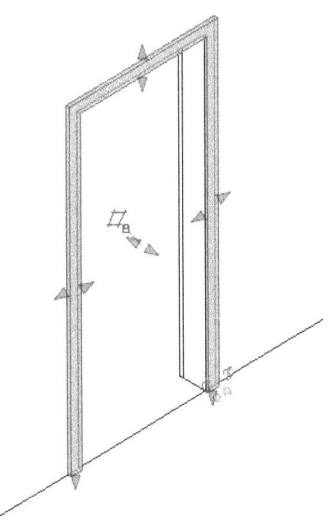

Switch the view to the other side.

Note that there is a frame/mullion already created for both sides.

Component Families

10. Select the **Family Types** tool.

11. Under Family Types:

    Select **New**.

    Family Types is where we define the different sizes of the door.

12. 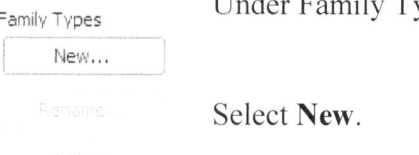 Enter the Door size.
    Press **OK**.

13. Enter the correct Height and Width.

    Press **Apply**.

    You should see the door opening flex in the display window.

14. Under Family Types:

    Select **New**.

15. 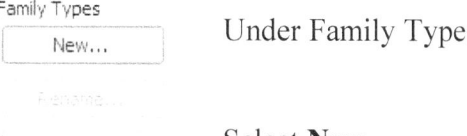 Enter the Door size.
    Press **OK**.

16. Enter the correct Height and Width.

    Press **Apply**.

    You should see the door opening flex in the display window.

17. Under Family Types:

    Select **New**.

3-41

18. Name: Exterior - 7' 2" x 3' 10"

    Enter the Door size.
    Press **OK**.

19. Dimensions
    Height   7' 2"
    Width   3' 10"
    Rough Width

    Enter the correct Height and Width.

    Press **Apply**.

    You should see the door opening flex in the display window.

20. Press **OK** to close the Family Types dialog.

21. Floor Plans
    Ref. Level

    Activate the **Ref Level** view.

22. Activate the **Create** ribbon.

    Select the **Reference Line** tool on the Datum panel.

    The reference line can be used to host inserted elements.

23. Draw an angled line above the door opening.

24. We need to align the ENDPOINT of the reference line with the horizontal plane and the left vertical plane.

    Select the **ALIGN** tool.

25. Select the horizontal plane then select the ENDPOINT of the reference line.

    Use the TAB key if needed to cycle through the selections.

26. Lock the alignment.

27. Select the Right reference plane and the ENDPOINT of the reference line to align.

    Use the TAB key if needed to cycle through the selections.

    You may need to move the reference line into position before setting the alignment lock.

Component Families

28.		Lock the alignment.
29.		Select the **Angular Dimension** tool.
30.	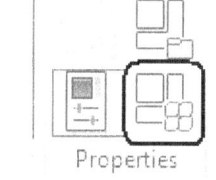	Place an angular dimension between the reference line and the exterior plane.
31.		Select the angular dimension and select Add parameter from the Options bar.
32.		Type **Door Swing** for the parameter name. Enable **Instance**.
		Users can control the door swing value if it is an instance value.
		Press **OK**.
33.		Select **Family Types**.
34.		Change the Door Swing to **90 degrees**. Press **Apply**.
		The reference line should flex.
		Try a few different angle values to verify that the reference line shifts properly.
		If the reference line does not flex properly, delete it and try to lock it into place again.
		Press **OK** to close the dialog.

3-43

35.  Activate the Insert ribbon.

    Select **Load Family.**

    Select the door panel family created at the start of the exercise.

36.  You should see the **Door Panel** in the Families area of the Project Browser.

37. 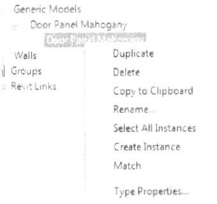 Right click on the Door Panel and select **Type Properties**.

38.  Click on the = next to **Width**.

39. 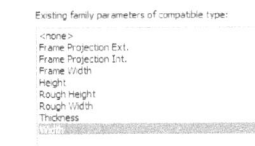 Select **Width**.

    Press **OK**.

    This links the panel width to the width of the door opening.

40.  Select the = next to **Height**.

41. 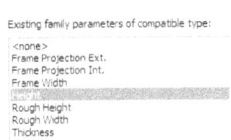 Select **Height**.

    Press **OK**.

42. 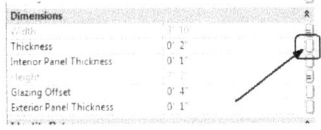 Select the = next to Thickness and link it to the Thickness parameter.

    Width, Thickness, and Height should all be grayed out.

| 43. |  | *The size of the door panel is now controlled by the Family Type parameters in the door family- the host family.* |

Press **OK**.

| 44. | 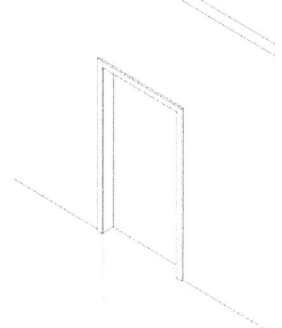 | Activate View 1. |

| 45. | | Orient the view so you are looking at the side of the door where the reference line controlling the door swing is located. |

| 46. |  | Activate the Create ribbon.

Select **Set Work Plane.** |

| 47. |  | Enable **Pick a plane**. |

| 48. | 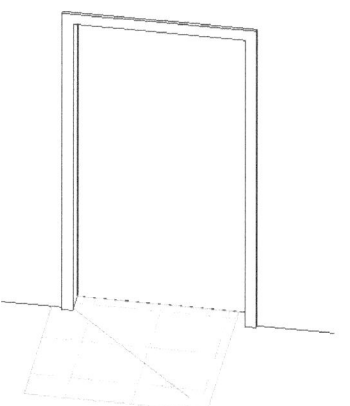 | Select the horizontal plane hosted by the reference line.

Use the TAB key to cycle through the selections until the horizontal plane is highlighted. |

49. 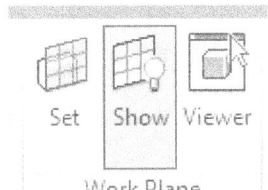 Select **Show Work Plane** to verify that the correct work plane is selected.

50. 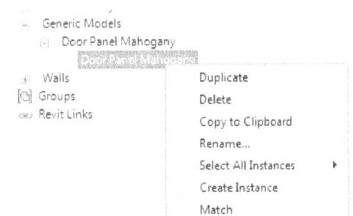 Right click on the Door Panel in the Project Browser.

    Select **Create Instance**.

51. 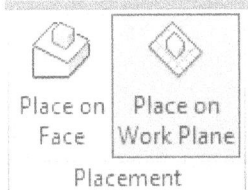 On the ribbon, select **Place on Work Plane**.

52. 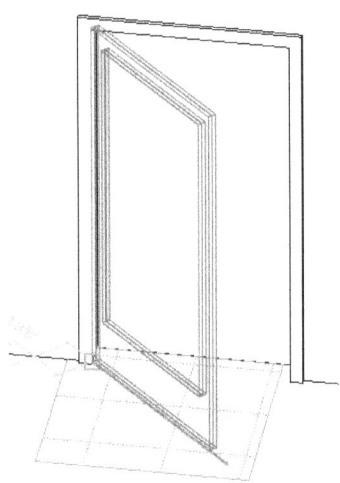 Press the SPACE BAR to rotate the door panel so it is aligned with the reference line.

    Locate the ENDPOINT of the reference line to insert the door. This ensures that the door stays constrained to that hinge point.

    Right click and select CANCEL to complete placing the door.

53. 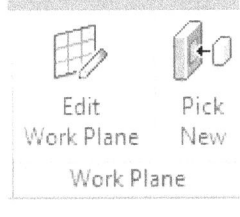 Select the door panel.
    Select **Edit Work Plane** on the ribbon.

54.  The door panel should be hosted by the Reference Line.

    Press **OK**.

Component Families

If the door panel is not hosted by the reference line, then delete the door panel and repeat the steps of setting the work plane and placing the door panel.

55. 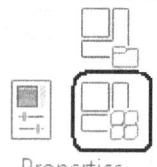 Select **Family Types**.

56. 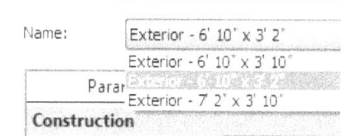 *Adjust the door swing and verify that the door moves.*

    Change the value of the door swing and press **Apply**.

    Do this for two to three values to test the family.

57. Flex the door sizes and verify that the door panel size changes.

    Press **OK**.

58. 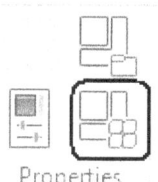 Launch the **Family Types** dialog.

59.  Under Parameters:

    Select **Add**.

60.  For Name, type **Exterior Mullion**.
    For Type of Parameter, select **Material**.
    Group parameter under: **Materials and Finishes**.
    Enable Type.

    Press **OK**.

3-47

61. Under Materials and Finishes:
Locate the **Exterior Mullion** parameter.
Click in the column to assign the material.

62. Select the **Mahogany** material.

Press **OK**.

63. You should see the Mahogany material listed.

64. Under Parameters:

Select **Add**.

65. For Name, type **Interior Mullion**.
For Type of Parameter, select **Material.**
Group parameter under: **Materials and Finishes**.
Enable **Type**.

Press **OK**.

66. Under Materials and Finishes:
Locate the **Interior Mullion** parameter.
Click in the column to assign the material.

67. Select the **Paint -Canvas** material.

Press **OK**.

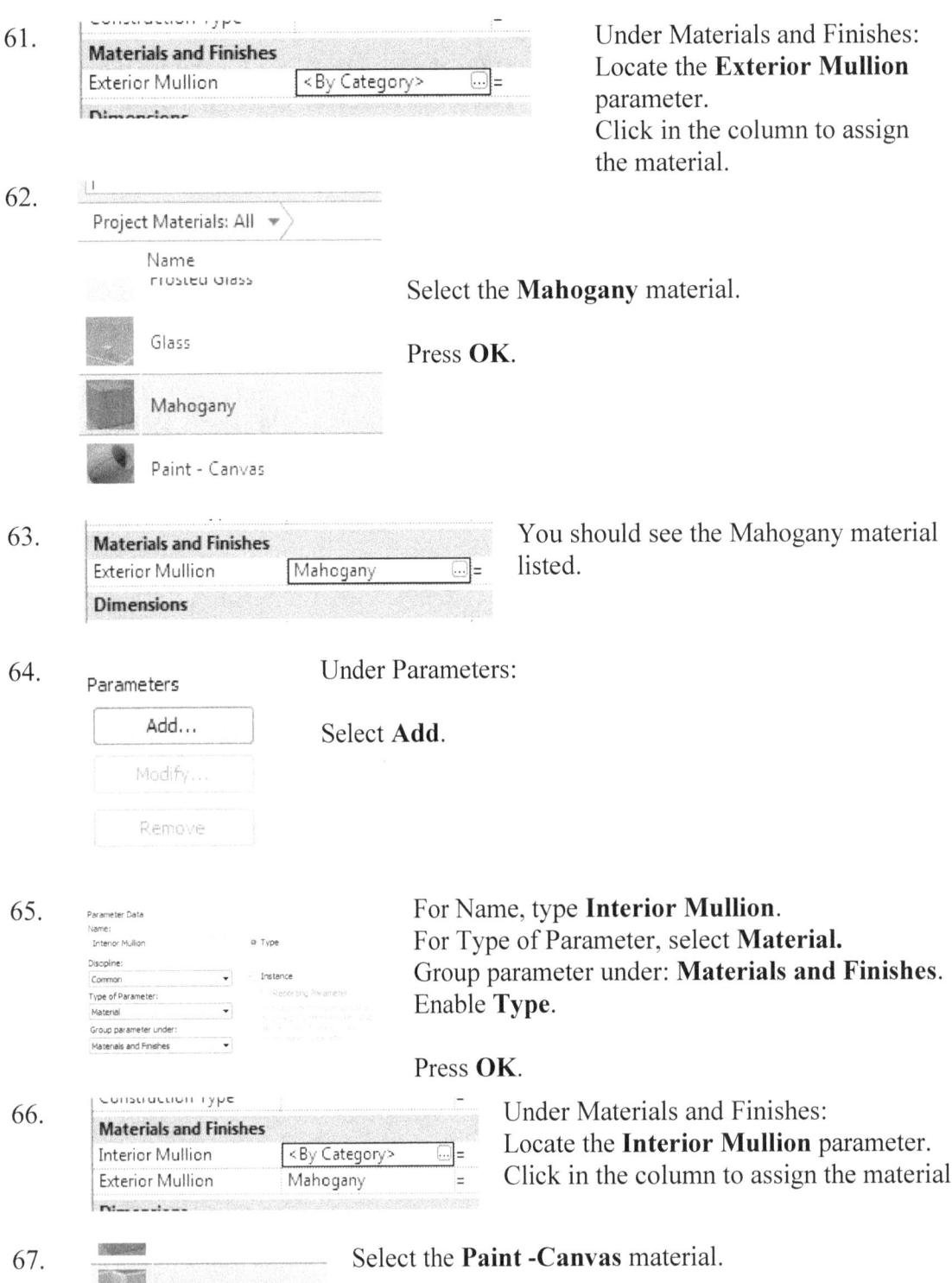

Component Families

68. You should see the Paint-Canvas material listed.
Switch to the other family type sizes and assign the materials.

Close the dialog.

69.  Select the Frame on the Exterior Side.

70. Set the Material to the **Exterior Mullion**.

    Press **OK**.

71.  Orbit the view to the interior side.

    Select the **Interior Mullion**.

72. Set the Material to the **Interior Mullion**.

    Press **OK**.

73. Materials have to be assigned for each family type. So select each family door size and verify the material has been assigned.
74. Save your model as *Door Exterior- Mahogany.rfa*.
75. Open a project file and insert the door and test it.

3-49

Command Exercise
Exercise 3-5 – Indentifying a Family

Drawing Name: **i_firestation_elem.rvt**
Estimated Time to Completion: 5 Minutes

Scope

Identify different elements and their families.

Solution

1. Open *i_firestation_elem.rvt*.

2. Activate the **South** elevation.

3. Select the second window from the left.

4. Select the **30" W** from the window type list in the Properties pane.

5. Use the Measure tool from the Quick Access toolbar to check the distance between the outside edges of the two left windows.

6. 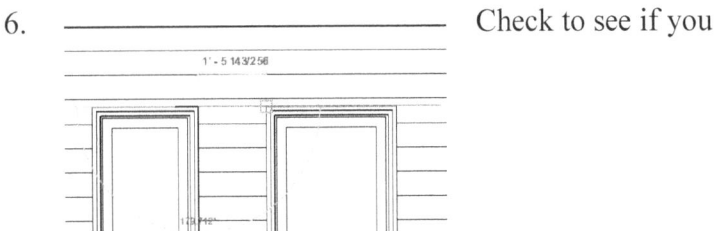 Check to see if you got the same value.

7. Close without saving.

Command Exercise

Exercise 3-6 – Assigning OMNI Class Numbers

Drawing Name: **Door-Exterior Mahogany.rvt, Park Bench**
Estimated Time to Completion: 10 Minutes

Scope

Assign an OMNI Class Number to your door family and park bench family

Solution

1. Open the **Door-Exterior Mahogany** family.

2. 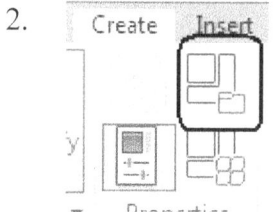 Select the **Family Categories** tool from the Create ribbon.

3. 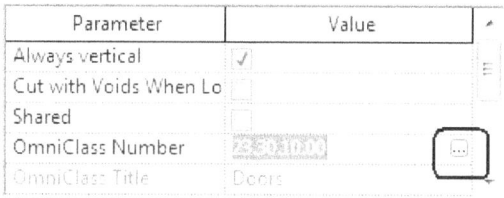 Click on the ... button in the Omni Class Number field.

   Left click in the column to access the ... button.

4. Locate the **Glazed Passage Doors** OMNI class.

   Highlight and press **OK**.

5. Family Parameters

Parameter	Value
Always vertical	✓
Cut with Voids When Lo	
Shared	
OmniClass Number	23.30.10.14.24
OmniClass Title	Glazed Passage Doors

The number updates and the OmniClass Title updates.

Press **OK**.

6. Save your model.

OMNI Classes are used for scheduling and keynotes.

Component Families

Command Exercise

Exercise 3-7 – Room Calculation Point

Drawing Name: **Door-Exterior Mahogany.rvt, Room Calculation**
Estimated Time to Completion: 10 Minutes

Scope

Enable Room Calculation Point for a Family

Solution

1. Open the **Door-Exterior Mahogany** family.

2. Activate the **Ref. Level** floor plan view.

3. Place a check for Room Calculation Point.

   This associates the placement of the door with a room.

4. You should see a squiggly line with two arrows.

   The direction of the arrows indicate which side of the door will be used to determine which room the door is placed in.

5. Open the Room Calculation project.

   There are three rooms and two doors.

6. Open the Door Schedule view.

7. You see the two doors listed and their associated rooms.

3-53

8. Return to the *Door – Exterior Mahogony* file.
Load into Project to add it to the Room Calculation project.

9.  Place the door in the center of the south exterior wall.

10. Note that when you select the door you can see the Room Calculation indicator.

11. Open the Door Schedule view.

12. Note how the schedule has updated.

13. Close without saving.

Practice Exam

1. If a level-based component is moved from Level 1 to Level 2 by changing the Element Properties, the new location is:
    A. At the origin.
    B. Corresponds to the location on the previous level.
    C. Where the user selects it to be.
    D. None of the above.

2. Families in the Project Browser are organized by:

    A. Instance, then Category, then Family
    B. Family, then Instance
    C. Category, then Family, then Type
    D. Family, then Type

3. A _____ is a group of elements with a common set of properties - called parameters - and a related graphical representation.

    A. view
    B. family
    C. project
    D. model
    E. mass

4. Select TWO characteristics you can specify for a family type:

    A. Category
    B. Materials
    C. Levels
    D. Views
    E. Dimensions

5. Select the THREE types of families:

    A. Datum
    B. View
    C. System
    D. Model
    E. Mass

6. To change or assign the OMNI class of a model family, select:

    A. Family Types
    B. Family Category and Parameters
    C. Properties
    D. Materials

7. To create a new system family you:

    A. Use a Revit family template
    B. Modify the instance parameters of an existing similar family
    C. Duplicate a similar system family and modify the type parameters
    D. Modify the type parameters of a similar system family.

8. In the Generic Model family template, you are provided with:

    A. Levels
    B. Reference Planes
    C. Grids
    D. Grid Guides

9. To create an opening in an extrusion when defining a model family, you need to place a:

    A. Opening
    B. Void
    C. Hole
    D. Cut

10. A(n) _____ family exists only in the project and can not be loaded into other projects.

    A. massing
    B. system
    C. loadable
    D. in-place
    E. model

Answers:
1) B; 2) C; 3) B; 4) B & E; 5) C, D, & E; 6) B or C; 7) C; 8) B; 9) B; 10) B

Lesson Four

View Properties

This lesson addresses the following User and Professional exam questions:

- View Properties
- Object Visibility Settings
- Section Views
- Elevation Views
- View Templates
- Scope Boxes

The Project Browser lists all the views available in the project. Any view can be dragged and dropped onto a sheet. Once a view is used or consumed on a sheet, it cannot be placed a second time on a sheet – even on a different sheet. Instead, you must create a duplicate view. You can create as many duplicate views as you like. Each duplicate view may have different annotations, line weight settings, detail levels, etc. Annotations are Userd to a view. If a view is deleted, any annotations are also deleted.

Revit has bidirectional associativity. This means that changes in one view are automatically reflected in all Userd views. For example, if you modify the dimensions or locations of a window in one view, the change is reflected in all the Userd views, including the 3D view.

You can control the appearance of Revit elements using Object Visibility Settings. These settings control line color, line type, and line weight. You can create templates which have different Object Visibility Settings for different project types.

Command Exercise

Exercise 4-1 – Creating a Level

Drawing Name: **i_levels.rvt**
Estimated Time to Completion: 5 Minutes

Scope

Placing a level.

Solution

1. Activate the **South Elevation**.

   The level names have been turned off.

2. 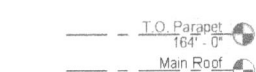 Select each level and place a check in the square that appears. This will turn on visibility of the level name.

3. You should be able to identify the names for each level.

   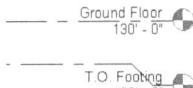

4. Select the **Level** tool from the Architecture ribbon.

5. 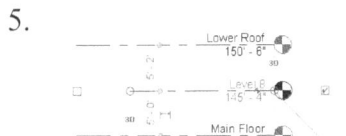 Place a level **5'-0"** above the Main Floor.

6.  Note the elevation value for the new level.

7. Close without saving.

Command Exercise

Exercise 4-2 – Story vs. Non-Story Levels

Drawing Name: **story_levels.rvt**
Estimated Time to Completion: 15 Minutes

Scope

Understanding the difference between story and non-story levels
Converting a non-story level to a story level.

Solution

1. Activate the **South Elevation**.

2. Select each level and place a check in the square that appears. This will turn on visibility of the level name.

3. Study the Main Floor level.

   Notice that it is the color black while all the other levels are blue.

   The Main Floor level is a non-story or reference level. It does not have a view associated with it.

4. Activate the Architecture ribbon.

   Select the **Level** tool on the Datum panel.

5. On the Options bar:
   Uncheck **Make Plan View**.
   Set the Offset to **8' 0"**.

6.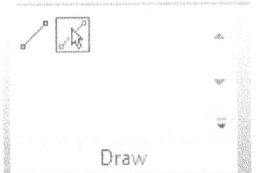

   Select the **Pick** tool on the Draw panel.

View Properties

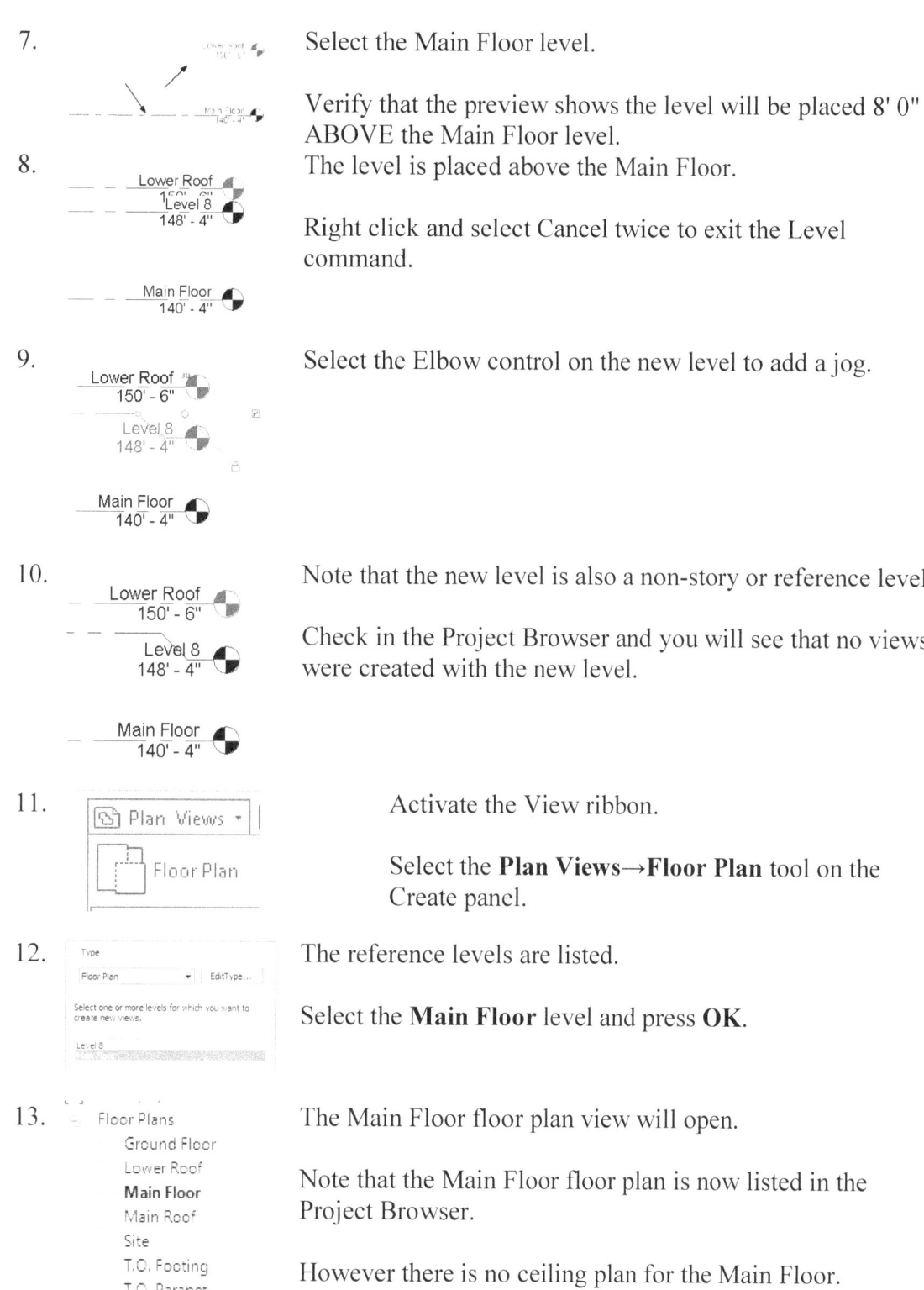

7. Select the Main Floor level.

Verify that the preview shows the level will be placed 8' 0" ABOVE the Main Floor level.

8. The level is placed above the Main Floor.

Right click and select Cancel twice to exit the Level command.

9. Select the Elbow control on the new level to add a jog.

10. Note that the new level is also a non-story or reference level.

Check in the Project Browser and you will see that no views were created with the new level.

11. Activate the View ribbon.

Select the **Plan Views→Floor Plan** tool on the Create panel.

12. The reference levels are listed.

Select the **Main Floor** level and press **OK**.

13. The Main Floor floor plan view will open.

Note that the Main Floor floor plan is now listed in the Project Browser.

However there is no ceiling plan for the Main Floor.

4-5

14. Activate the View ribbon.

    Select the **Plan Views→Reflected Ceiling Plan** tool on the Create panel.

15. The reference levels are listed.

    Select the **Main Floor** level and press **OK**.

16. The Main Floor ceiling plan view will open.

17. Activate the **South Elevation**.

18. The Main Floor level is now the color Blue to indicate it has Userd views.

19. Close the file without saving.

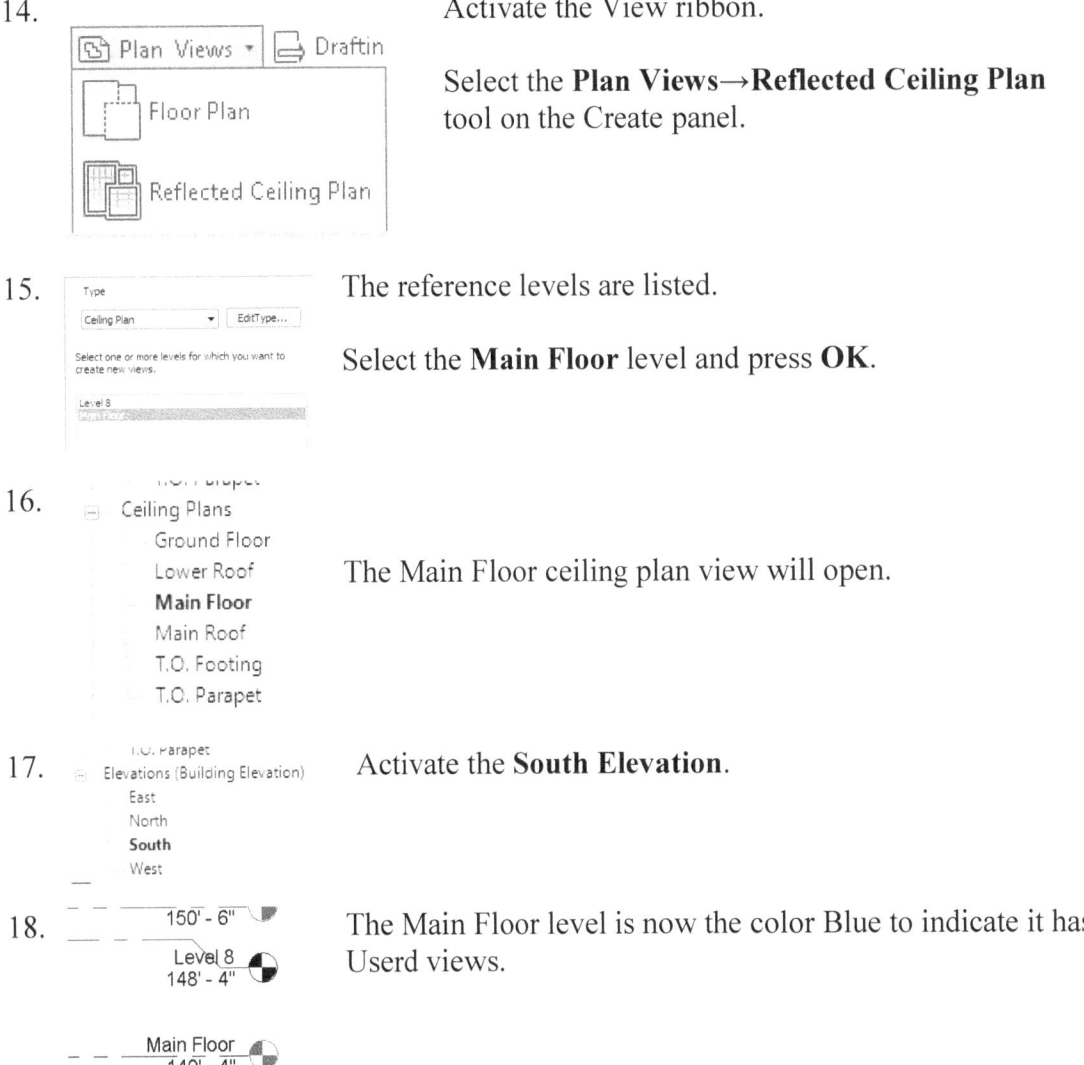

Command Exercise

Exercise 4-3 – Creating Column Grids

Drawing Name: **i_grids.rvt**
Estimated Time to Completion: 10 Minutes

Scope

Placing column grids.

Solution

1. Activate the **Ground Floor** view.

2. Zoom into the building area displayed.

3. Select the **Grid** tool on the Datum panel from the Architecture ribbon.

4. Select the **Pick Lines** mode from the Draw panel.

5. 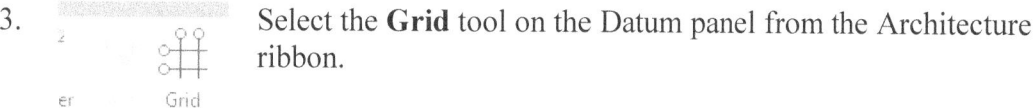 Set the Offset to **2'-0" [600 mm]** on the Options bar.

6. 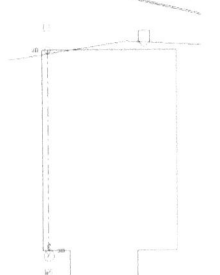 Click to place a vertical grid line as shown.

7. 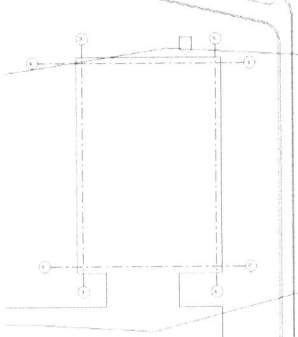 Place gridlines as shown.

   Use the grips by the heads to drag the grid bubbles into position.

8. 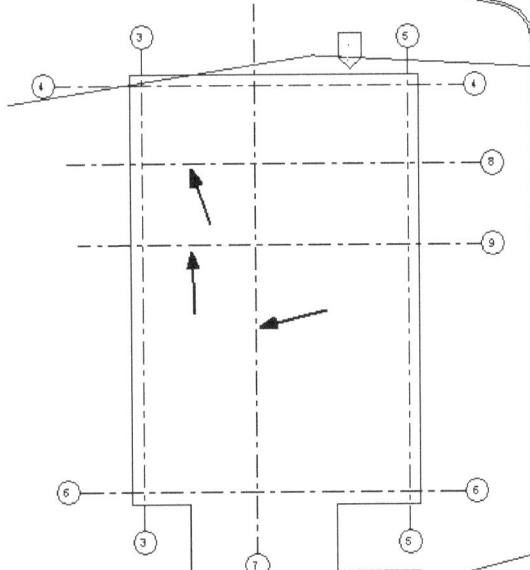 Switch to draw grid mode.

   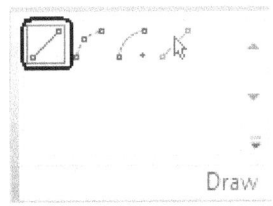

   Add a vertical grid line as shown.

   Add two horizontal grid lines as shown.

9. 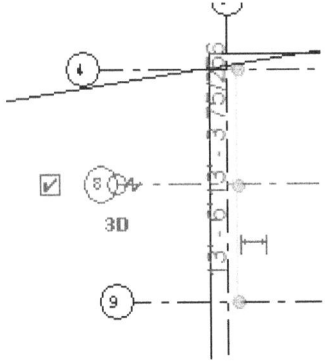 Place a check on both rectangles to make bubbles visible on both sides of the grid line.

10. Label the horizontal grid lines A-D.
The alpha labeled grids are incremented from bottom to top.

Label the vertical grid lines 1-3.
The number labeled grids are incremented from left to right.

Enable the grid bubbles on both ends.

Turn off the visibility of elevations.

11. Activate the **Annotate** ribbon.

Select the **ALIGNED** dimension tool from the Dimension panel.

12. Place a multi-segmented dimension between the three vertical grid lines.

Enable the EQ toggle.

13. 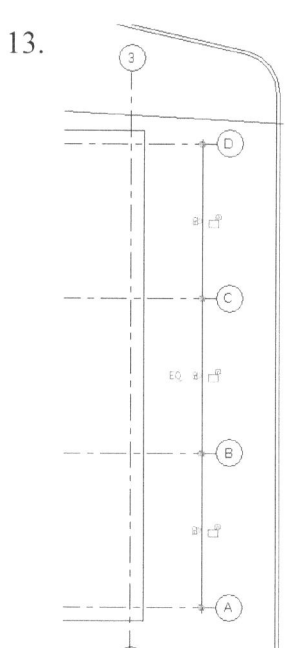 Place a multi-segmented dimension between the horizontal grid lines.

Enable the EQ toggle.

To place a continuous dimension, select each grid line in order, then left click above (if the last grid line is the top one) or below (if the last grid line is the bottom one).

Exit the command.

14. Select the vertical dimensions.
    Right click and toggle **EQ Display** off.

15. The dimension is now visible.

16. Select the horizontal dimensions.
    Right click and toggle **EQ Display** off.

    The horizontal dimensions are now visible.

17. Activate the Architecture ribbon.

    Select the **Structural Column** tool from the Build panel.

18. Select the **24 x 24 Concrete Square Column [600 x 600 mm]** from the Type Selector list on the Properties pane.

19. Enable the **At Grids** option to place columns at grid intersections.

20. Hold down the CONTROL key.
    Select each grid line.

    A column will be placed at each grid intersection.

21.  Select the Green Check when you see a column at every grid intersection.

Right click and select CANCEL to exit the command.

22. 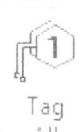 Activate the Annotate ribbon.

Select **Tag All** from the Tag panel.

23.  Highlight the **Structural Column Tag**.

Press **OK**.

24. 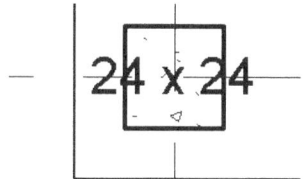 Zoom into a column to read the tag.

The column is labeled with the Column Type.

25. 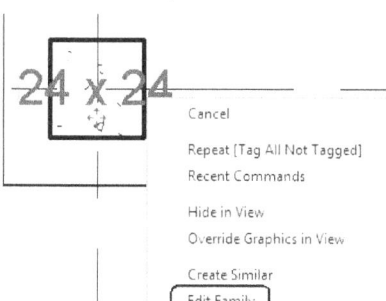 Select one of the structural column tags.

Right click and select **Edit Family**.

Make sure you select the tag and not the column.

26. Go to **File→Save As→Family**.

27. Rename the tag *Structural Column Location Tag.rfa*.

28. 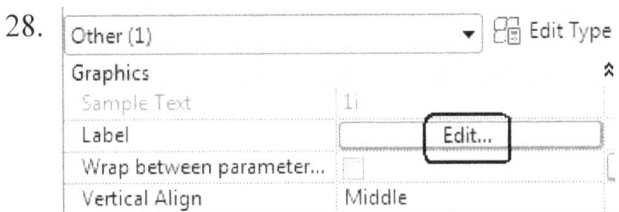 Select the text.

    Select the **Edit** button in the Label field on the Properties pane.

29. Use the Add and Remove tools in the middle of the dialog to remove the Type Name and add the Column Location Mark.

    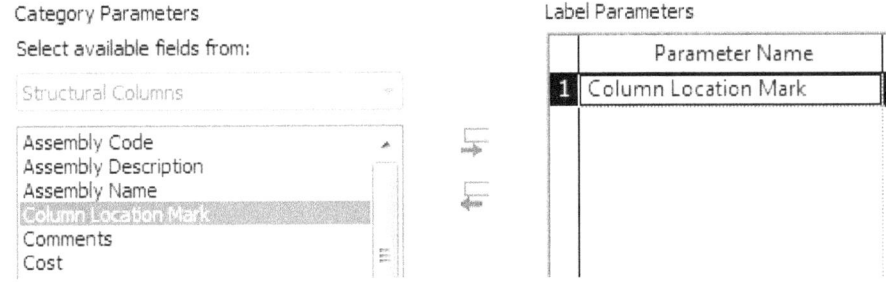

30. Press **OK**.

31. Save the file.

32. Select the **Load into Project** tool.

33. 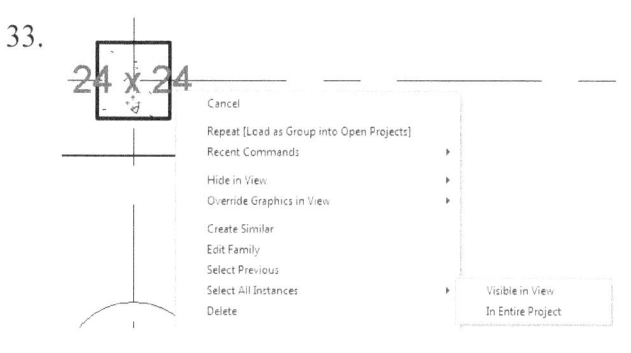 Select one of the column tags.

Right click and select **Select All Instances→ Visible in View**.

34. 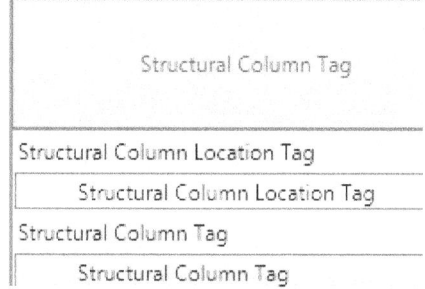 Select the **Structural Column Location Tag** from the Type Selector on the Properties pane.

35. The tags update with the location.

36. Close without saving.

Command Exercise

Exercise 4-4 – Setting View Depth

Drawing Name: **i_firestation_basic_plan.rvt**
Estimated Time to Completion: 5 Minutes

Scope

Determine the view depth of a view

Solution

1. Activate the **Site** view.

2.  In the Properties pane:

   Scroll down to **View Range** located under the Extents category.

   Select the **Edit** button.

3. Determine the **View Depth**.

4. Press **OK**.

5. Close without saving.

Command Exercise

Exercise 4-5 – Create a Cropped View

Drawing Name: **i_firestation_managing_views.rvt**
Estimated Time to Completion: 10 Minutes

Scope

Create a cropped view

Solution

1. Activate the **Main Floor – Furniture Plan** view.

2. In the Properties Pane:

   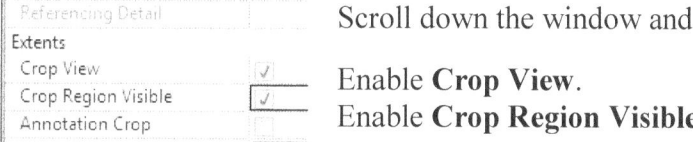

   Scroll down the window and:

   Enable **Crop View**.
   Enable **Crop Region Visible**.

4.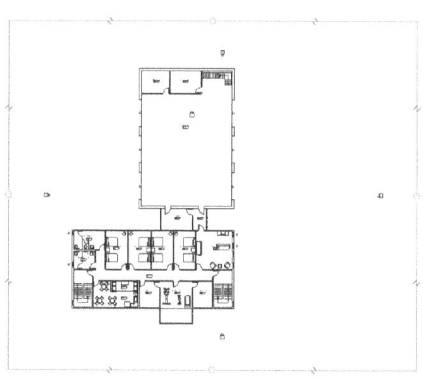

   Zoom out.

   Select the viewport rectangle.

4-15

5. 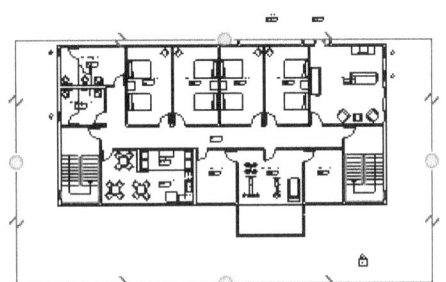   Use the bubbles to position the viewport so that only the furniture floor plan is visible.

6.

   Select **Hide Crop Region** using the tool in the View Control bar.
   Press **OK**.

7. Close without saving.

View Properties

Command Exercise
Exercise 4-6 – Change View Display

Drawing Name: **i_firestation_managing_views.rvt**
Estimated Time to Completion: 15 Minutes

Scope

*Use Temporary Hide/Isolate to control visibility of elements.
Change Line Width Display
Change Object Display Settings*

Solution

1. Activate the **Main Floor** floor plan.

   Floor Plans
   Ground Floor
   Lower Roof
   Main Floor
   Main Floor- Furniture Plan
   Main Roof
   Site

2. Select one of the exterior walls so it is highlighted.

3. 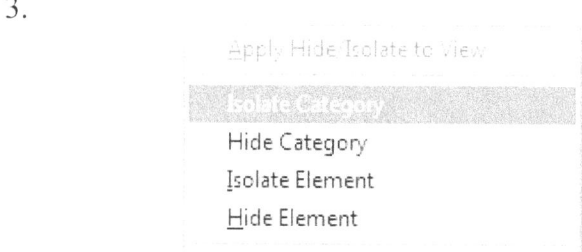 Select the **Temporary Hide/Isolate** tool.
   Right click and select **Isolate Category**.

   Apply Hide/Isolate to View
   Isolate Category
   Hide Category
   Isolate Element
   Hide Element
   Reset Temporary Hide/Isolate

4-17

4.	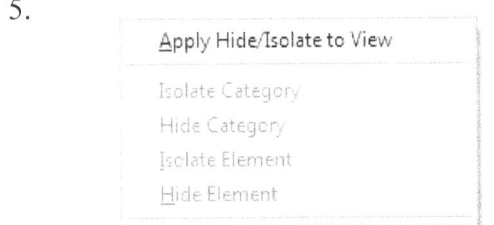	Only the exterior walls are visible.
5.	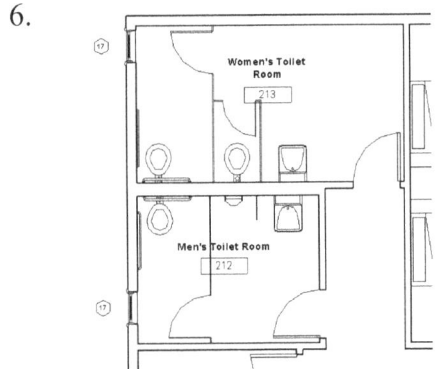	Select the **Temporary Hide/Isolate** tool. Right click and select **Reset Temporary Hide/Isolate**.
6.		Zoom into the region where the lavatories are located.

7. 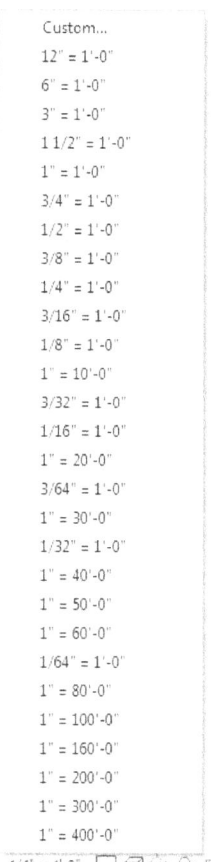 Change the view scale to **1/8″ = 1′-0″**.

8. 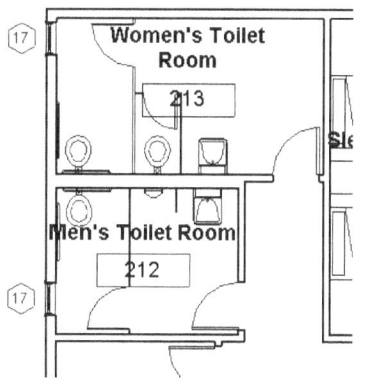 Note that the room tags scale to the view.

9. Activate the **Modify** ribbon.

10. 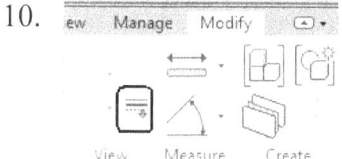 Select the **Linework** tool on the View panel.

11. Set the Line Style to **Overhead**.

12. 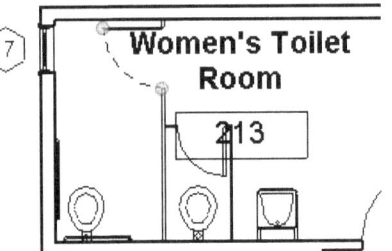 Select the door swing on the toilet cubicle.

    Note that the door swing's appearance changes.

13. Activate **Section 1**.

14. Activate the Manage ribbon.

    Select **Settings→Object Styles**.

15. Expand the **Doors** category on the Model Objects tab.
    Change the Line Weight, Line Color, and Line Pattern for the Panel and Frame.
    Press **Apply** to see the changes.

16. 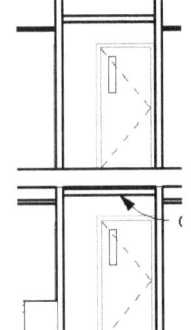 You can move the dialog over so you can see how the display is changed.

    Press OK to close the dialog.

    Note that linework changes are specific to the view, but object settings changes affect all views.

17. Close without saving.

Command Exercise

Exercise 4-7 – Reveal Hidden Elements

Drawing Name: **i_visibility.rvt**
Estimated Time to Completion: 5 Minutes

Scope

Turn on the display of hidden elements

Solution

1. Activate **the Ground Floor Admin Wing** floor plan.

2. Select the **Reveal Hidden Elements** tool.

3. Items highlighted in magenta are hidden.

   Window around the two tables while holding down the CONTROL key to select them.

4. Select **Unhide element** from the ribbon.

5. The tables will no longer be displayed as magenta (hidden elements).

6. Select the **Close Hidden Elements** tool.

7. The view will be restored.

8. Select the **Measure** tool from the Quick Access toolbar.

9. Determine the distance between the center of the two tables.

   Did you get 48' 2"?

   If you didn't get that measurement, check that you selected the midpoint or center of the two tables.

10. Close without saving.

Command Exercise

Exercise 4-8 – Create a View Template

Drawing Name: **view_templates.rvt**
Estimated Time to Completion: 30 Minutes

Scope

Apply a wall tag
Create a view template
Create a view filter
Apply view settings to a view

Solution

1. Activate Level 1.

2. Activate the **Annotate** ribbon.

3. Select **Tag All**.

4. 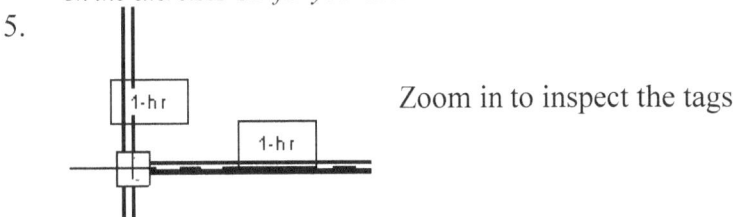 Highlight the **Wall tag - fire rating** as the tag to be used and press **OK**.

   The wall tag - fire rating is a custom family. It was pre-loaded into this exercise, but is included on the exercises CD for your use.

5. Zoom in to inspect the tags.

View Properties

6.  In the Properties pane:

   Click on the **<None>** button next to View Template.

7. 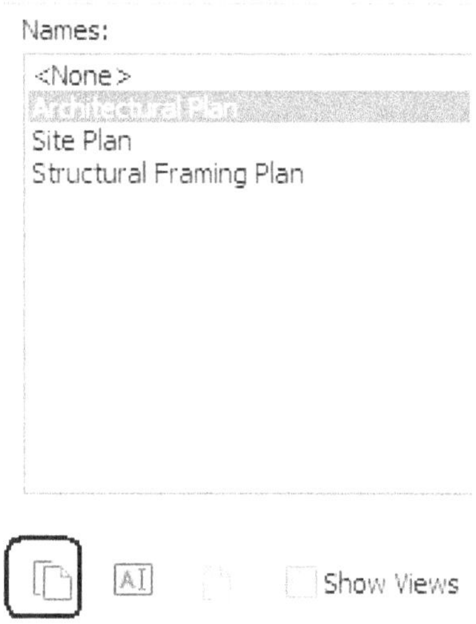 Highlight **Architectural Plan** and select the **Duplicate** button at the bottom of the dialog.

8.  Type **Fire Rating** in the Name field.

   Press **OK**.

9. 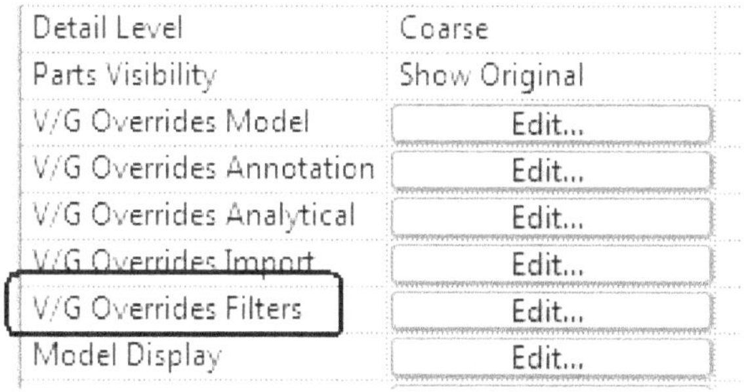 Select the **Edit** button next to **V/G Overrides Filters**.

10.  Select **Edit/New** at the bottom of the dialog.

11.  Select **New**.

12. 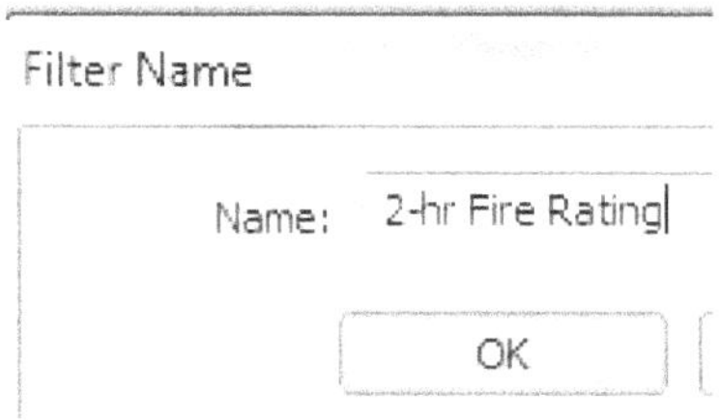 Type **2-hr Fire Rating** in the Name field.

    Press **OK**.

4-23

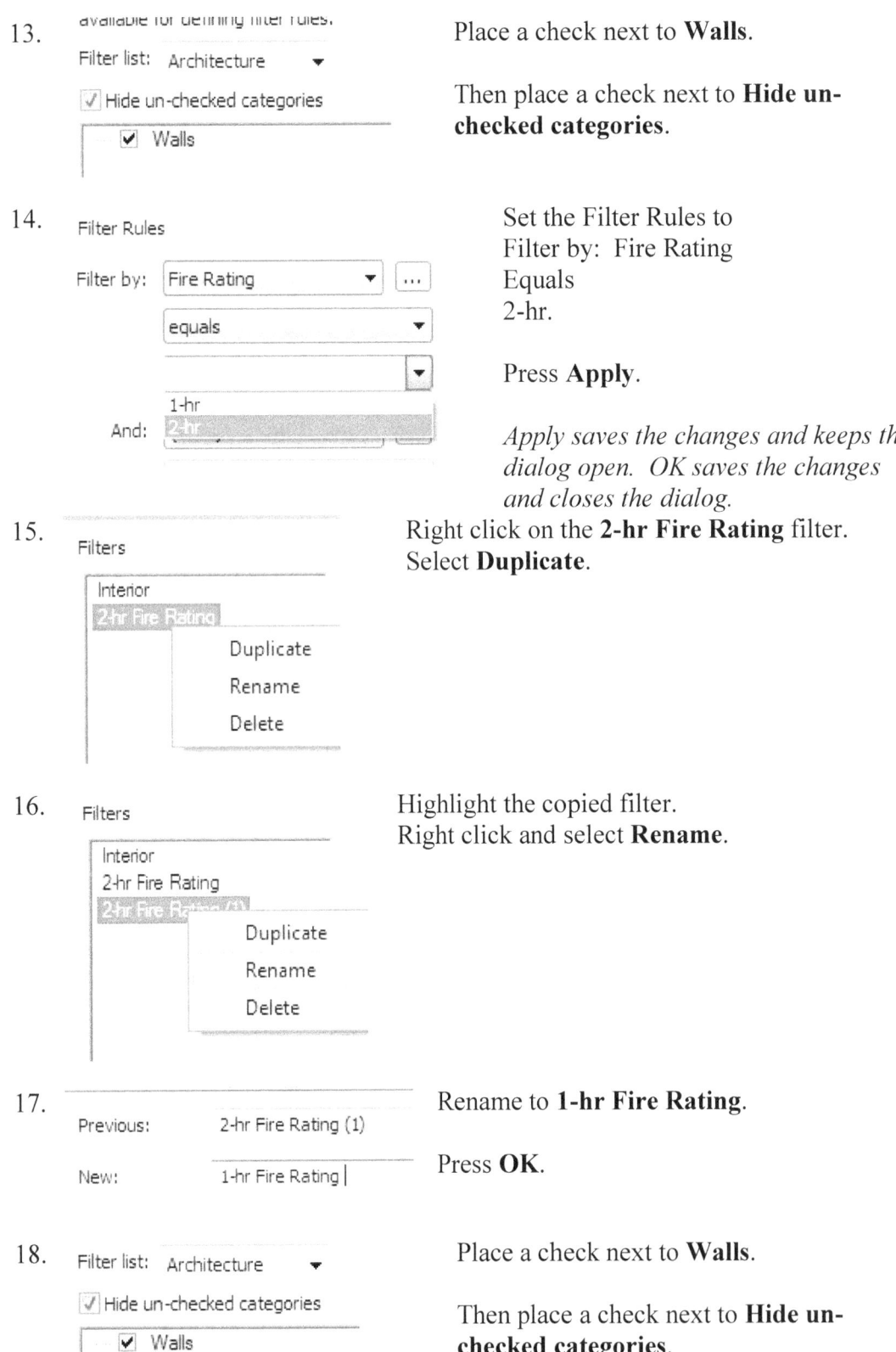

13. Place a check next to **Walls**.

    Then place a check next to **Hide un-checked categories**.

14. Set the Filter Rules to
    Filter by: Fire Rating
    Equals
    2-hr.

    Press **Apply**.

    Apply saves the changes and keeps the dialog open. OK saves the changes and closes the dialog.

15. Right click on the **2-hr Fire Rating** filter. Select **Duplicate**.

16. Highlight the copied filter. Right click and select **Rename**.

17. Rename to **1-hr Fire Rating**.

    Press **OK**.

18. Place a check next to **Walls**.

    Then place a check next to **Hide un-checked categories**.

19.  Set the Filter Rules to
Filter by: Fire Rating
Equals
1-hr.

Press **Apply**.

Press **OK**.

20. Select the **Add** button at the bottom of the Filters tab.

21. 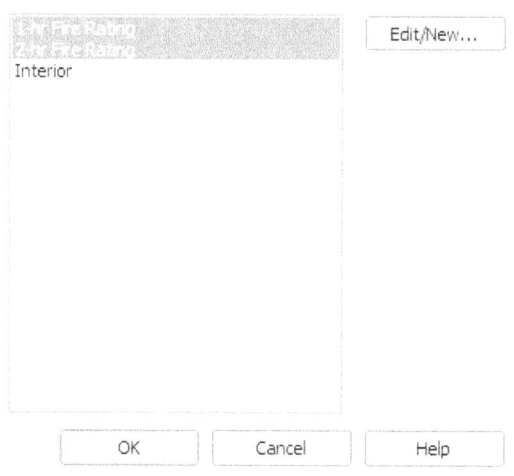 Hold down the Control key. Highlight the 1-hr and 2-hr fire rating filters and press **OK**.

22. You should see the two fire rating filters listed.

If the interior filter was accidentally added, simply highlight it and select Remove to delete it.

23. Highlight the **2-hr Fire Rating** filter.

24. Select the **Pattern Override** under Projection/Surface.

25. Select the **2 Hour** fill pattern.

Press **OK**.

26. Set the Color to **Blue.**

Press **OK.**

27. Highlight the **1-hr Fire Rating** filter

28. Select the **Pattern Override** under Projection/Surface.

29. Set the fill pattern to **1 Hour** and the Color to **Magenta**.

Press **OK**.

30. Apply the same settings to the Cut Overrides.

Press **OK**.

View Properties

31. Highlight the **Fire Rating** View Template.

    Press **Apply**.

    Press **OK**.

32. The view updates.

33. Activate **Level 2**.

34. In the Properties pane:

    Click on the **<None>** button next to View Template.

35. Highlight the **Fire Rating** View Template.

    Press **Apply**.

    Press **OK**.

36. Activate **Level 3**.

37. In the Properties pane:

    Click on the **<None>** button next to View Template.

38. Highlight the **Fire Rating** View Template.

    Press **Apply**.

    Press **OK**.

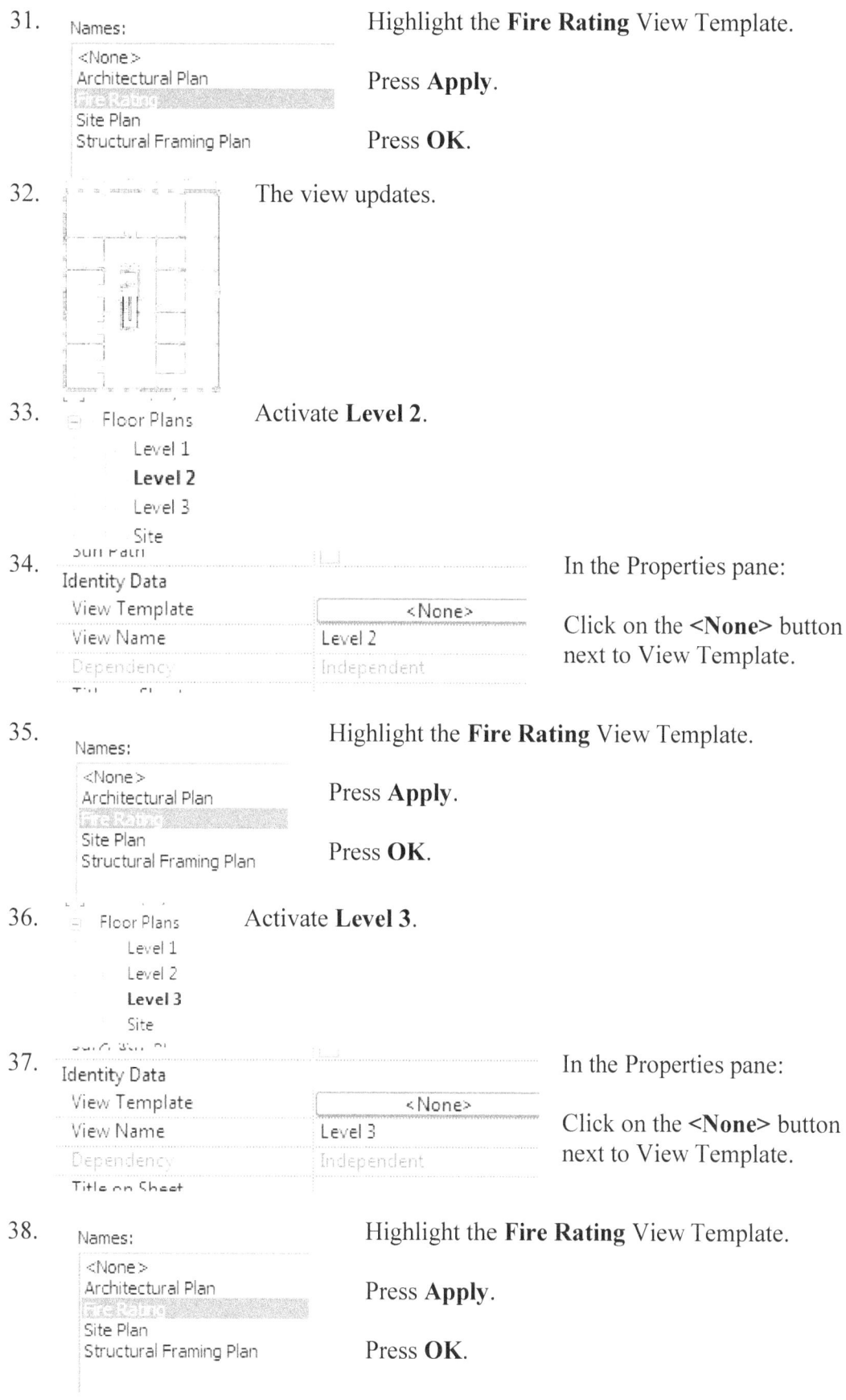

4-27

39. Zoom in to see the hatch patterns.

40. 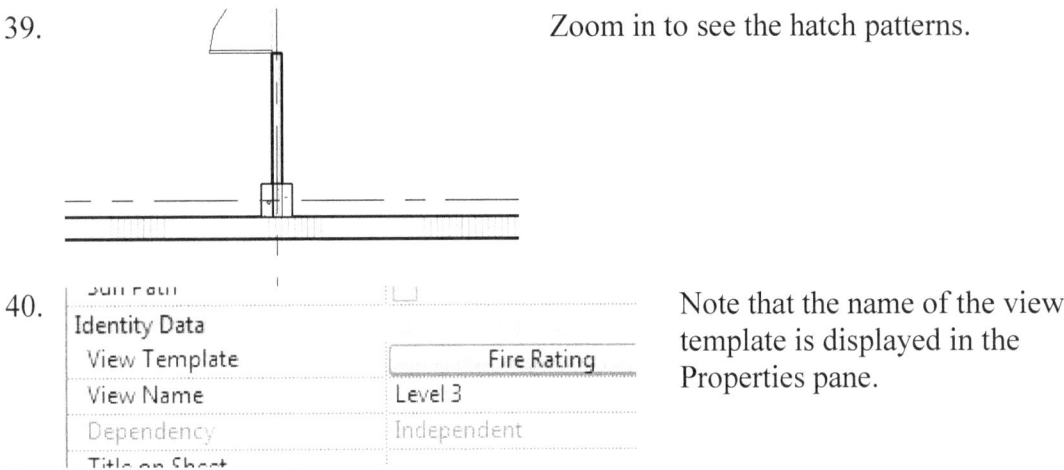 Note that the name of the view template is displayed in the Properties pane.

41. Close without saving.

The 1-hr, 2-hr, and 3-hr hatch patterns are custom fill patterns provided inside this exercise file. The *.pat files are included on the exercise CD for your use.

Command Exercise

Exercise 4-9 – Create a Scope Box

Drawing Name: **i_scope_box.rvt**
Estimated Time to Completion: 15 Minutes

Scope

Create and apply a scope box.
Scope boxes are used to control the visibility of grid lines and levels in views.

Solution

1. Activate the **Level 1** floor plan.

2. Activate the **Architecture** ribbon.
   Select the **Grid** tool from the Datum panel.

3. Set the Offset to **2' 0"** on the Options bar.

4. Select the **Pick Lines** tool from the Draw panel.

5. Place three grid lines using the exterior side of the walls to offset.

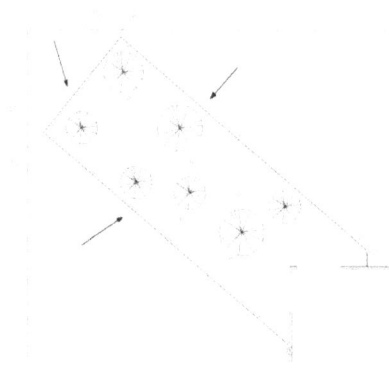

6. 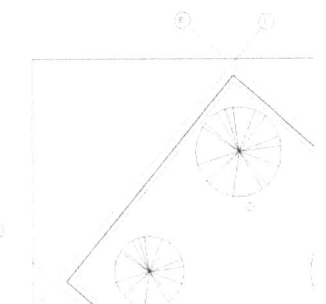 Re-label the grid bubbles so that the two long grid lines are A and B and the short grid line is 1.

7. Select the **Measure** tool from the Quick Access toolbar.

8. 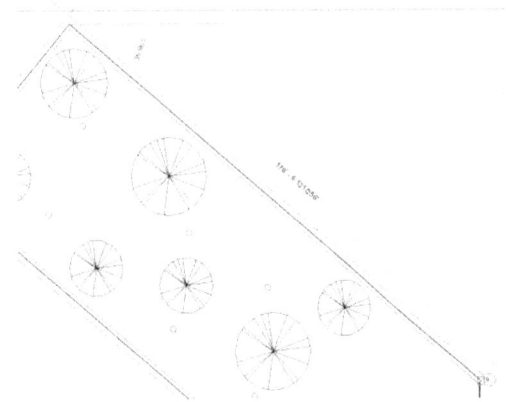 Measure one side of the yard area.

9. 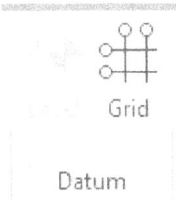 Activate the **Architecture** ribbon.

    Select the **Grid** tool from the Datum panel.

10. Offset: 180' 0"   Set the Offset to **180' 0"** on the Options bar.

11. 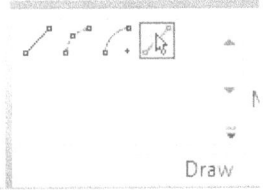 Select the **Pick Lines** tool from the Draw panel.

View Properties

12. 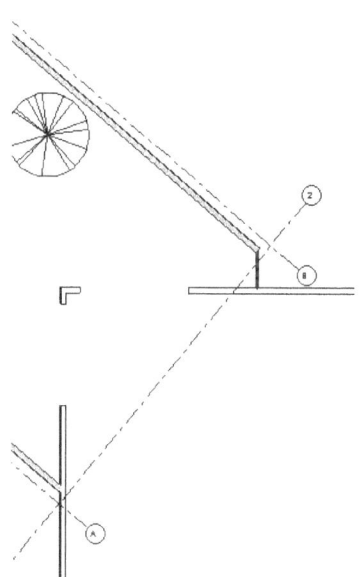 Place the lower grid line by selecting the upper wall and offsetting 180′.

Re-label the grid line **2**.

13. Activate the **View** ribbon.

Select the **Scope Box** tool from the Create panel.

14. 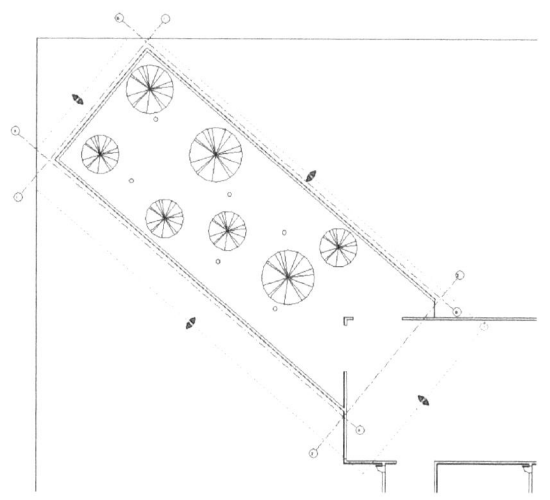 Place the scope box.

Use the Rotate icon on the corner to rotate the scope box into position.

Use the blue grips to control the size of the scope box.

15. Select the grid line labeled **B**.

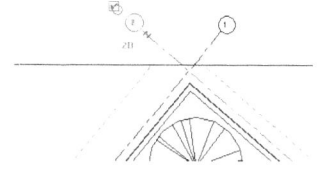

4-31

16. 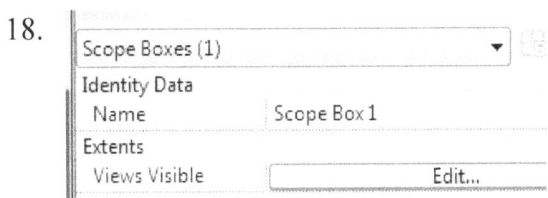 In the Properties pane:

    Set the Scope Box to Scope Box 1, the scope box which was just placed.

17. Repeat for the other three grid lines.

18. 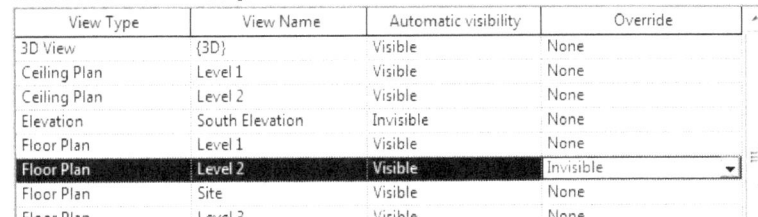 Select the Scope Box.

    Select **Edit** next to Views Visible in the Properties pane.

19.  Set the Level 2 Floor Plan Override Invisible.

    Press **OK**.

20. Select the scope box.
    Right click and select **Hide in View→ Elements**.

    The scope box is no longer visible in the view.

21. Activate Level 2.

    The grid lines and scope box are not visible.

22. Close the file without saving.

Command Exercise

Exercise 4-10 – Duplicating Views

Drawing Name: **duplicating_views.rvt**
Estimated Time to Completion: 20 Minutes

Scope

Duplicate view with Detailing
Duplicate view as Dependent
Duplicate view

Understand the difference between the different duplicating views options

Solution

1. Activate the **Level 1** floor plan.

   Note that the doors all have door tags.

2. Highlight Level 1.

   Select **Duplicate View→Duplicate**.

   When you select Duplicate View→Duplicate, any annotations, such as tags and dimensions, are not duplicated.

4-33

3.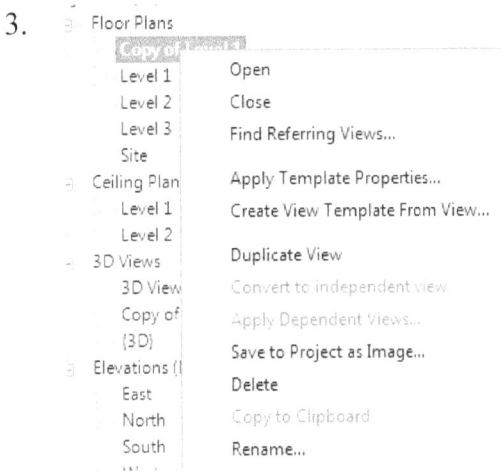

    Highlight the copied level 1.

    Right click and select **Rename**.

    Name: Level 1 - No Annotations

    Type **Level 1- No Annotations**.

    Press **OK**.

4.

    Highlight Level 1.

    Right click and select **Duplicate View→Duplicate with Detailing**.

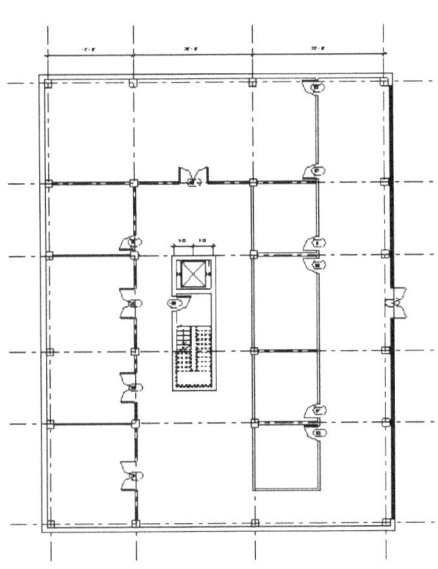

Note that when you select Duplicate with Detailing the new view includes annotations, such as tags and dimensions.

View Properties

5.  Highlight the copied level 1.

   Right click and select **Rename**.

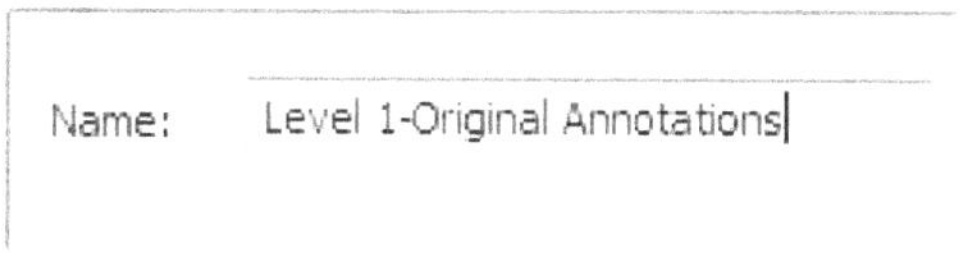

   Type **Level 1- Original Annotations**.

   Press **OK**.

6. 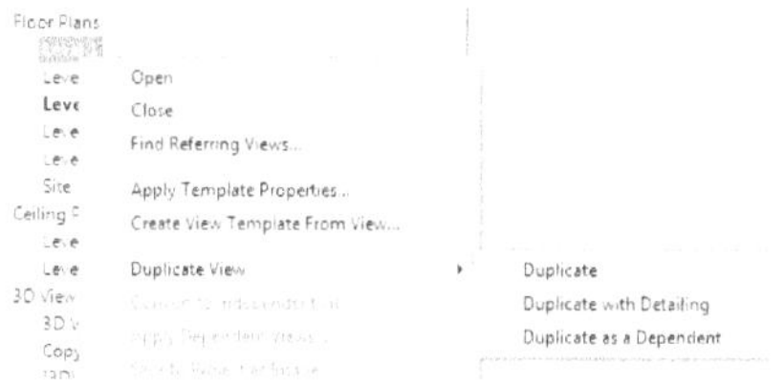 Highlight Level 1.
   Right click and select **Duplicate View→Duplicate as a Dependent**.

7. 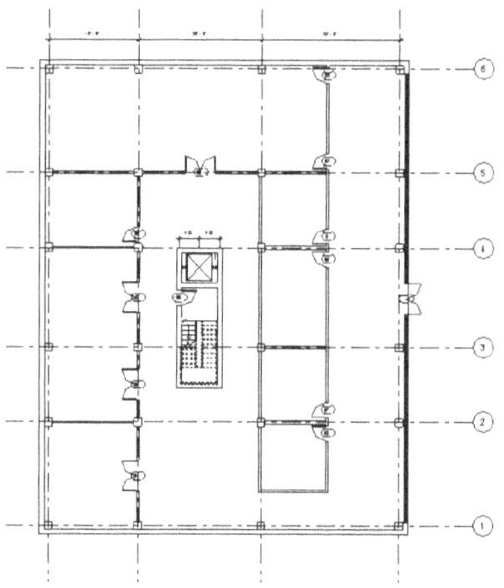 *Notice that the duplicated view includes the annotations.*

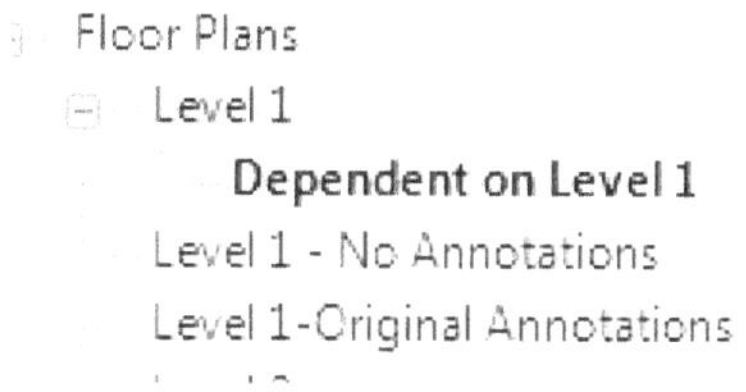

   In the Project Browser, the dependent view is listed underneath the parent view.

8.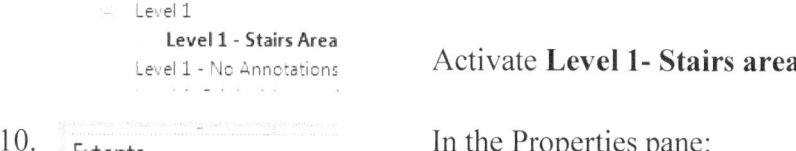

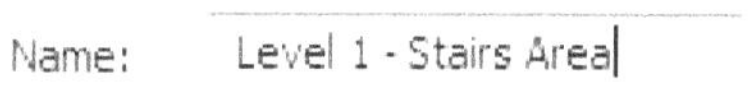

   Rename the dependent view **Level 1- Stairs area**.

9. Activate **Level 1- Stairs area**.

10. 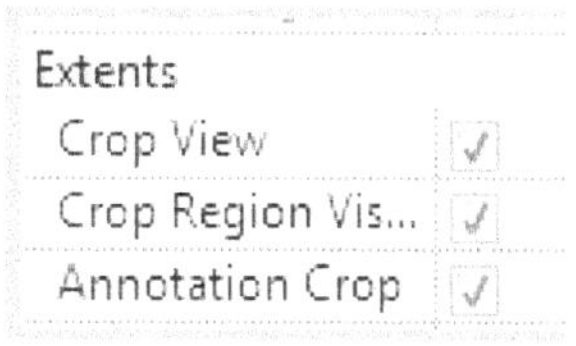 In the Properties pane:

    Enable **Crop View**.
    Enable **Crop Region Visible**.

4-35

11. 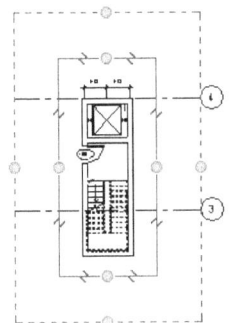 Adjust the crop region to focus the view on the stairs area.

12. Floor Plans
    Level 1
      Level 1 - Stairs Area
      Level 1 - No Annotations
      Level 1 - Original Annotations

    Activate **Level 1**.

13. 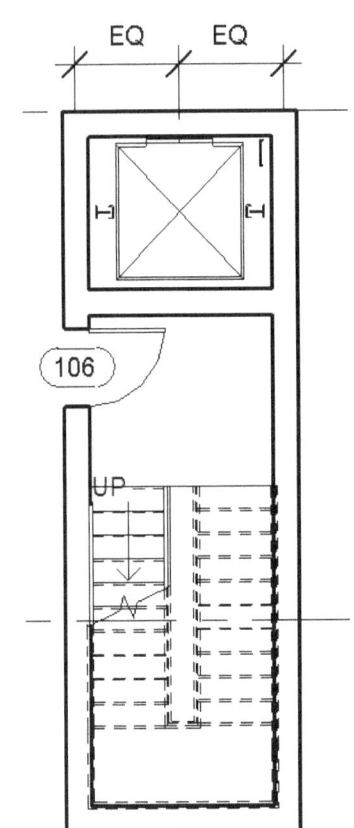 Zoom into the stairs area.

14. Annotate — Activate the **Annotate** ribbon.

15. Tread Number — Select the **Tread Number** tool on the Tag panel.

    Tread Numbers can only be applied to Component-based stairs.

16. Select the middle line that highlights when your mouse hovers over the left side of the stairs.

17. Select the middle line on the right side of the stairs.

    Right click and select CANCEL to exit the command.

18. Activate **Level 1- Original Annotations**.

    Notice that the new annotations - the tread numbers - are not visible in this view.

19. Activate **Level 1 - No Annotations**.

    Notice that the new annotations - the tread numbers - are not visible in this view.

20. Activate the **Level 1- Stairs area** dependent view.

    Notice that any annotations added to the parent view are added to the dependent view.

21. Close without saving.

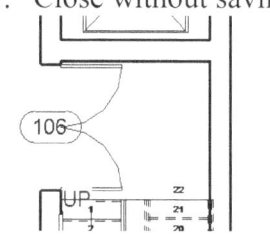

Extra: Change the door labeled 106 to a double flush door. Which Level 1 views display the new door type?

Command Exercise

Exercise 4-11 – Segmented Views

Drawing Name: **segmented views.rvt**
Estimated Time to Completion: 10 Minutes

Scope

Modify a section view into segments

Solution

1.  Activate the **A101- Segmented Elevation** sheet.

   There are two views on the sheet. One is the floor plan and the other is a section view defined in the floor plan.

2. 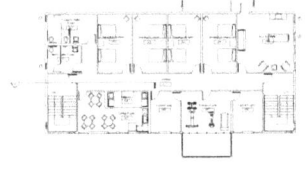 Activate the **Main Floor** floor plan.

   This is the top view on the sheet.

3. 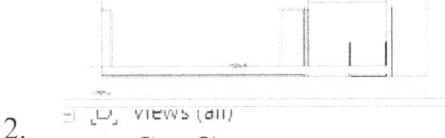 Select the section line.

4. 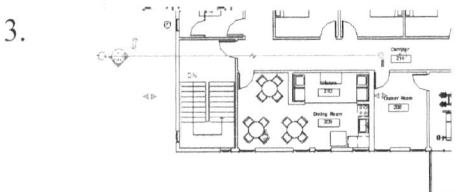 Select **Split Segment** on the ribbon.

4-38

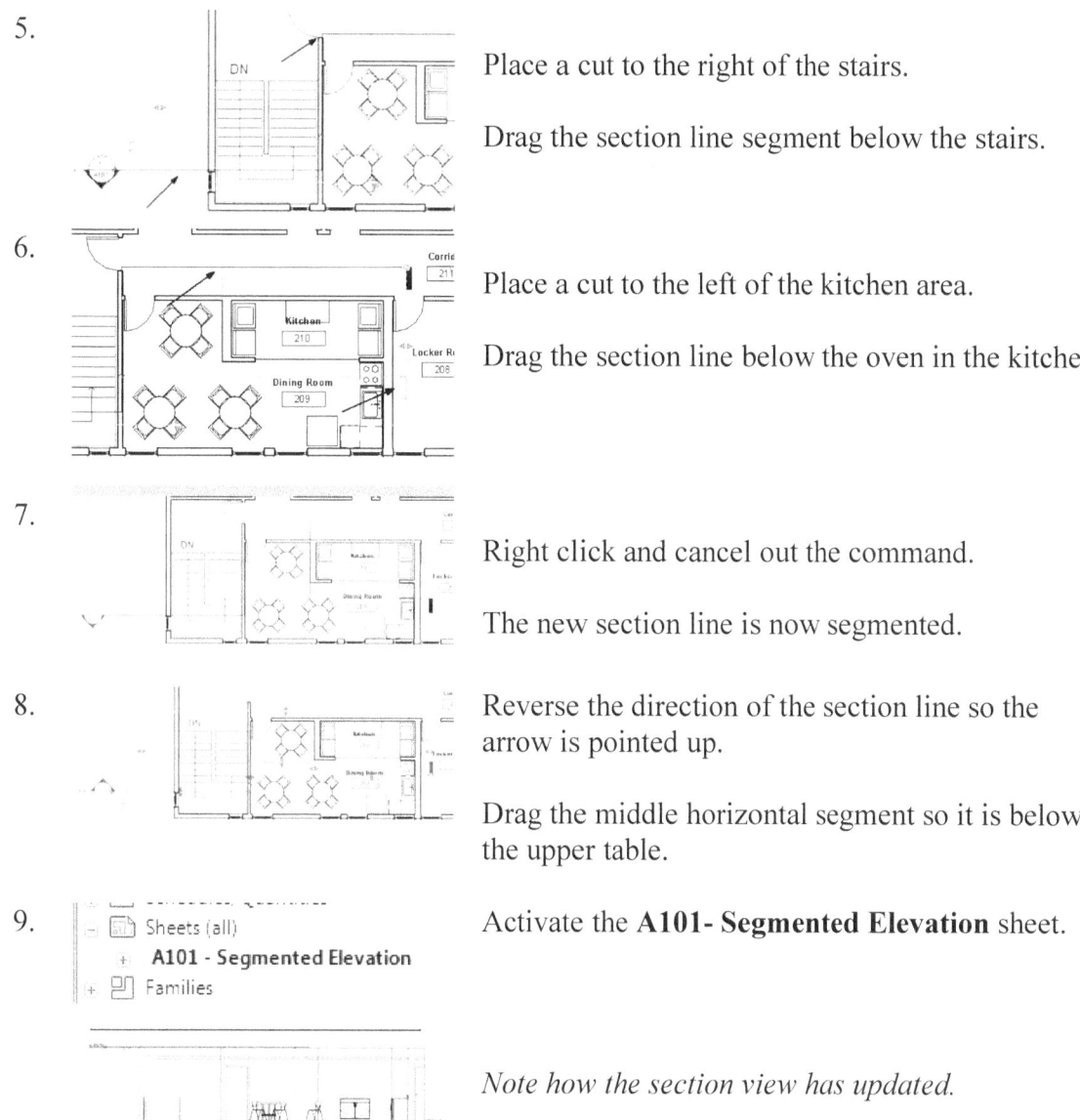

5. Place a cut to the right of the stairs.

   Drag the section line segment below the stairs.

6. Place a cut to the left of the kitchen area.

   Drag the section line below the oven in the kitchen.

7. Right click and cancel out the command.

   The new section line is now segmented.

8. Reverse the direction of the section line so the arrow is pointed up.

   Drag the middle horizontal segment so it is below the upper table.

9. Activate the **A101- Segmented Elevation** sheet.

   Note how the section view has updated.

10. Close without saving.

Practice Exam

1. Straight grid lines are visible in the following view types:
   - A. ELEVATION
   - B. PLAN
   - C. 3D
   - D. SECTION
   - E. DETAIL

2. Which two shortcut keys launch the Visibility/Graphics dialog?
   - A. VG
   - B. VV
   - C. VE
   - D. VW
   - E. F5

3. True or False: Objects hidden using the Temporary Hide/Isolate tool are not visible, but they are still printed.

4. Select the THREE options for Detail Level for a view:
   - A. COARSE
   - B. SHADED
   - C. FINE
   - D. HIDDEN
   - E. WIREFRAME
   - F. MEDIUM

5. True or False: If you delete a view, the annotations placed in the view are also deleted.

6. True or False: A camera can not be placed in an elevation view.

7. Scope boxes control the visibility of:
   - A. Elements
   - B. Plumbing Fixtures
   - C. Object Styles
   - D. Grid lines and levels

8. To change the graphic appearance of your model from Hidden Line to Realistic, you:
   - A. Modify the Rendering Settings in the View Control Bar
   - B. Edit Visibility/Graphics Overrides
   - C. Change graphic display options in View Properties
   - D. Click Visual Styles on the View Control Bar

9. A story level is the color:
    A. Yellow
    B. Blue
    C. Black
    D. Green

10. If you create a level using the COPY or ARRAY tool, what type of level is created?
    A. Story
    B. Non-Story or Reference
    C. Elevation
    D. Plan

11. This type of level does not have a PLAN view associated to it:
    A. Story
    B. Non-Story or Reference
    C. Elevation
    D. Plan

12. In which view type can you place a level?
    A. PLAN
    B. LEGEND
    C. 3D
    D. CONSTRUCTION
    E. ELEVATION

Answers:
1) A, B & D; 2) A & B; 3) True; 4) A, C & F; 5) True; 6) False; 7) D; 8) D; 9) B; 10) B; 11) B; 12) E

Lesson Five

Dimensions and Constraints

This lesson addresses the following exam questions:

- Constraints
- Temporary and Permanent Dimensions
- Multisegmented Dimensions

Dimensions are system families. They are in the Annotation category. They have type and instance properties.

Revit has three types of dimensions: listening, temporary and permanent.

A listening dimension is the dimension that is displayed as you are drawing, modifying, or moving an element.

A temporary dimension is displayed when an element is placed or selected. In order to modify a temporary dimension, you must select the element.

A permanent dimension is placed using the Dimension tool. In order to modify a permanent dimension, you must move the element to a new position. The permanent dimension will automatically update. To reposition an element, you can modify the temporary dimension or move the element using listening dimensions.

When you enter dimension values using feet and inches, you do not have to enter the units. You can separate the feet and inches values with a space and Revit will fill in the units. If a single unit is entered, for example '10', Revit assumes that value is 10 feet, not 10 inches.

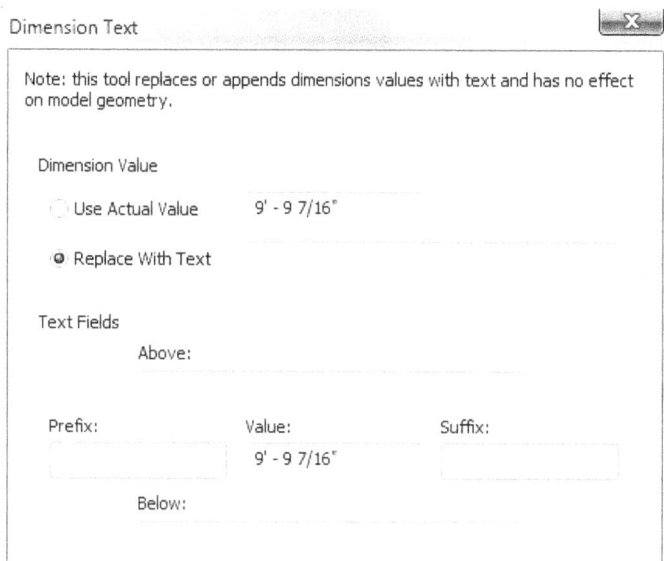

To override a dimension value, select the permanent dimension and enable Replace with Text and put in the desired text.

You can also add additional notes using the Text Fields for Above, Prefix, Suffix, or Below.

Command Exercise

Exercise 5-1 – Placing Permanent Dimensions

Drawing Name: **i_dimensions.rvt**
Estimated Time to Completion: 20 Minutes

Scope

Placing dimensions

Solution

1. Activate the **Ground Floor Admin Wing** floor plan.

   Floor Plans
   Ground Floor
   Ground Floor Admin Wing
   Lower Roof
   Main Floor
   Main Floor Admin Wing

2. Activate the Modify ribbon.

   Select the **Match Properties** tool from the Clipboard panel.

3. Select the wall indicated as the source object.

4. 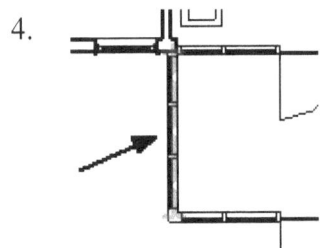 Select the wall indicated as the target object.

5-3

5. 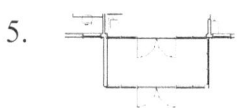 Note that the curtain wall changed to an Exterior-Siding wall.

   Cancel out of the Match Properties command.

6. Activate the Annotate ribbon.

   Select the **Aligned Dimension** tool.

7.

   On the Options bar, set the dimensions to select the **Wall faces**.

8. On the Options bar, set the dimensions to **Pick Entire Walls**.

9. Pick the wall indicated.

   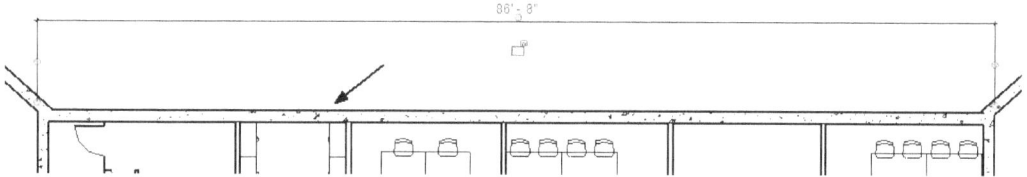

   Move the mouse above the selected wall to place the dimension.

10. Note the entire wall is selected and the dimension is located at the faces of the walls.
    Cancel or escape to end the Dimension command.

11. Select the dimension.

    Note that there are several grips available. The grips are the small blue bubbles.

Dimensions and Constraints

12. 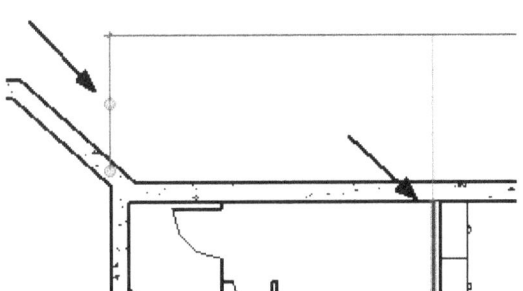 Select the second grip indicated on the left.

    Move the witness line to the wall face indicated by the arrow in the center of the image.

    Note that the witness line automatically snaps to the wall face.

13. Prefer: Wall centerlines ▼   With the dimension selected:

    On the Options bar, change the Prefer to **Wall centerlines**.

14. 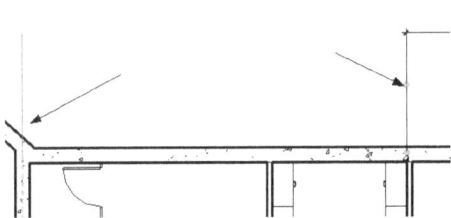 Select the dimension and activate the grip indicated.

    Drag the witness line to the center of the left wall.

15. 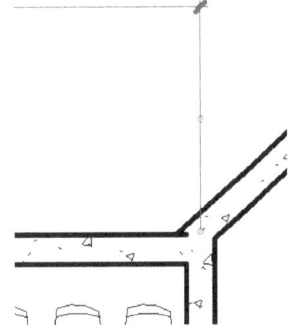 Left click once on the grip on the right side of the dimension to shift the witness line to the wall centerline.

16. Note how the dimension value updates.

17. Select the **Aligned** tool.

    Aligned

18. 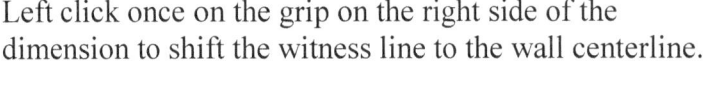 Set Pick to Individual References on the options bar.

5-5

19. 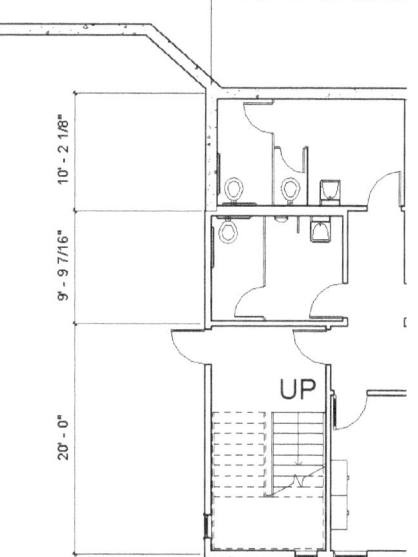 Select the centerlines of the walls indicated.

20. 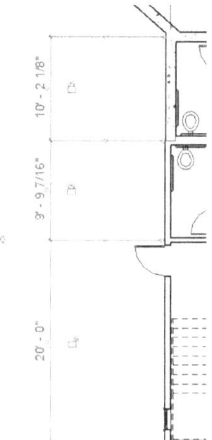 Select the dimension so it highlights.

   Select the two locks on the top dimensions to switch the permanent dimensions to *locked* dimensions. This means these distances will not be changed.

21. 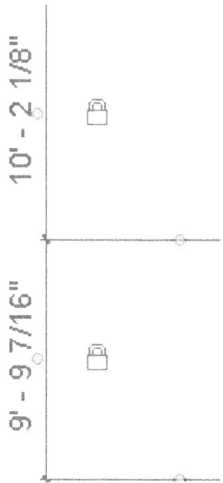 You should see two of the padlocks as closed and one as open.

Dimensions and Constraints

22. 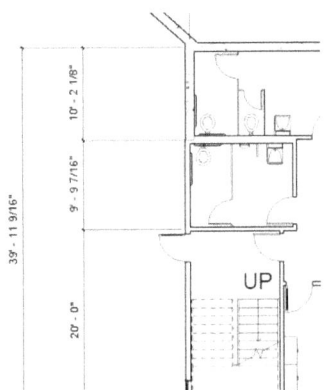 Place an overall dimension using the wall centerlines.

23. 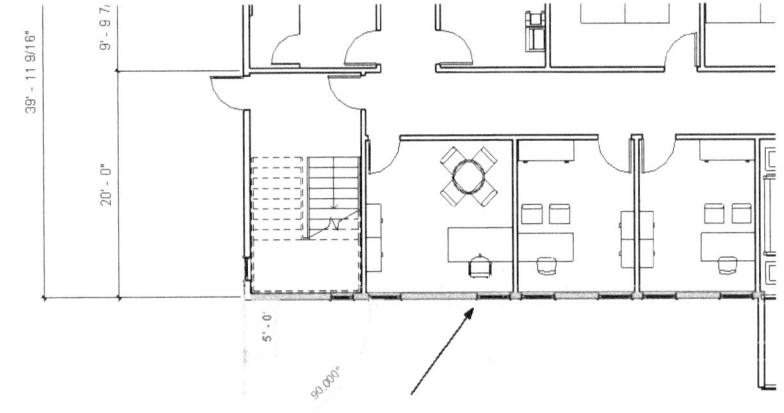 Drag the wall indicated down to a new position.

    Move down 5'-0".

24. If you get an error message, select **Unjoin Elements**.

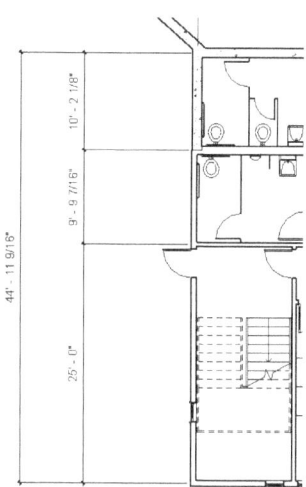

Note that the two locked dimensions did not change. Only the unlocked dimension updated.

25. Close without saving.

Dimensions and Constraints

Command Exercise
Exercise 5-2 – Modifying Dimension Text

Drawing Name: **dimtext.rvt**
Estimated Time to Completion: 15 Minutes

Scope

Replace Dimension Text
Restore Dimension Text
Modify Dimension Text

Solution

1. Activate the **Ground Floor Admin Wing** floor plan.

2. Select the dimension indicated.

   Double left click on the dimension text to bring up the dimension text dialog.

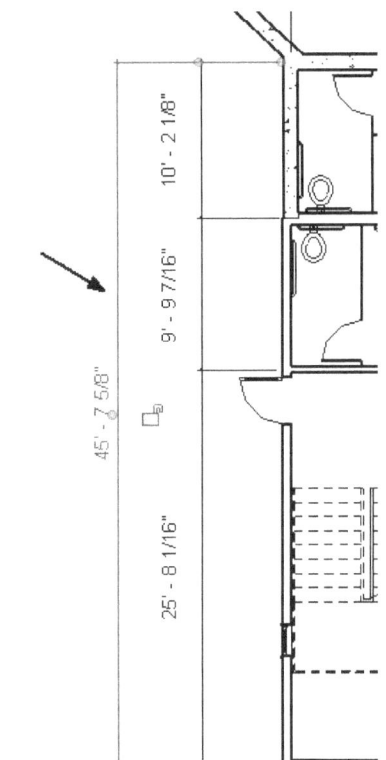

5-9

3. Enable **Replace with Text**.
   Enter **45' 8"**.

   Press **OK**.

4. You will get an error message stating that you cannot change the numeric value without moving the element to correspond with that value.

   Press **Close**

5. Enable **Replace with Text**.
   Enter **OVERALL**.

   Press **OK**.

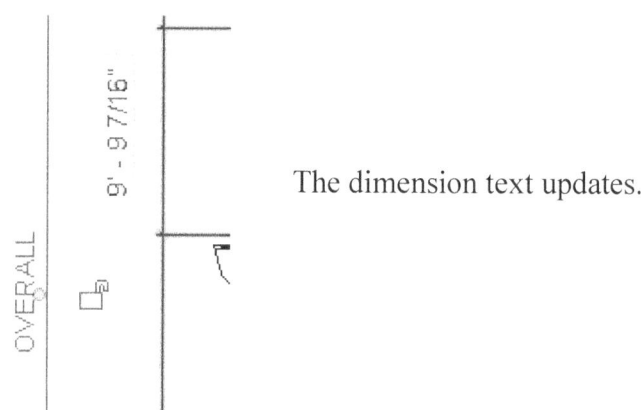

The dimension text updates.

6. Activate the **Manage** ribbon.

   Select **Project Units** on the Settings panel.

7.

   Select the **Length** button under the Format column.

8. Set the Rounding **To the nearest ½"**.

   Press **OK** until all dialogs are closed.

5-10

Dimensions and Constraints

9. Double left click on the 10′ 2″ dimension to be edited.

10. In the Below field, type: **WOMEN'S LAVATORY**.

   Press **OK**.

11. Note how the dimension updates.

12. Double left click on the dimension with **???**.

13. Enable **Use Actual Value**.

   Press **OK**.

14. Close without saving.

5-11

Command Exercise

Exercise 5-3 – Converting Temporary Dimensions to Permanent Dimensions

Drawing Name: **i_dimensions.rvt**
Estimated Time to Completion: 10 Minutes

Scope

Place dimensions using different options.
Convert temporary dimension to permanent

Solution

1. Activate the **Ground Floor Admin Wing** floor plan.

2. 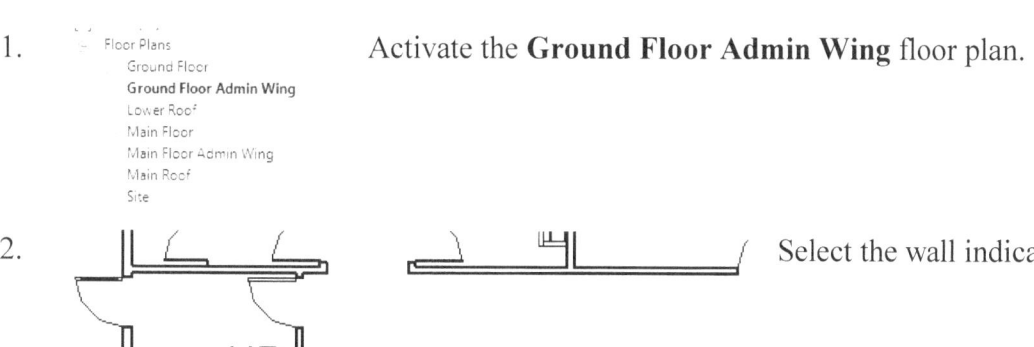 Select the wall indicated.

3. 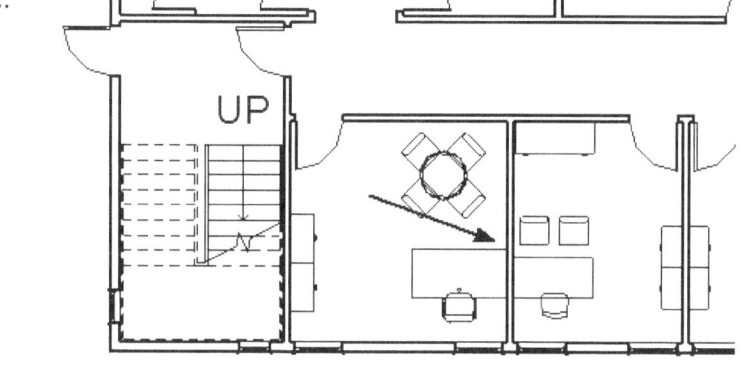 Two dimensions will appear.

   There is a small dimension icon visible.

   This icon converts a temporary dimension to a permanent dimension.

   Left click on this icon.

Dimensions and Constraints

4. 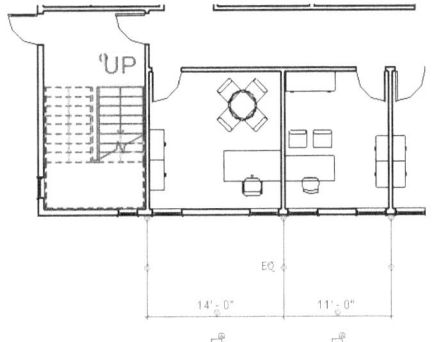 The dimensions are now converted to permanent dimensions.

   Drag the dimensions below the view.

5. Click on the dimensions so they highlight.

   Click on the witness line grip indicated.

   Notice how each time the grip is clicked on, the witness line shifts from the wall centerline to the face of the wall.

6. Select the grip at the endpoint of the witness line. Drag the endpoint to increase the gap between the wall and the witness line.

   Release the grip.

7. Select the witness line grip indicated.

   Drag the witness line to the outer wall. Note that if the witness line is set to the outer wall face it maintains that orientation.

   Release.

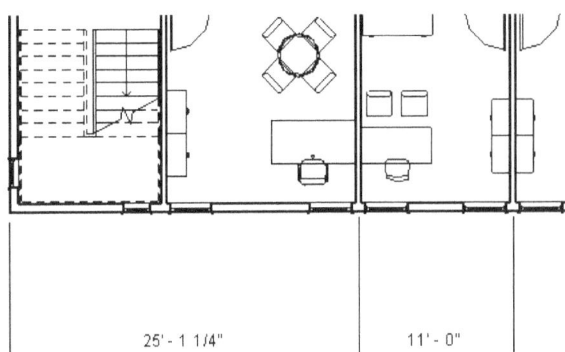 The dimension updates.

   Note the new dimension value.

8. Close without saving.

5-13

Command Exercise

Exercise 5-4 – Applying Constraints

Drawing Name: **i_constraints.rvt**
Estimated Time to Completion: 20 Minutes

Scope

Using the Align tool to constrain elements

Solution

1. 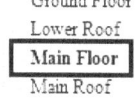 Activate the **Main Floor** floor plan.

2. 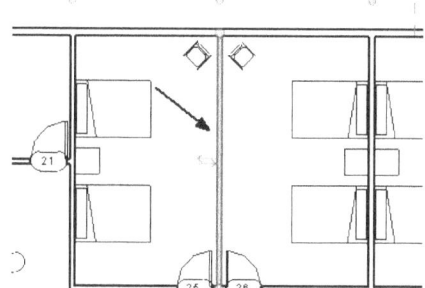 Select the interior wall located between Door 25 and Door 26 as indicated.

3. 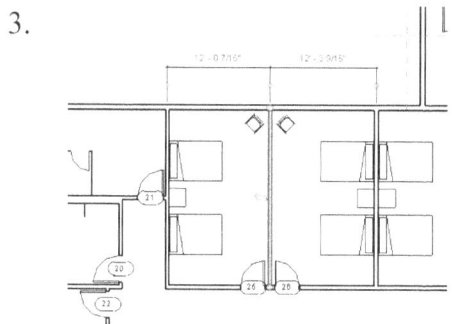 The associated temporary dimensions will become visible.

   Left click on the permanent dimension toggle to convert the dimensions to permanent dimensions.

4. Left click anywhere in the display window.

In the Properties pane:

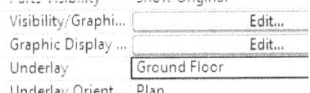

Scroll down to the **Underlay** parameter.

Set the Underlay to **Ground Floor**.

This will make the ground floor visible in the graphics window. Elements on the ground floor will be displayed in a lighter shade.

5. 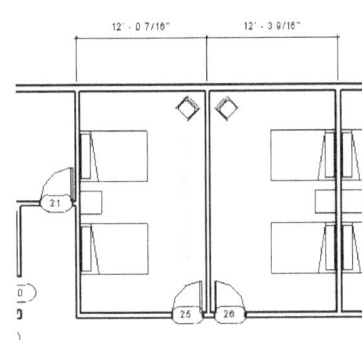 Note in the upper left corner the interior walls are not aligned.

6. Activate the **Modify** ribbon.

Select the **Align** tool from the Modify panel.

7. Prefer: Wall faces Set the preference to **Wall faces** on the Options bar.

8. Select the right side of the ground floor interior wall as the source for the alignment.

9. 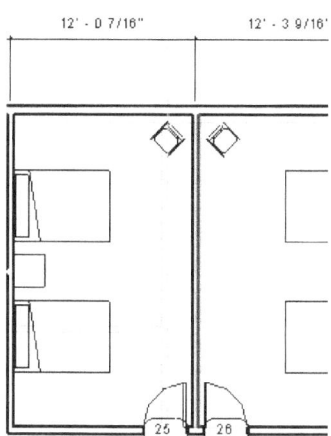 Select the right side of the main floor interior wall as the target for the alignment (this is the wall that will be shifted).

10. A warning dialog will appear due to the door location. Ignore the warning. ***Don't move the door or wall.***

11. 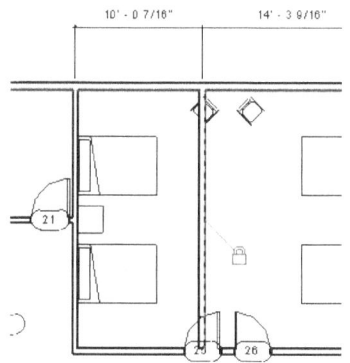 Lock the alignment in place so that the walls remain constrained together.

12. Right click and select Cancel to exit the Align mode.

13. Activate the **Ground Floor** floor plan.

Dimensions and Constraints

14.

Select the wall as well as the cabinets and copier next to the wall.

15. Select the **Move** tool on the Modify ribbon.

16.

Select a base point.

Move the selected elements to the right **1′ 11″**.

Left click in the window to release the selected elements.

17. Activate the **Main Floor** floor plan.

18. 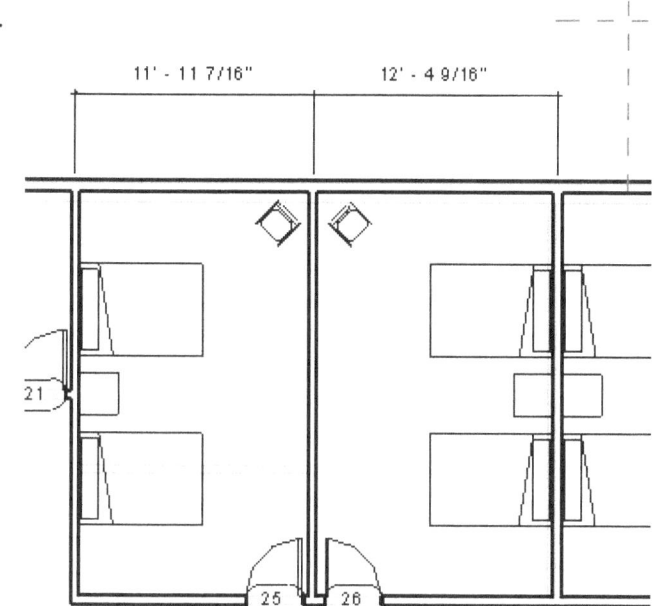 Note that the main floor interior wall has also shifted (the walls remained aligned because they are locked together) and the dimension has updated.

19. Close without saving.

Command Exercise

Exercise 5-5 – Equality Formula

Drawing Name: **equality_formula.rvt**
Estimated Time to Completion: 10 Minutes

Scope

Using an Equality Formula

Solution

1. Activate the **Main Floor** floor plan.

2. Select the EQ dimensions.
   Right click and disable EQ display.
   Note the value is 16' 4".

3. With the dimension still selected, go to the Properties pane.

   Under Equality Display:

   Select **Equality Formula**.

4. Select **Edit Type**.

5. Scroll down.
   Select the button next to Equality Formula.

6. Add Number of Segments and Length of Segment to Label Parameters.

   Organize so that Number of Segments is on Row 1; Length of Segment is on Row 2, and Total Length is on Row 3.

7.

	Parameter Name	Spac	Prefix
1	Number of Segments	1	@
2	Length of Segment	1	=
3	Total Length	0	

Add a Suffix for Number of Segments - @.
Add a Suffix for Length of Segment - =.
Place a space in each Prefix field.

This builds an equation that reads

<Number of Segments> @ <Length of Segment> = <Total Length>

Press **OK**.

8. Other
Equality Text — EQ
Equality Formula — Quantity @ Length= Total Length
Equality Witness Display — Tick and Line

The Equality Formula is updated.

Press **OK**.

9. 3 @ 16' - 4" = 49' - 0"

Left click in the graphics window to release the selection and update the dimension.

If you don't see spaces, go back and add spaces to the prefix fields.

10. Close without saving.

Dimensions and Constraints

Command Exercise

Exercise 5-6 – Alternate Dimensions

Drawing Name: **Alternate Dimensions.rvt**
Estimated Time to Completion: 10 Minutes

Scope

Apply Alternate Dimensions

Solution

1. Activate the **Main Floor** floor plan.

2. Zoom into the lower horizontal dimensions. Select the dimensions.

3. Select **Edit Type** in the Properties pane.

4. Select **Duplicate**.

5. Rename by adding **Alternate** to the name.

6. Press **OK**.
   Scroll down.
   Select the drop-down next to **Alternate Units** and select **Below**.

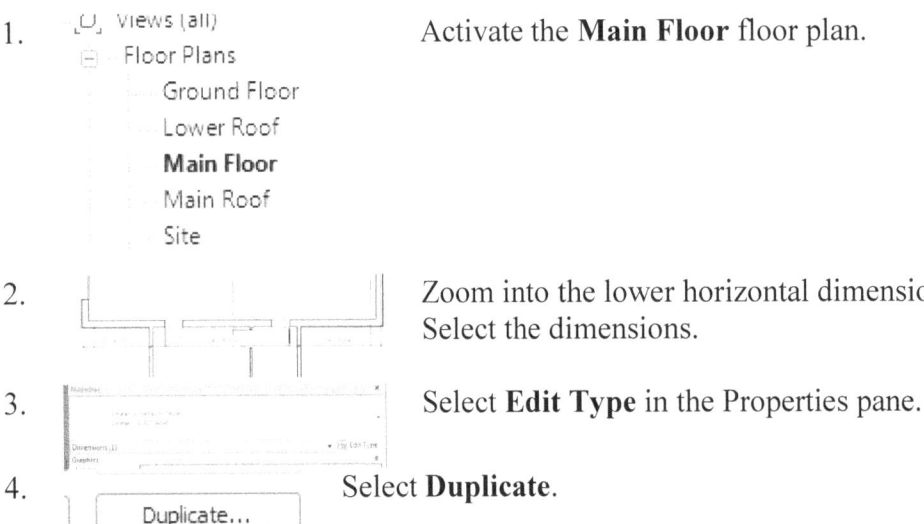

7. Type **mm** in the Alternate Units Suffix field.

   Press **OK**.

8. *The dimensions now display with both Imperial and millimeter units.*

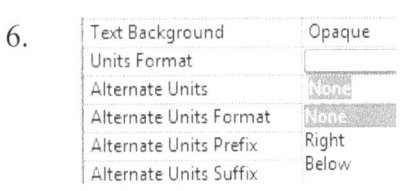

9. Close without saving.

5-21

Command Exercise

Exercise 5-7 – Multi-Segmented Dimensions

Drawing Name: **multisegment.rvt**
Estimated Time to Completion: 10 Minutes

Scope

Add and delete witness lines to a multi-segment dimension

Solution

1. Activate the **Main Floor** floor plan.

2. Select the lower dimension so it highlights.

3. On the ribbon, select **Edit Witness Lines**.

4. Select the center of each door indicated.

   Then, left pick below the building to accept the additions.

5. Drag the dimensions down so you can see them easily.

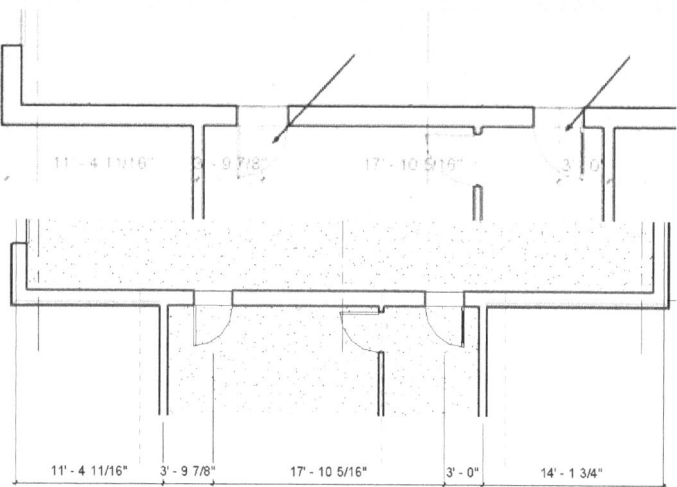

5-22

6. Select the dimension. Select **Edit Witness Lines**. Add the witness line at the wall indicated.

7. Select the dimension. Select **Edit Witness Lines**. Add the witness line at the two reference planes indicated.

8.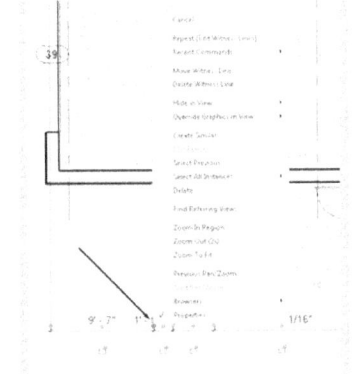

Select the dimension.

Select the grip on the witness line for the left reference plane.

Right click and select **Delete Witness Line**.

9. The dimension should update with the witness line removed.

10. 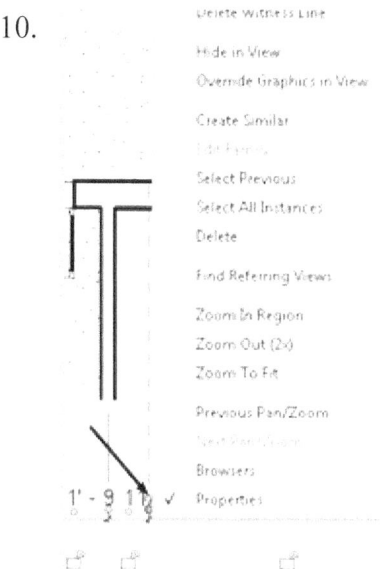 Repeat to delete the witness line on the right reference plane.

11. The dimension should update.

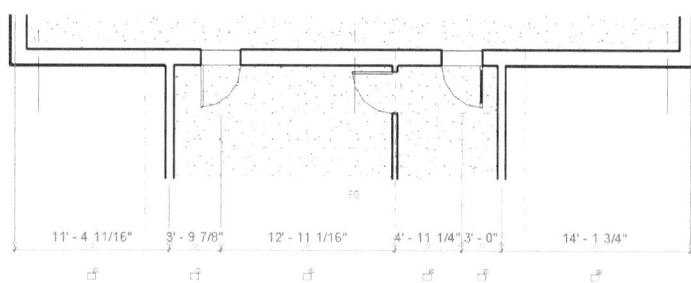

12. Close without saving.

Practice Exam

1. Constraints can be used to:
    A. Prevent other users from moving an element in a shared project.
    B. Lock two elements at a fixed distance.
    C. Keep two elements equally spaced.
    D. Alert the user when someone changes or moves an element.

2. To change a dimension's value:
    A. Use the Change Dimension tool.
    B. Click on Edit Witness Line.
    C. Select the element to be re-dimensioned.
    D. Click on the dimension.

3. True or False: You can override a dimension value with a new dimension value.

4. Select the TWO types of dimensions:
    A. PERMANENT
    B. TEMPORARY
    C. PERPETUAL
    D. INVISIBLE

5. To create a dimension using feet and inches, you can use this keystroke to separate the feet and inches values:
    A. COMMA
    B. SEMI-COLON
    C. SPACE
    D. COLON

6. A _____ allows you to locate elements equidistant from one another.
    A. multi-segmented dimension
    B. single-segmented dimension
    C. scope box
    D. reference plane

7. You are editing a dimension in Revit. You type 0 48. Revit will translate this to:
    A. 4' 0"
    B. 48' 0"
    C. 48"
    D. 4' 8"

8. You are placing a wall in Revit. You use the listening dimension to set the wall length. You enter '**5.5**'. Revit places the wall and it is _____ long.

    A. 5' 6"
    B. 5 1/2"
    C. 5' 0"
    D. 5' 5"

9. To change the position of an element, such as a door or wall"

    A. Place a permanent dimension, then modify the dimension.
    B. Select the element, then modify the temporary dimension
    C. Select the element and move it
    D. Use the Measure tool

10. The three types of dimensions used in Revit are:

    A. LINEAR
    B. ANGULAR
    C. TEMPORARY
    D. PERMANENT
    E. CONTINUOUS
    F. LISTENING

Answers:
1) B & C; 2) C; 3) False; 4) A & B; 5) C; 6) A; 7) A; 8) A; 9) B; 10) C, D, & F

Lesson Four

View Properties

This lesson addresses the following User and Professional exam questions:

- View Properties
- Object Visibility Settings
- Section Views
- Elevation Views
- View Templates
- Scope Boxes

The Project Browser lists all the views available in the project. Any view can be dragged and dropped onto a sheet. Once a view is used or consumed on a sheet, it cannot be placed a second time on a sheet – even on a different sheet. Instead, you must create a duplicate view. You can create as many duplicate views as you like. Each duplicate view may have different annotations, line weight settings, detail levels, etc. Annotations are Userd to a view. If a view is deleted, any annotations are also deleted.

Revit has bidirectional associativity. This means that changes in one view are automatically reflected in all Userd views. For example, if you modify the dimensions or locations of a window in one view, the change is reflected in all the Userd views, including the 3D view.

You can control the appearance of Revit elements using Object Visibility Settings. These settings control line color, line type, and line weight. You can create templates which have different Object Visibility Settings for different project types.

Command Exercise

Exercise 4-1 – Creating a Level

Drawing Name: **i_levels.rvt**
Estimated Time to Completion: 5 Minutes

Scope

Placing a level.

Solution

1. Activate the **South Elevation**.

    The level names have been turned off.

2. Select each level and place a check in the square that appears. This will turn on visibility of the level name.

3. 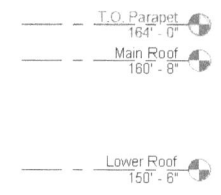 You should be able to identify the names for each level.

4.  Select the **Level** tool from the Architecture ribbon.

5. 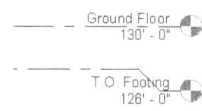 Place a level **5'-0"** above the Main Floor.

6. 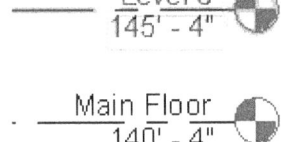   Note the elevation value for the new level.

7. Close without saving.

Command Exercise

Exercise 4-2 – Story vs. Non-Story Levels

Drawing Name: **story_levels.rvt**
Estimated Time to Completion: 15 Minutes

Scope

Understanding the difference between story and non-story levels
Converting a non-story level to a story level.

Solution

1. Activate the **South Elevation**.

2.  Select each level and place a check in the square that appears. This will turn on visibility of the level name.

3. Study the Main Floor level.

   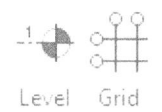 Notice that it is the color black while all the other levels are blue.

   The Main Floor level is a non-story or reference level. It does not have a view associated with it.

4. Activate the Architecture ribbon.

   Select the **Level** tool on the Datum panel.

5. On the Options bar:
   Uncheck **Make Plan View**.
   Set the Offset to **8' 0"**.

6. 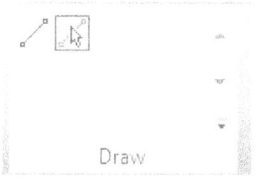 Select the **Pick** tool on the Draw panel.

4-4

View Properties

7. Select the Main Floor level.

Verify that the preview shows the level will be placed 8' 0" ABOVE the Main Floor level.

8. The level is placed above the Main Floor.

Right click and select Cancel twice to exit the Level command.

9. Select the Elbow control on the new level to add a jog.

10. Note that the new level is also a non-story or reference level.

Check in the Project Browser and you will see that no views were created with the new level.

11. Activate the View ribbon.

Select the **Plan Views→Floor Plan** tool on the Create panel.

12. The reference levels are listed.

Select the **Main Floor** level and press **OK**.

13. The Main Floor floor plan view will open.

Note that the Main Floor floor plan is now listed in the Project Browser.

However there is no ceiling plan for the Main Floor.

4-5

14. 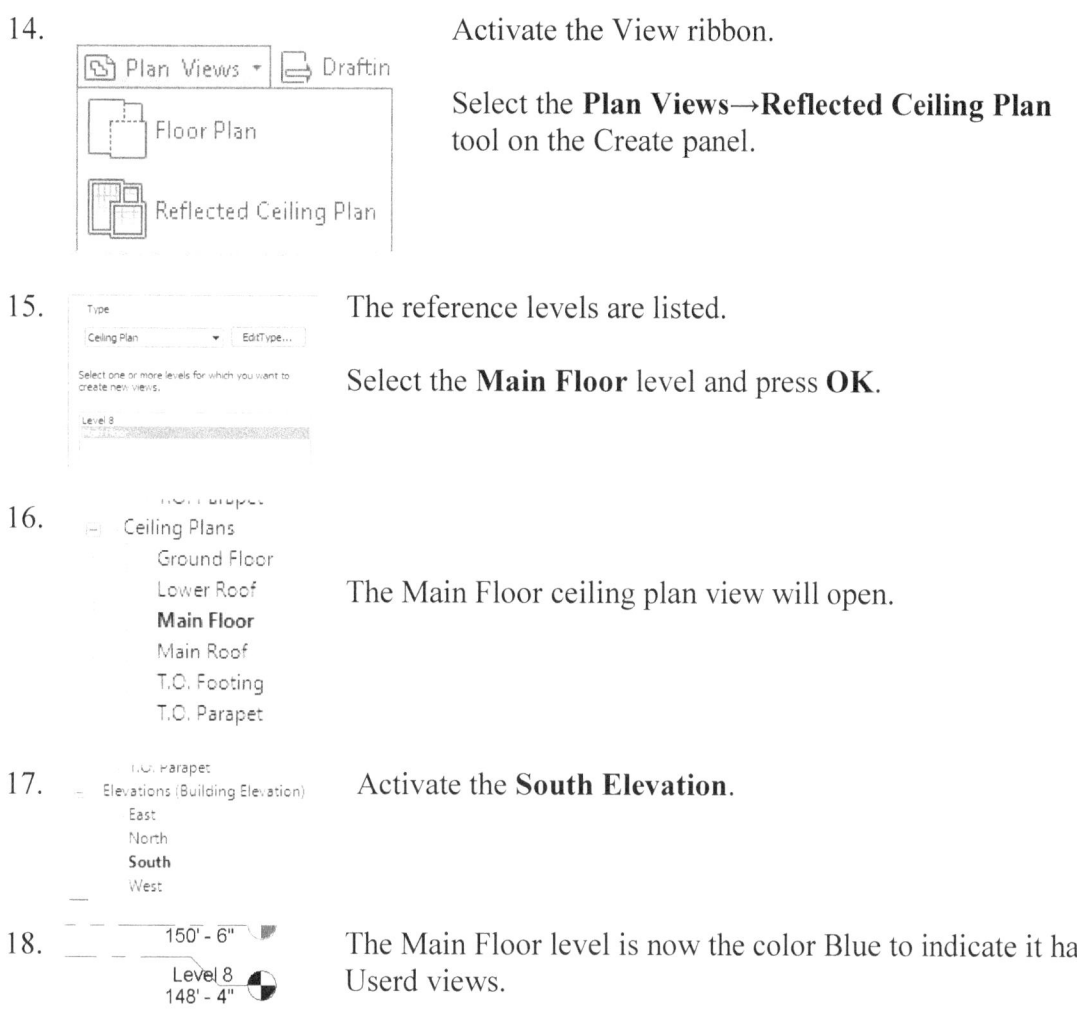 Activate the View ribbon.

    Select the **Plan Views→Reflected Ceiling Plan** tool on the Create panel.

15. The reference levels are listed.

    Select the **Main Floor** level and press **OK**.

16. The Main Floor ceiling plan view will open.

17. Activate the **South Elevation**.

18. The Main Floor level is now the color Blue to indicate it has Userd views.

19. Close the file without saving.

View Properties

Command Exercise

Exercise 4-3 – Creating Column Grids

Drawing Name: **i_grids.rvt**
Estimated Time to Completion: 10 Minutes

Scope

Placing column grids.

Solution

1. Floor Plans
   Ground Floor
   Lower Roof
   Main Floor
   Main Roof
   Site
   T.O. Footing
   T.O. Parapet

   Activate the **Ground Floor** view.

2. 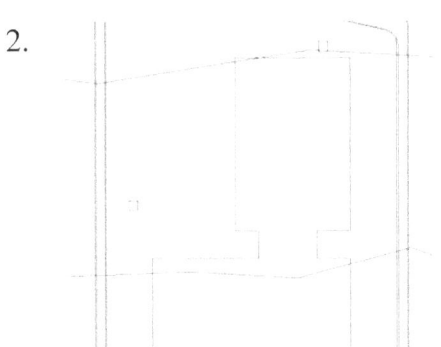 Zoom into the building area displayed.

3. Select the **Grid** tool on the Datum panel from the Architecture ribbon.
   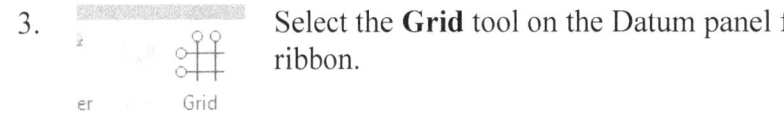

4. Select the **Pick Lines** mode from the Draw panel.
   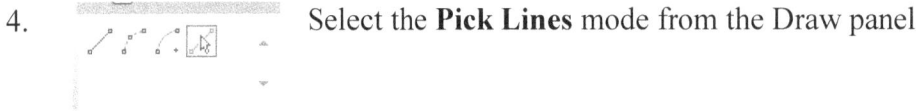

5.  Set the Offset to **2'-0" [600 mm]** on the Options bar.

4-7

6.  Click to place a vertical grid line as shown.

7. 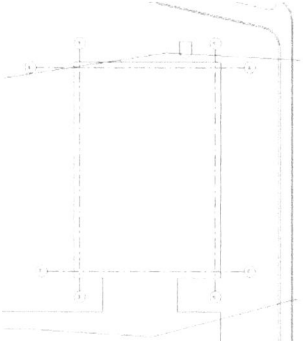 Place gridlines as shown.

   Use the grips by the heads to drag the grid bubbles into position.

8. 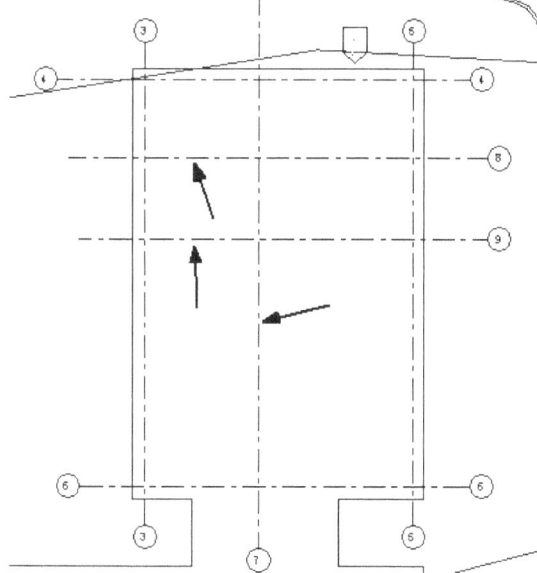 Switch to draw grid mode.

   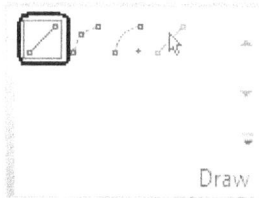

   Add a vertical grid line as shown.

   Add two horizontal grid lines as shown.

9. 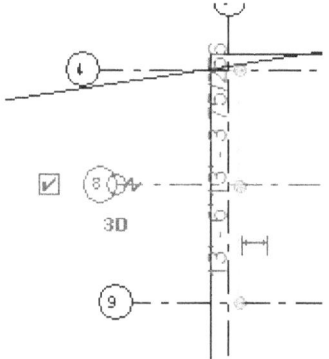 Place a check on both rectangles to make bubbles visible on both sides of the grid line.

10. 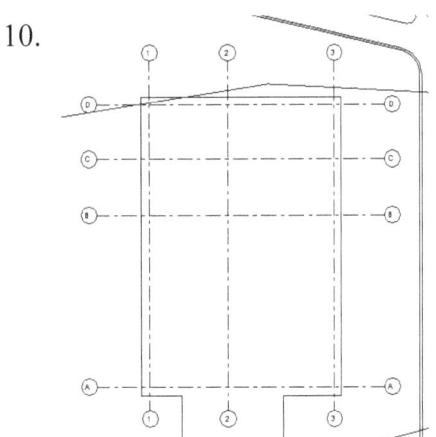 Label the horizontal grid lines A-D.
The alpha labeled grids are incremented from bottom to top.

Label the vertical grid lines 1-3.
The number labeled grids are incremented from left to right.

Enable the grid bubbles on both ends.

Turn off the visibility of elevations.

11. 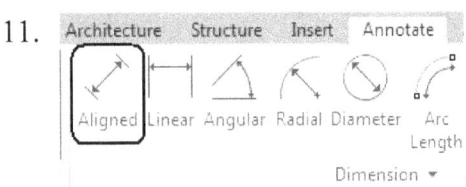 Activate the **Annotate** ribbon.

Select the **ALIGNED** dimension tool from the Dimension panel.

12. Place a multi-segmented dimension between the three vertical grid lines.

Enable the EQ toggle.

13. 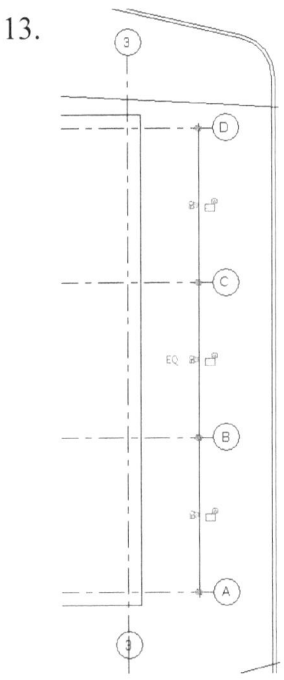 Place a multi-segmented dimension between the horizontal grid lines.

Enable the EQ toggle.

To place a continuous dimension, select each grid line in order, then left click above (if the last grid line is the top one) or below (if the last grid line is the bottom one).

Exit the command.

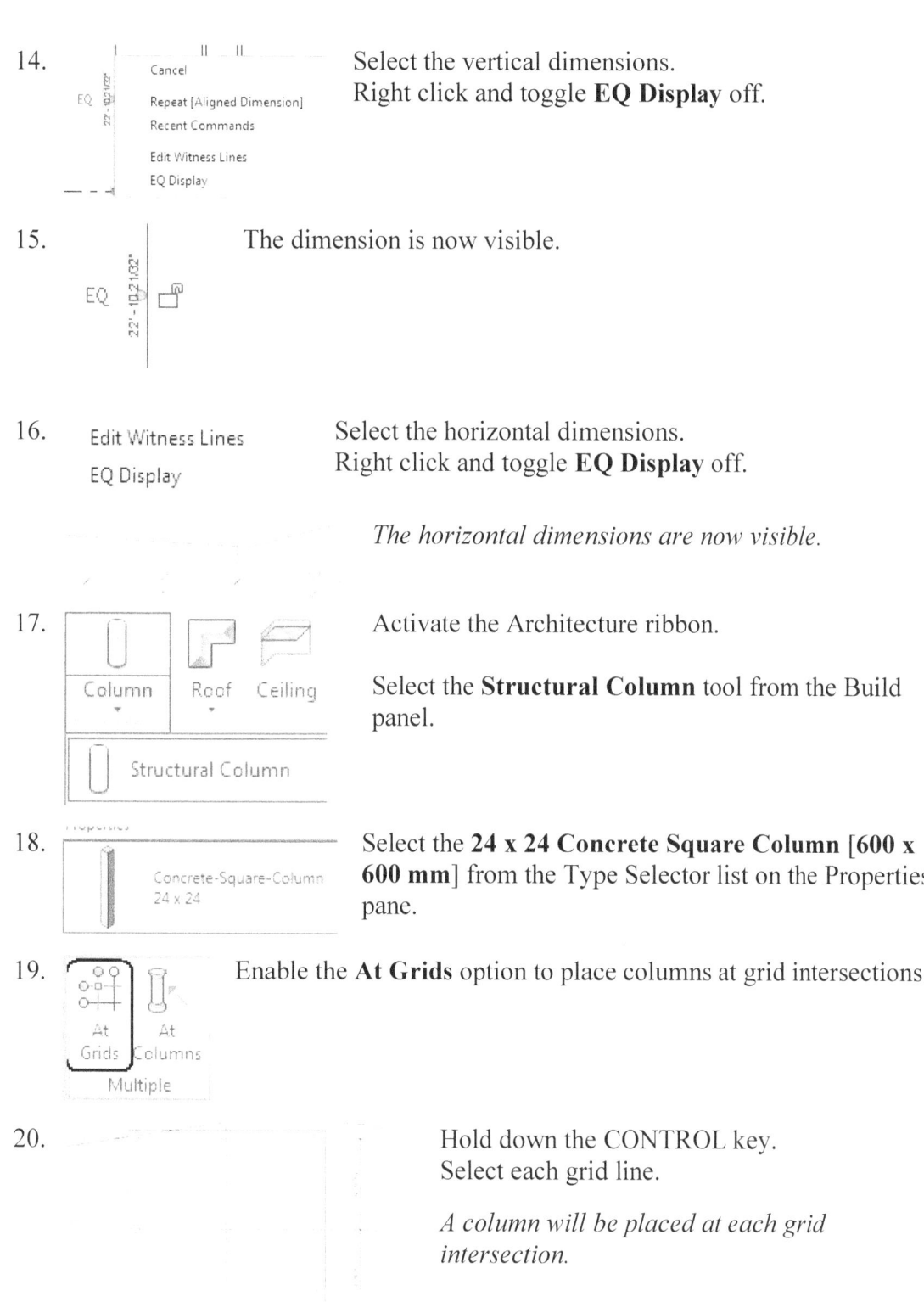

14. Select the vertical dimensions.
Right click and toggle **EQ Display** off.

15. The dimension is now visible.

16. Select the horizontal dimensions.
Right click and toggle **EQ Display** off.

The horizontal dimensions are now visible.

17. Activate the Architecture ribbon.

Select the **Structural Column** tool from the Build panel.

18. Select the **24 x 24 Concrete Square Column [600 x 600 mm]** from the Type Selector list on the Properties pane.

19. Enable the **At Grids** option to place columns at grid intersections.

20. Hold down the CONTROL key.
Select each grid line.

A column will be placed at each grid intersection.

View Properties

21. 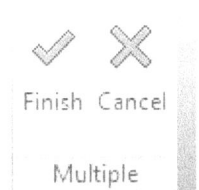 Select the Green Check when you see a column at every grid intersection.

    Right click and select CANCEL to exit the command.

22.  Activate the Annotate ribbon.

    Select **Tag All** from the Tag panel.

23.  Highlight the **Structural Column Tag**.

    Press **OK**.

24. 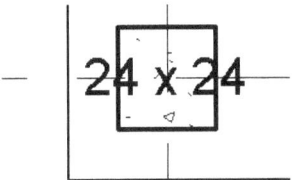 Zoom into a column to read the tag.

    The column is labeled with the Column Type.

25. 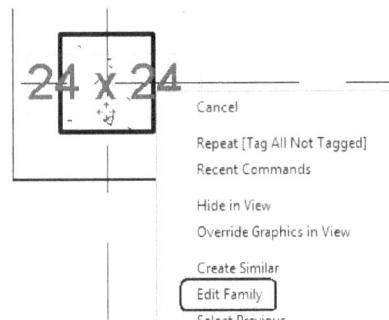 Select one of the structural column tags.

    Right click and select **Edit Family**.

    Make sure you select the tag and not the column.

4-11

26. 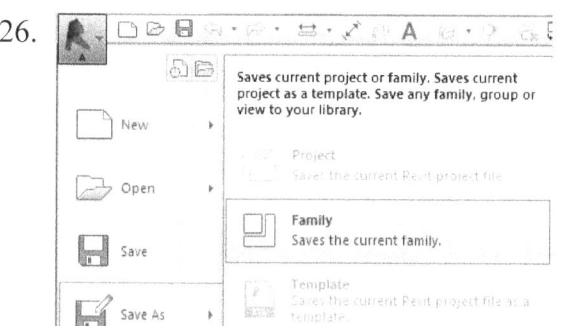 Go to **File→Save As→Family**.

27. 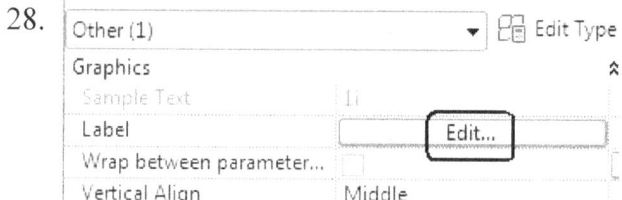 Rename the tag *Structural Column Location Tag.rfa*.

28. 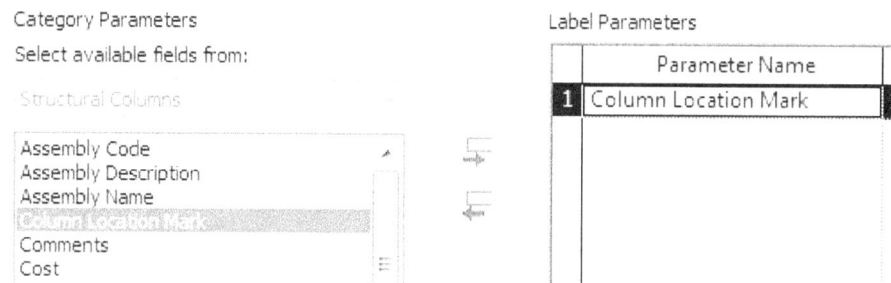 Select the text.

    Select the **Edit** button in the Label field on the Properties pane.

29. Use the Add and Remove tools in the middle of the dialog to remove the Type Name and add the Column Location Mark.

30. Press **OK**.

31. Save the file.

32. Select the **Load into Project** tool.

33. 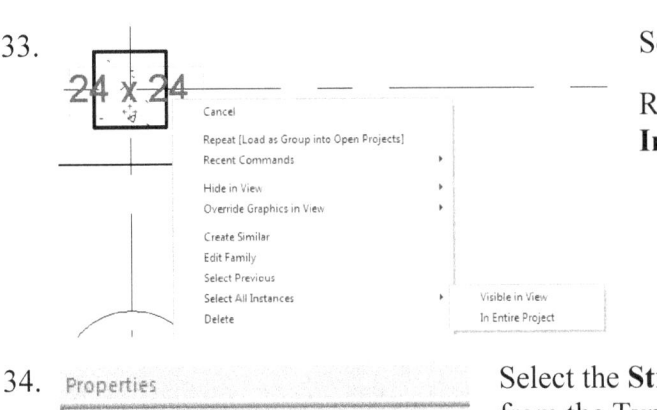 Select one of the column tags.

Right click and select **Select All Instances**→ **Visible in View**.

34. 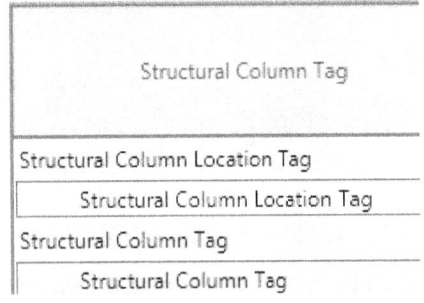 Select the **Structural Column Location Tag** from the Type Selector on the Properties pane.

35. The tags update with the location.

36. Close without saving.

Command Exercise

Exercise 4-4 – Setting View Depth

Drawing Name: **i_firestation_basic_plan.rvt**
Estimated Time to Completion: 5 Minutes

Scope

Determine the view depth of a view

Solution

1. Activate the **Site** view.

2.  In the Properties pane:

   Scroll down to **View Range** located under the Extents category.

   Select the **Edit** button.

3. Determine the **View Depth**.

4. Press **OK**.

5. Close without saving.

Command Exercise

Exercise 4-5 – Create a Cropped View

Drawing Name: **i_firestation_managing_views.rvt**
Estimated Time to Completion: 10 Minutes

Scope

Create a cropped view

Solution

1. Activate the **Main Floor – Furniture Plan** view.

   Floor Plans
      Ground Floor
      Lower Roof
      Main Floor
      Main Floor- Furniture Plan
      Main Roof
      Site
      T. O. Footing
      T. O. Parapet

2. In the Properties Pane:

   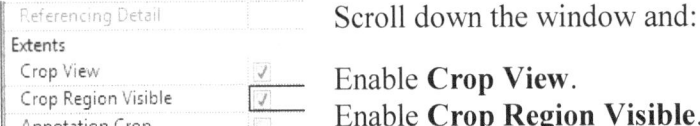

   Scroll down the window and:

   Enable **Crop View**.
   Enable **Crop Region Visible**.

4.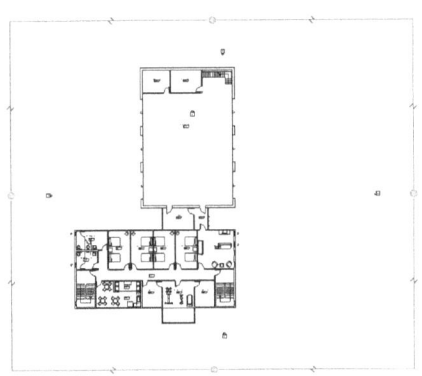

   Zoom out.

   Select the viewport rectangle.

5. 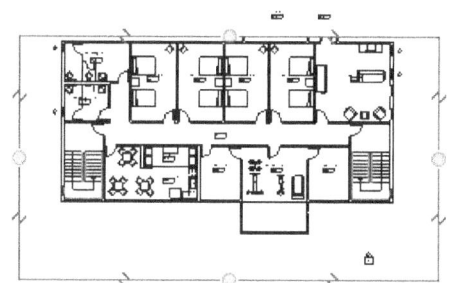 Use the bubbles to position the viewport so that only the furniture floor plan is visible.

6.

   Select **Hide Crop Region** using the tool in the View Control bar.
   Press **OK**.

7. Close without saving.

View Properties

Command Exercise
Exercise 4-6 – Change View Display

Drawing Name: **i_firestation_managing_views.rvt**
Estimated Time to Completion: 15 Minutes

Scope

*Use Temporary Hide/Isolate to control visibility of elements.
Change Line Width Display
Change Object Display Settings*

Solution

1. Activate the **Main Floor** floor plan.

2. 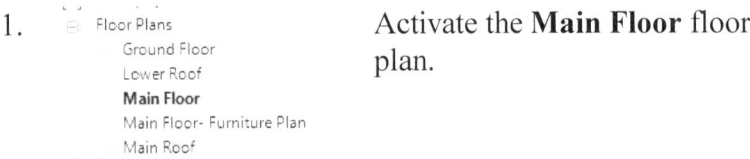 Select one of the exterior walls so it is highlighted.

3.  Select the **Temporary Hide/Isolate** tool.
Right click and select **Isolate Category**.

4-17

4. 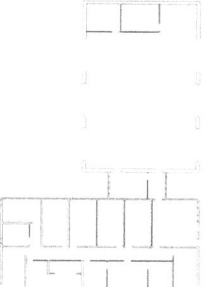 Only the exterior walls are visible.

5. 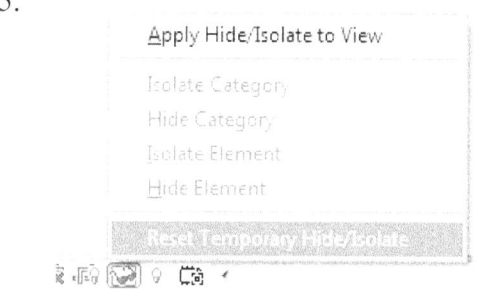 Select the **Temporary Hide/Isolate** tool. Right click and select **Reset Temporary Hide/Isolate**.

6.  Zoom into the region where the lavatories are located.

7. 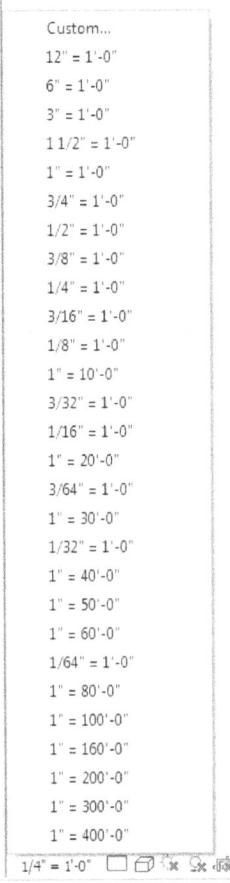 Change the view scale to **1/8″ = 1′-0″**.

8. 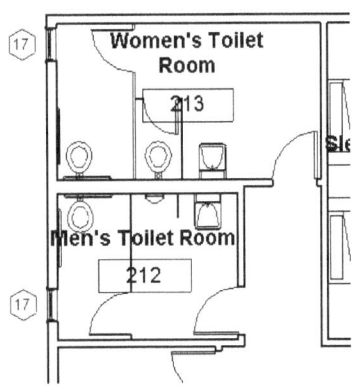 Note that the room tags scale to the view.

9. `Modify` Activate the **Modify** ribbon.

10. 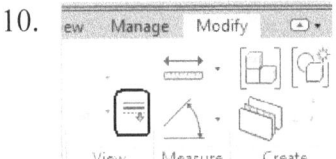 Select the **Linework** tool on the View panel.

11. Set the Line Style to **Overhead**.

12. 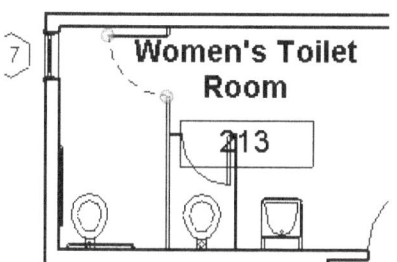 Select the door swing on the toilet cubicle.

    Note that the door swing's appearance changes.

13. Activate **Section 1**.

14. Activate the Manage ribbon.

    Select **Settings→Object Styles**.

15. Expand the **Doors** category on the Model Objects tab.
    Change the Line Weight, Line Color, and Line Pattern for the Panel and Frame.
    Press **Apply** to see the changes.

16. 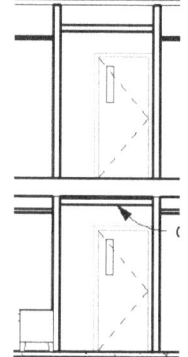 You can move the dialog over so you can see how the display is changed.

    Press OK to close the dialog.

    Note that linework changes are specific to the view, but object settings changes affect all views.

17. Close without saving.

View Properties

Command Exercise

Exercise 4-7 – Reveal Hidden Elements

Drawing Name: **i_visibility.rvt**
Estimated Time to Completion: 5 Minutes

Scope

Turn on the display of hidden elements

Solution

1. Activate **the Ground Floor Admin Wing** floor plan.

2. Select the **Reveal Hidden Elements** tool.

3. Items highlighted in magenta are hidden.

   Window around the two tables while holding down the CONTROL key to select them.

4. Select **Unhide element** from the ribbon.

5. The tables will no longer be displayed as magenta (hidden elements).

6. Select the **Close Hidden Elements** tool.

7. The view will be restored.

8. Select the **Measure** tool from the Quick Access toolbar.

9. Determine the distance between the center of the two tables.

   Did you get 48' 2"?

   If you didn't get that measurement, check that you selected the midpoint or center of the two tables.

10. Close without saving.

4-21

Command Exercise

Exercise 4-8 – Create a View Template

Drawing Name: **view_templates.rvt**
Estimated Time to Completion: 30 Minutes

Scope

Apply a wall tag
Create a view template
Create a view filter
Apply view settings to a view

Solution

1. Activate Level 1.

2. Activate the **Annotate** ribbon.

3. Select **Tag All**.

4. 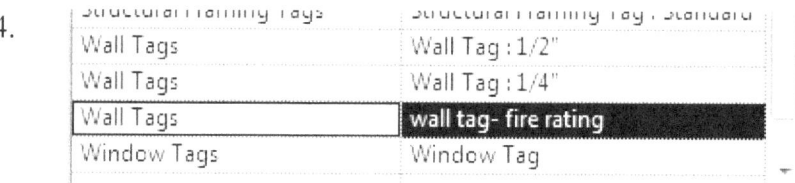 Highlight the **Wall tag - fire rating** as the tag to be used and press **OK**.

   The wall tag - fire rating is a custom family. It was pre-loaded into this exercise, but is included on the exercises CD for your use.

5. 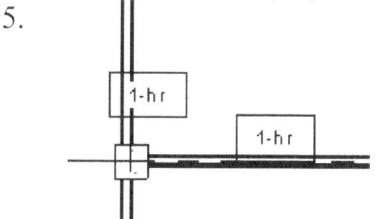 Zoom in to inspect the tags.

4-22

View Properties

6. In the Properties pane:

   Click on the **<None>** button next to View Template.

7. Highlight **Architectural Plan** and select the **Duplicate** button at the bottom of the dialog.

8. Type **Fire Rating** in the Name field.

   Press **OK**.

9. Select the **Edit** button next to **V/G Overrides Filters**.

10. Select **Edit/New** at the bottom of the dialog.

11. Select **New**.

12. Type **2-hr Fire Rating** in the Name field.

    Press **OK**.

4-23

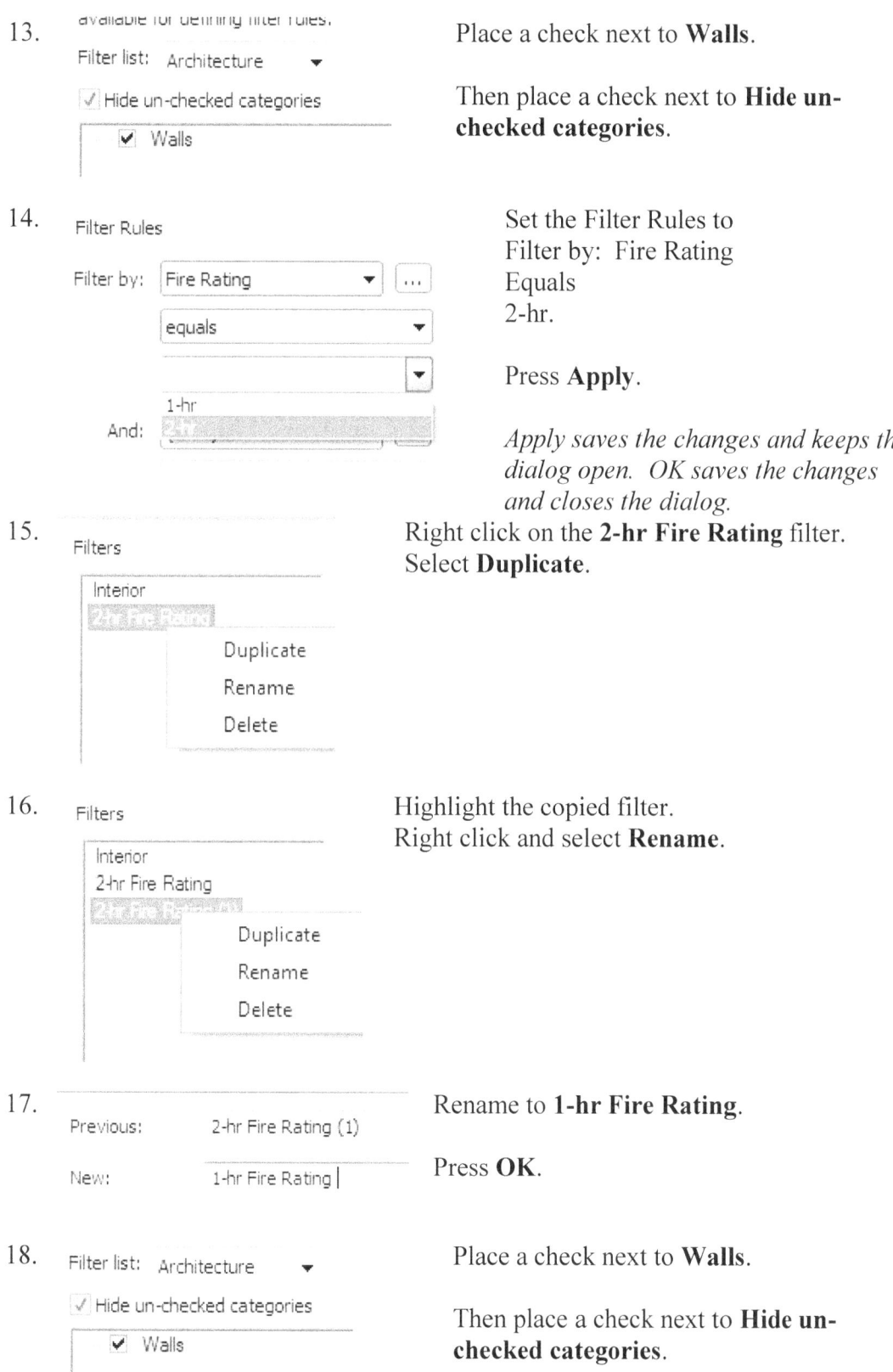

13. Place a check next to **Walls**.

    Then place a check next to **Hide un-checked categories**.

14. Set the Filter Rules to
    Filter by: Fire Rating
    Equals
    2-hr.

    Press **Apply**.

    Apply saves the changes and keeps the dialog open. OK saves the changes and closes the dialog.

15. Right click on the **2-hr Fire Rating** filter. Select **Duplicate**.

16. Highlight the copied filter.
    Right click and select **Rename**.

17. Rename to **1-hr Fire Rating**.

    Press **OK**.

18. Place a check next to **Walls**.

    Then place a check next to **Hide un-checked categories**.

19. Set the Filter Rules to
     Filter by: Fire Rating
     Equals
     1-hr.

     Press **Apply**.

     Press **OK**.

20. Select the **Add** button at the bottom of the Filters tab.

21. 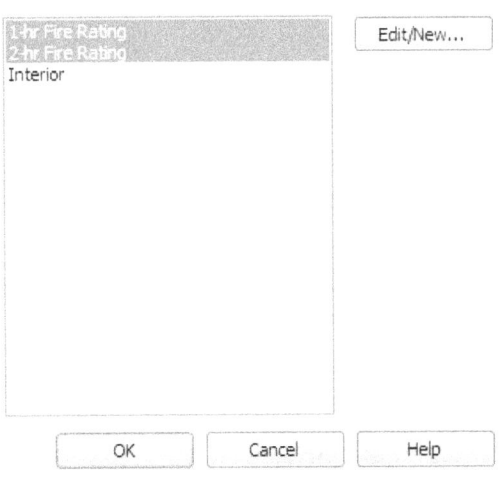 Hold down the Control key. Highlight the 1-hr and 2-hr fire rating filters and press **OK**.

4-25

22. You should see the two fire rating filters listed.

If the interior filter was accidentally added, simply highlight it and select Remove to delete it.

23. Highlight the **2-hr Fire Rating** filter.

24. Select the **Pattern Override** under Projection/Surface.

25. Select the **2 Hour** fill pattern.

Press **OK**.

26. Set the Color to **Blue.**

Press **OK.**

27. Highlight the **1-hr Fire Rating** filter

28. Select the **Pattern Override** under Projection/Surface.

29. Set the fill pattern to **1 Hour** and the Color to **Magenta**.

Press **OK.**

30. Apply the same settings to the Cut Overrides.

Press **OK**.

View Properties

31. Highlight the **Fire Rating** View Template.

    Press **Apply**.

    Press **OK**.

32. The view updates.

33. Activate **Level 2**.

34. In the Properties pane:

    Click on the **<None>** button next to View Template.

35. Highlight the **Fire Rating** View Template.

    Press **Apply**.

    Press **OK**.

36. Activate **Level 3**.

37. In the Properties pane:

    Click on the **<None>** button next to View Template.

38. Highlight the **Fire Rating** View Template.

    Press **Apply**.

    Press **OK**.

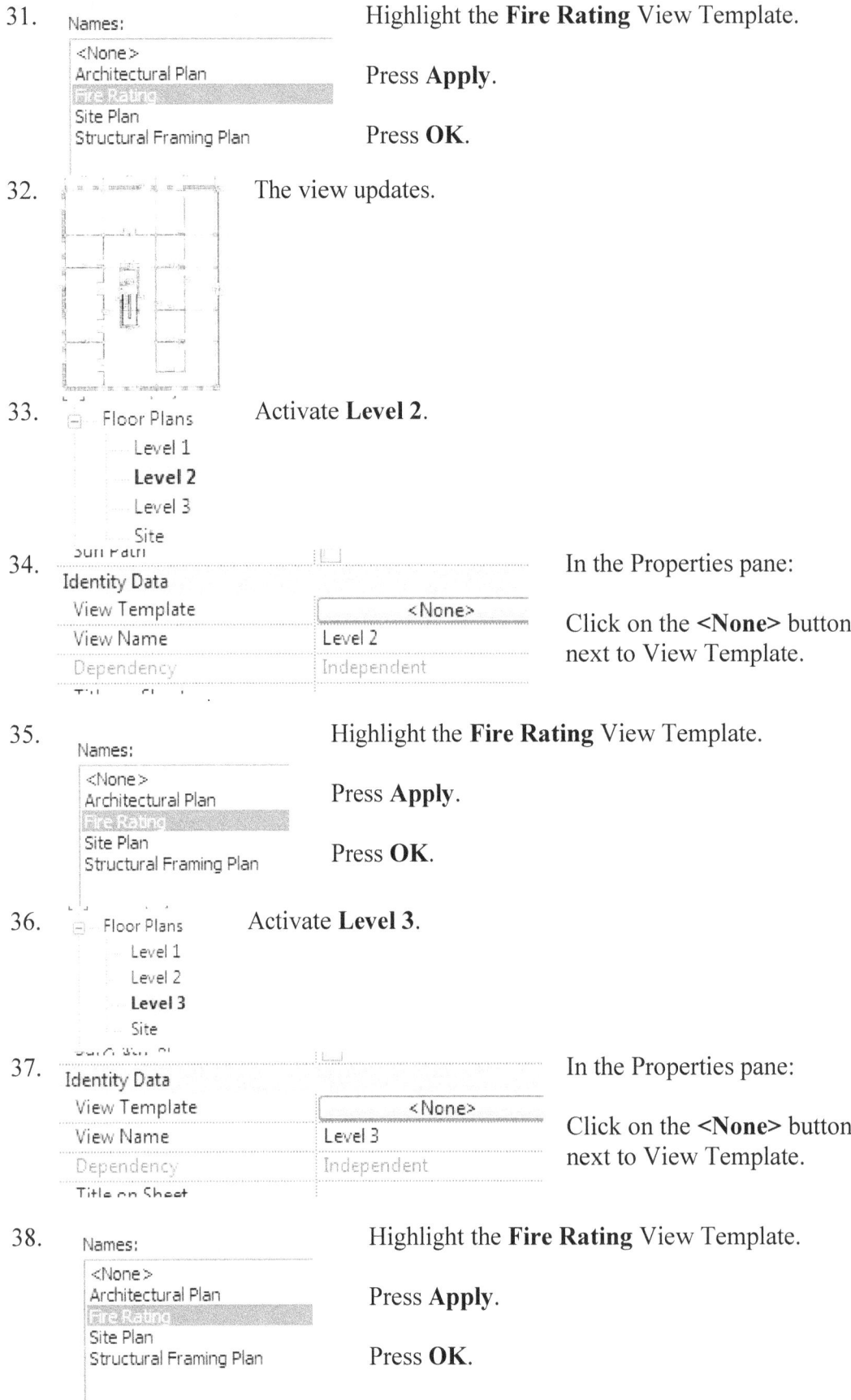

4-27

39. Zoom in to see the hatch patterns.

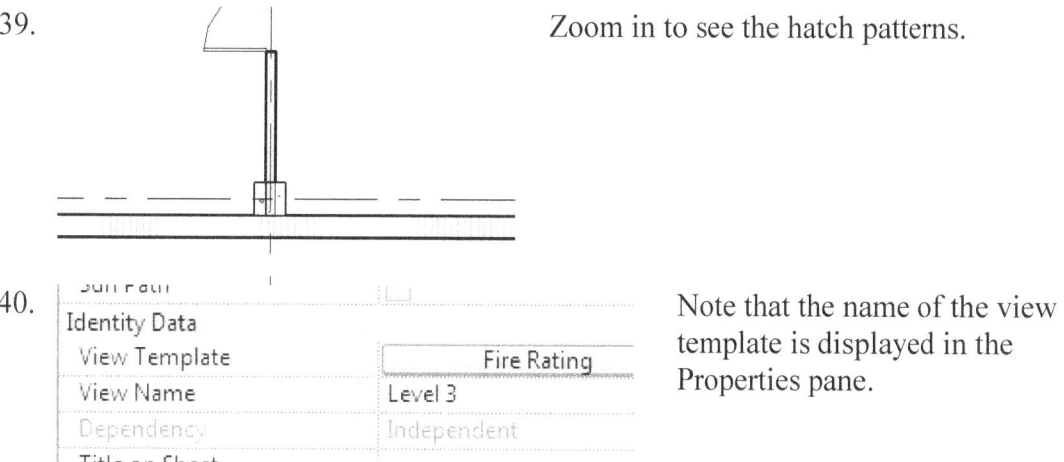

40. Note that the name of the view template is displayed in the Properties pane.

41. Close without saving.

The 1-hr, 2-hr, and 3-hr hatch patterns are custom fill patterns provided inside this exercise file. The *.pat files are included on the exercise CD for your use.

Command Exercise

Exercise 4-9 – Create a Scope Box

Drawing Name: **i_scope_box.rvt**
Estimated Time to Completion: 15 Minutes

Scope

Create and apply a scope box.
Scope boxes are used to control the visibility of grid lines and levels in views.

Solution

1. Activate the **Level 1** floor plan.

2. Activate the **Architecture** ribbon.
   Select the **Grid** tool from the Datum panel.

3. Set the Offset to **2' 0"** on the Options bar.

4. Select the **Pick Lines** tool from the Draw panel.

5. Place three grid lines using the exterior side of the walls to offset.

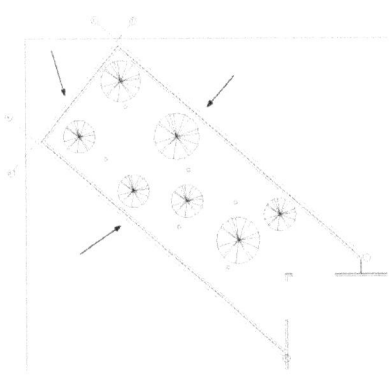

4-29

6. 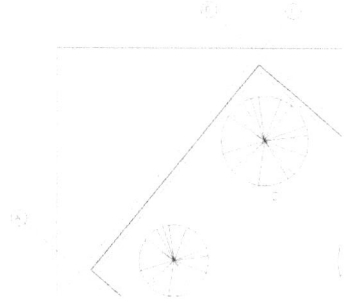 Re-label the grid bubbles so that the two long grid lines are A and B and the short grid line is 1.

7. 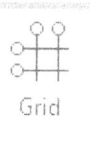 Select the **Measure** tool from the Quick Access toolbar.

8. 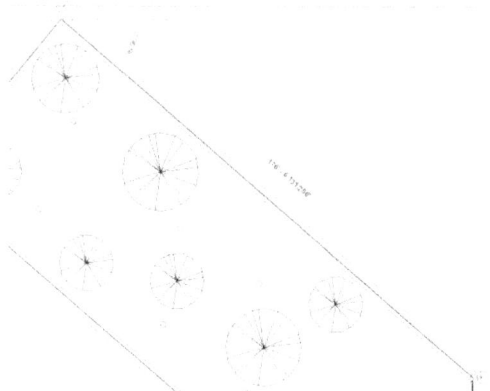 Measure one side of the yard area.

9. Activate the **Architecture** ribbon.

   Select the **Grid** tool from the Datum panel.

10. Set the Offset to **180′ 0″** on the Options bar.

11. 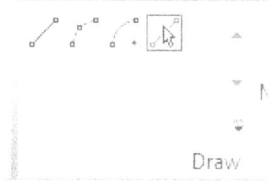 Select the **Pick Lines** tool from the Draw panel.

View Properties

12. Place the lower grid line by selecting the upper wall and offsetting 180'.

    Re-label the grid line **2**.

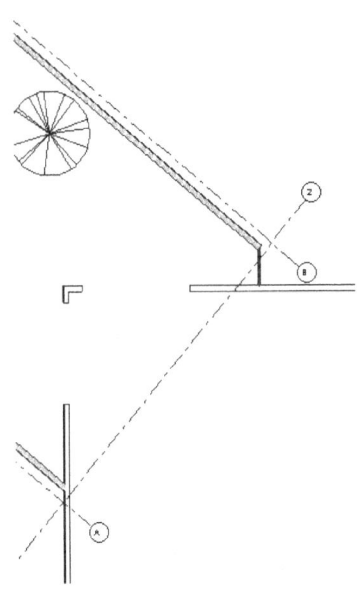

13. Activate the **View** ribbon.

    Select the **Scope Box** tool from the Create panel.

14. Place the scope box.

    Use the Rotate icon on the corner to rotate the scope box into position.

    Use the blue grips to control the size of the scope box.

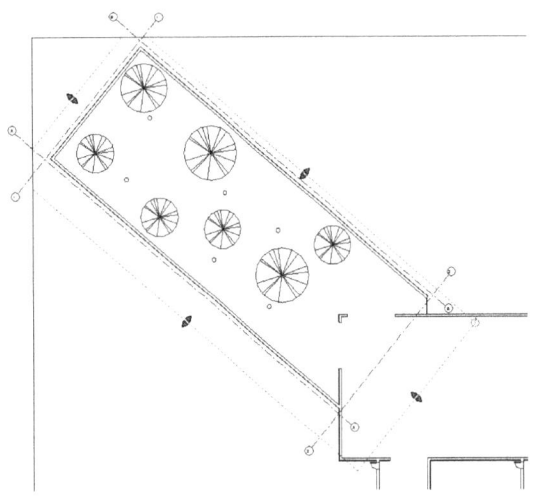

15. Select the grid line labeled **B**.

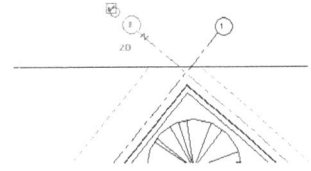

4-31

16. 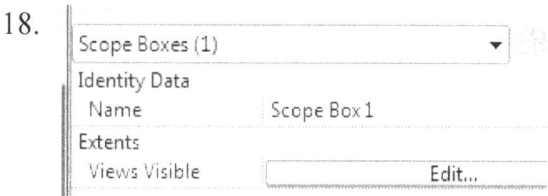 In the Properties pane:

    Set the Scope Box to Scope Box 1, the scope box which was just placed.

17. Repeat for the other three grid lines.

18. Select the Scope Box.

    Select **Edit** next to Views Visible in the Properties pane.

19. 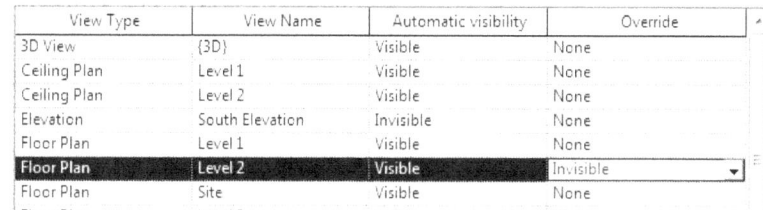 Set the Level 2 Floor Plan Override Invisible.

    Press **OK**.

20. Select the scope box.
    Right click and select **Hide in View→ Elements**.

    The scope box is no longer visible in the view.

21. Activate Level 2. The grid lines and scope box are not visible.

22. Close the file without saving.

4-32

Command Exercise
Exercise 4-10 – Duplicating Views

Drawing Name: **duplicating_views.rvt**
Estimated Time to Completion: 20 Minutes

Scope

Duplicate view with Detailing
Duplicate view as Dependent
Duplicate view

Understand the difference between the different duplicating views options

Solution

1. Activate the **Level 1** floor plan.

   Note that the doors all have door tags.

2. Highlight Level 1.

   Select **Duplicate View→Duplicate**.

   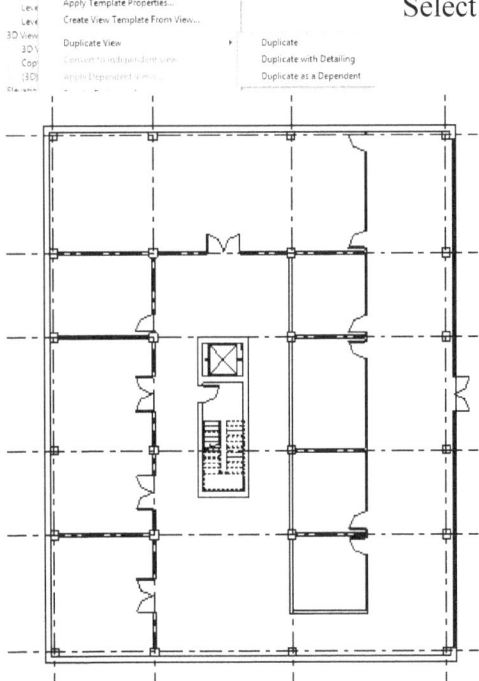

   When you select Duplicate View→Duplicate, any annotations, such as tags and dimensions, are not duplicated.

4-33

3. Highlight the copied level 1.

    Right click and select **Rename**.

    Name: Level 1 - No Annotations

    Type **Level 1- No Annotations**.

    Press **OK**.

4. Highlight Level 1.

    Right click and select **Duplicate View→Duplicate with Detailing**.

    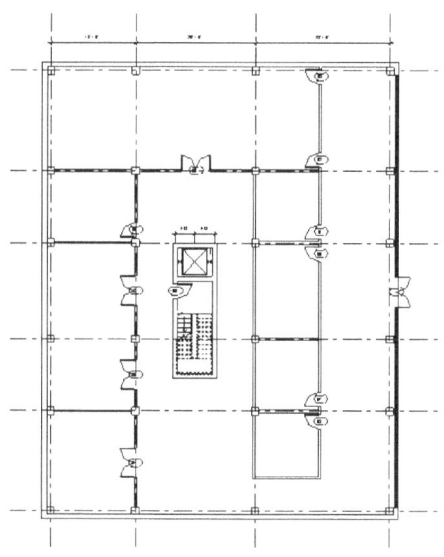

    Note that when you select Duplicate with Detailing the new view includes annotations, such as tags and dimensions.

View Properties

5. Highlight the copied level 1.

   Right click and select **Rename**.

   Name: Level 1-Original Annotations

   Type **Level 1- Original Annotations**.

   Press **OK**.

6. Highlight Level 1.
   Right click and select **Duplicate View→Duplicate as a Dependent**.

7. *Notice that the duplicated view includes the annotations.*

   Floor Plans
      Level 1
         Dependent on Level 1
      Level 1 - No Annotations
      Level 1-Original Annotations

   In the Project Browser, the dependent view is listed underneath the parent view.

8. Name: Level 1 - Stairs Area

   Rename the dependent view **Level 1- Stairs area**.

9. Floor Plans
      Level 1
         Level 1 - Stairs Area
      Level 1 - No Annotations

   Activate **Level 1- Stairs area**.

10. In the Properties pane:

    Enable **Crop View**.
    Enable **Crop Region Visible**.

4-35

11. 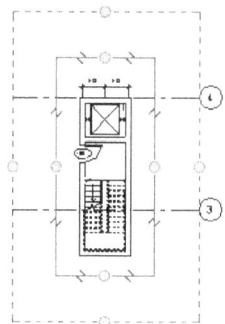 Adjust the crop region to focus the view on the stairs area.

12. Activate **Level 1**.

13. 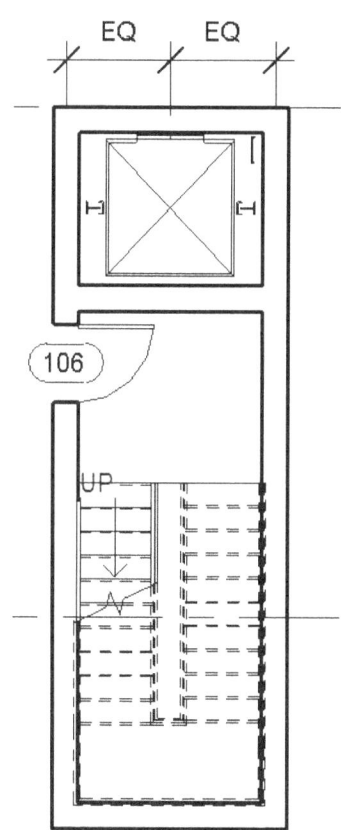 Zoom into the stairs area.

14. Activate the **Annotate** ribbon.
15. Select the **Tread Number** tool on the Tag panel.

Tread Numbers can only be applied to Component-based stairs.

View Properties

16. Select the middle line that highlights when your mouse hovers over the left side of the stairs.

17. Select the middle line on the right side of the stairs.

Right click and select CANCEL to exit the command.

18. 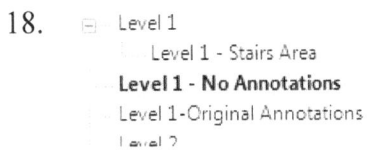 Activate **Level 1- Original Annotations**.

   Notice that the new annotations - the tread numbers - are not visible in this view.

19. Activate **Level 1 - No Annotations**.

   Notice that the new annotations - the tread numbers - are not visible in this view.

20. Activate the **Level 1- Stairs area** dependent view.

   Notice that any annotations added to the parent view are added to the dependent view.

21. Close without saving.

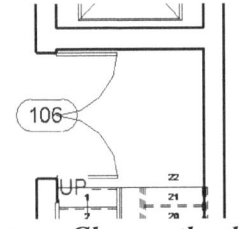

Extra: Change the door labeled 106 to a double flush door. Which Level 1 views display the new door type?

4-37

Command Exercise
Exercise 4-11 – Segmented Views

Drawing Name: **segmented views.rvt**
Estimated Time to Completion: 10 Minutes

Scope

Modify a section view into segments

Solution

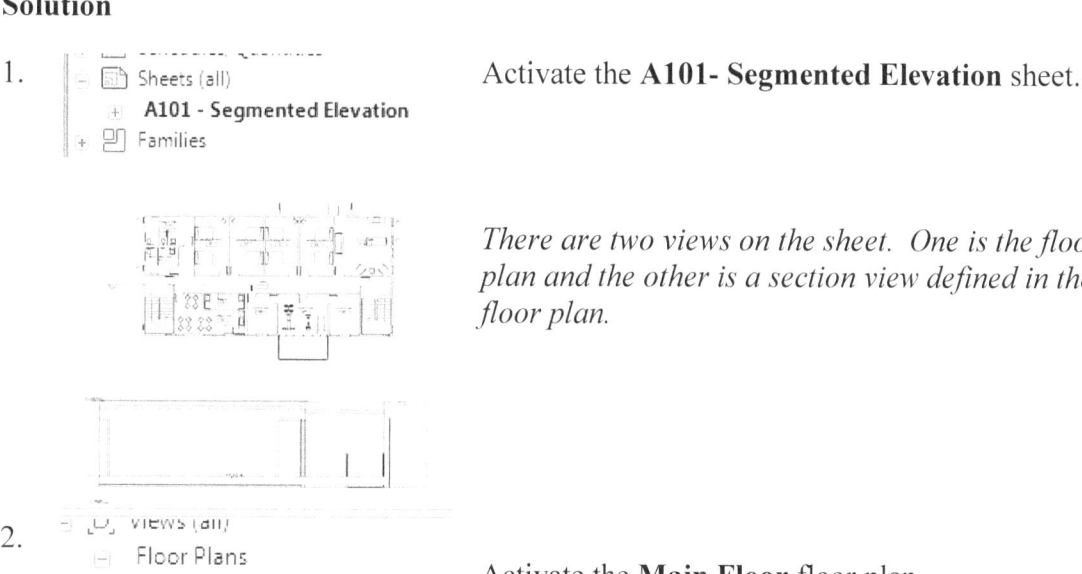

1. Activate the **A101- Segmented Elevation** sheet.

   There are two views on the sheet. One is the floor plan and the other is a section view defined in the floor plan.

2. Activate the **Main Floor** floor plan.

   This is the top view on the sheet.

3. Select the section line.

4. Select **Split Segment** on the ribbon.

4-38

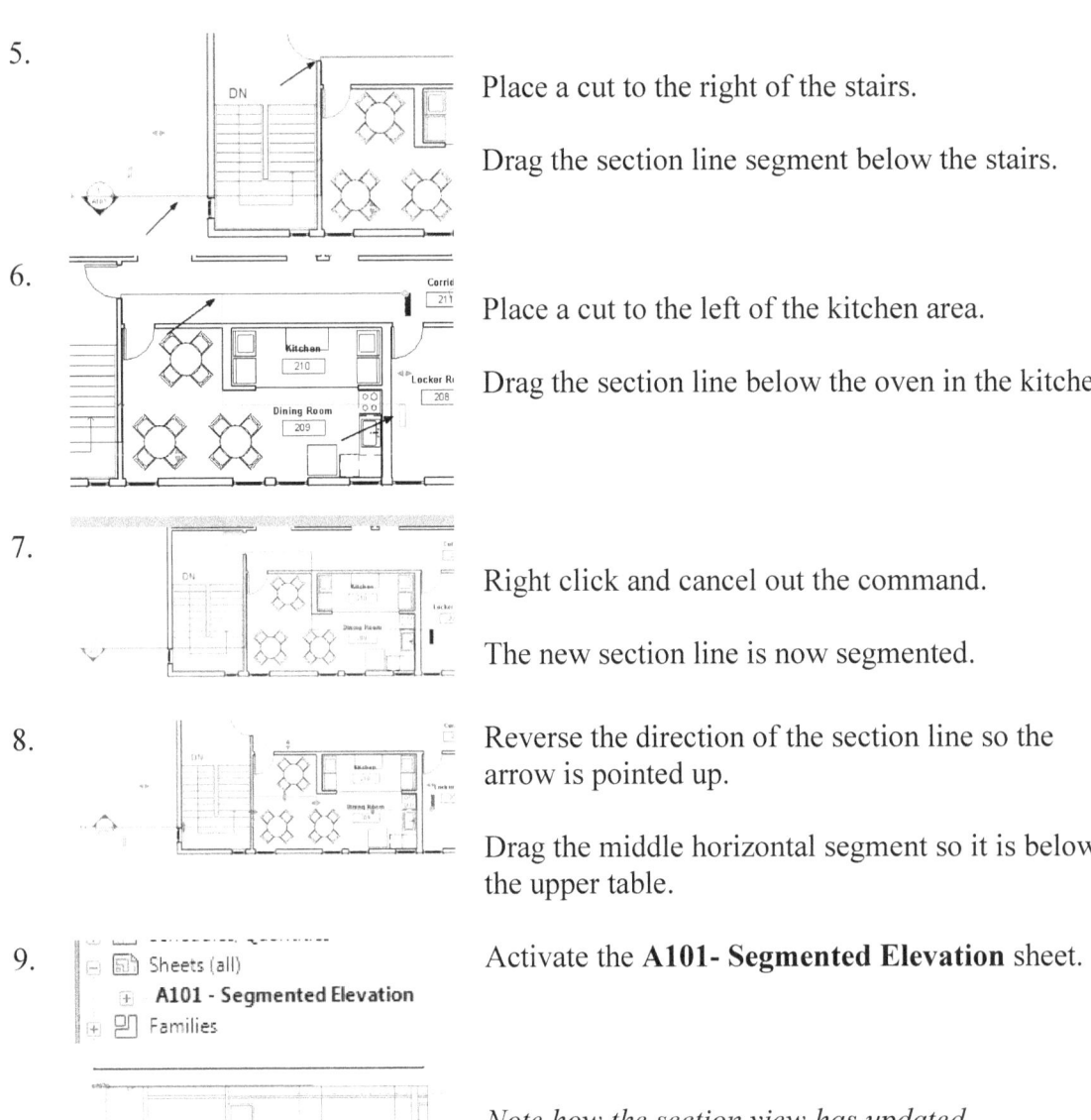

5. Place a cut to the right of the stairs.

   Drag the section line segment below the stairs.

6. Place a cut to the left of the kitchen area.

   Drag the section line below the oven in the kitchen.

7. Right click and cancel out the command.

   The new section line is now segmented.

8. Reverse the direction of the section line so the arrow is pointed up.

   Drag the middle horizontal segment so it is below the upper table.

9. Activate the **A101- Segmented Elevation** sheet.

   Note how the section view has updated.

10. Close without saving.

Practice Exam

1. Straight grid lines are visible in the following view types:
   A. ELEVATION
   B. PLAN
   C. 3D
   D. SECTION
   E. DETAIL

2. Which two shortcut keys launch the Visibility/Graphics dialog?
   A. VG
   B. VV
   C. VE
   D. VW
   E. F5

3. True or False: Objects hidden using the Temporary Hide/Isolate tool are not visible, but they are still printed.

4. Select the THREE options for Detail Level for a view:
   A. COARSE
   B. SHADED
   C. FINE
   D. HIDDEN
   E. WIREFRAME
   F. MEDIUM

5. True or False: If you delete a view, the annotations placed in the view are also deleted.

6. True or False: A camera can not be placed in an elevation view.

7. Scope boxes control the visibility of:
   A. Elements
   B. Plumbing Fixtures
   C. Object Styles
   D. Grid lines and levels

8. To change the graphic appearance of your model from Hidden Line to Realistic, you:
   A. Modify the Rendering Settings in the View Control Bar
   B. Edit Visibility/Graphics Overrides
   C. Change graphic display options in View Properties
   D. Click Visual Styles on the View Control Bar

9. A story level is the color:
    A. Yellow
    B. Blue
    C. Black
    D. Green

10. If you create a level using the COPY or ARRAY tool, what type of level is created?
    A. Story
    B. Non-Story or Reference
    C. Elevation
    D. Plan

11. This type of level does not have a PLAN view associated to it:
    A. Story
    B. Non-Story or Reference
    C. Elevation
    D. Plan

12. In which view type can you place a level?
    A. PLAN
    B. LEGEND
    C. 3D
    D. CONSTRUCTION
    E. ELEVATION

Answers:
1) A, B & D; 2) A & B; 3) True; 4) A, C & F; 5) True; 6) False; 7) D; 8) D; 9) B; 10) B; 11) B; 12) E

Lesson Five

Dimensions and Constraints

This lesson addresses the following exam questions:

- Constraints
- Temporary and Permanent Dimensions
- Multisegmented Dimensions

Dimensions are system families. They are in the Annotation category. They have type and instance properties.

Revit has three types of dimensions: listening, temporary and permanent.

A listening dimension is the dimension that is displayed as you are drawing, modifying, or moving an element.

A temporary dimension is displayed when an element is placed or selected. In order to modify a temporary dimension, you must select the element.

A permanent dimension is placed using the Dimension tool. In order to modify a permanent dimension, you must move the element to a new position. The permanent dimension will automatically update. To reposition an element, you can modify the temporary dimension or move the element using listening dimensions.

When you enter dimension values using feet and inches, you do not have to enter the units. You can separate the feet and inches values with a space and Revit will fill in the units. If a single unit is entered, for example '10', Revit assumes that value is 10 feet, not 10 inches.

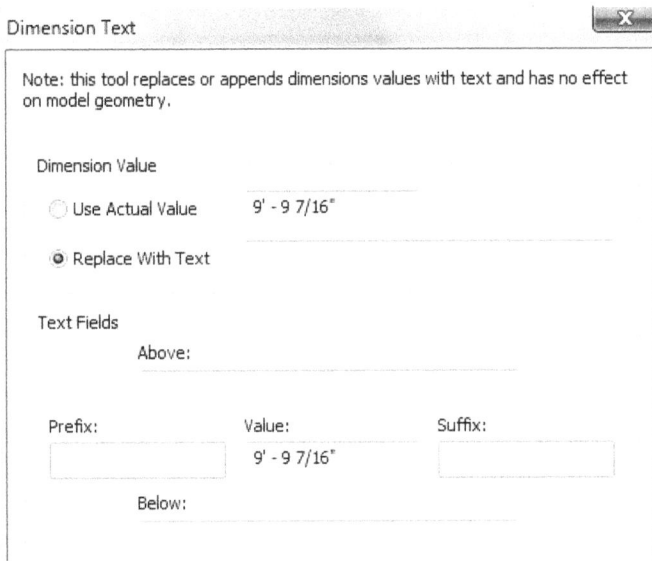

To override a dimension value, select the permanent dimension and enable Replace with Text and put in the desired text.

You can also add additional notes using the Text Fields for Above, Prefix, Suffix, or Below.

Dimensions and Constraints

Command Exercise

Exercise 5-1 – Placing Permanent Dimensions

Drawing Name: **i_dimensions.rvt**
Estimated Time to Completion: 20 Minutes

Scope

Placing dimensions

Solution

1. Floor Plans
   Ground Floor
   Ground Floor Admin Wing
   Lower Roof
   Main Floor
   Main Floor Admin Wing

   Activate the **Ground Floor Admin Wing** floor plan.

2.  Activate the Modify ribbon.

   Select the **Match Properties** tool from the Clipboard panel.

3. Select the wall indicated as the source object.

4. 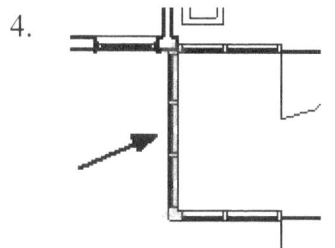 Select the wall indicated as the target object.

5-3

5. Note that the curtain wall changed to an Exterior-Siding wall.

   Cancel out of the Match Properties command.

6. Activate the Annotate ribbon.

   Select the **Aligned Dimension** tool.

7.

   On the Options bar, set the dimensions to select the **Wall faces**.

8. On the Options bar, set the dimensions to **Pick Entire Walls**.

9. Pick the wall indicated.

   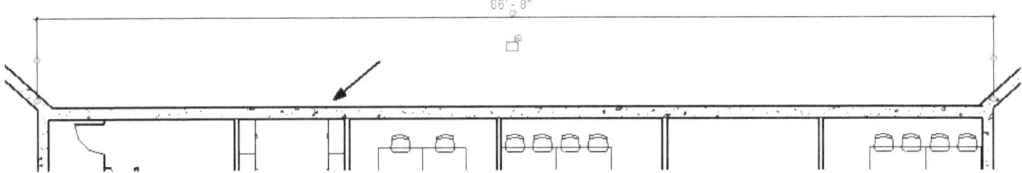

   Move the mouse above the selected wall to place the dimension.

10. Note the entire wall is selected and the dimension is located at the faces of the walls.
    Cancel or escape to end the Dimension command.

11. Select the dimension.

    Note that there are several grips available. The grips are the small blue bubbles.

12. 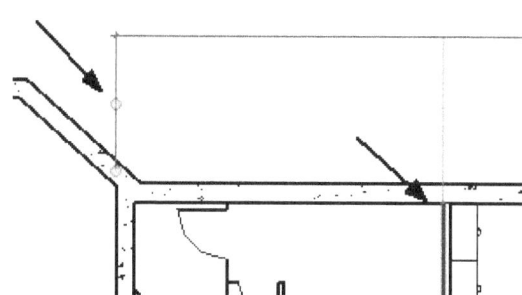 Select the second grip indicated on the left.

Move the witness line to the wall face indicated by the arrow in the center of the image.

Note that the witness line automatically snaps to the wall face.

13. Prefer: Wall centerlines   With the dimension selected:

On the Options bar, change the Prefer to **Wall centerlines**.

14. 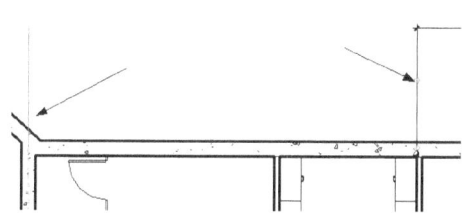 Select the dimension and activate the grip indicated.

Drag the witness line to the center of the left wall.

15. 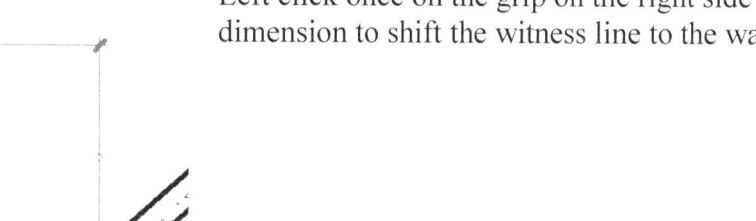 Left click once on the grip on the right side of the dimension to shift the witness line to the wall centerline.

16. Note how the dimension value updates.

17. Select the **Aligned** tool.

18. 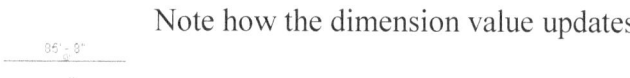 Set Pick to Individual References on the options bar.

19. 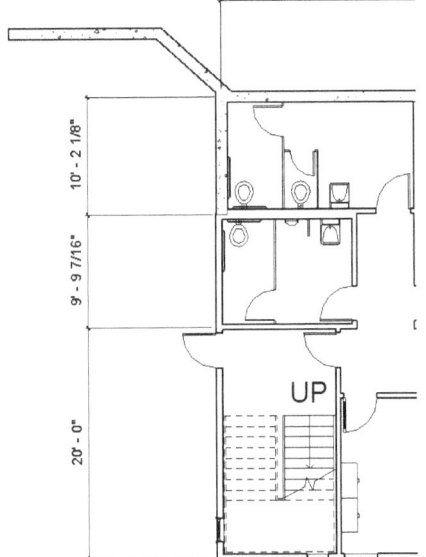 Select the centerlines of the walls indicated.

20. 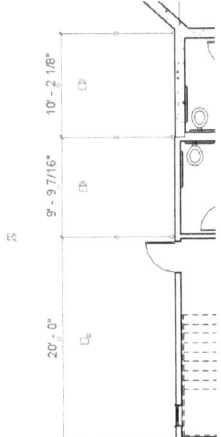 Select the dimension so it highlights.

    Select the two locks on the top dimensions to switch the permanent dimensions to *locked* dimensions. This means these distances will not be changed.

21. 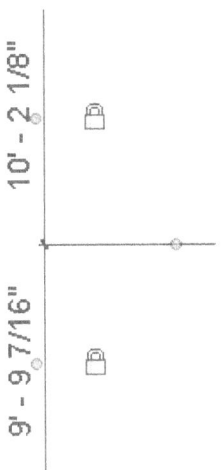 You should see two of the padlocks as closed and one as open.

22. 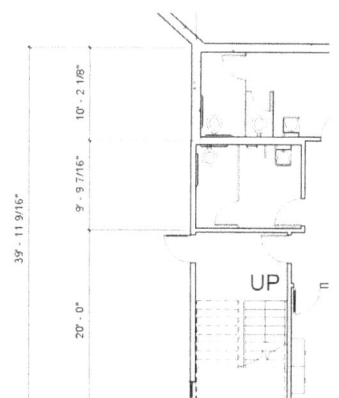 Place an overall dimension using the wall centerlines.

23. 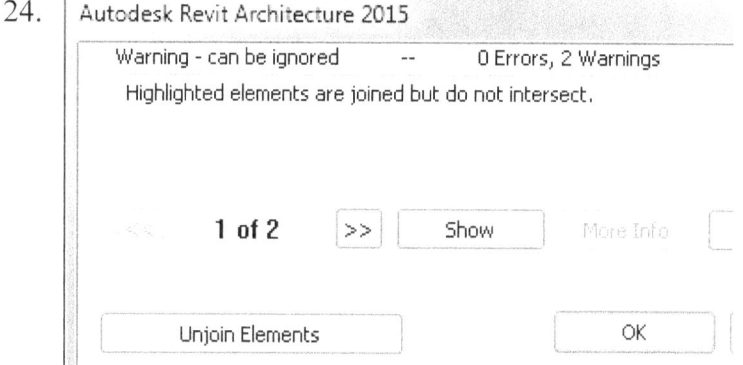 Drag the wall indicated down to a new position.

    Move down 5'-0".

24. If you get an error message, select **Unjoin Elements**.

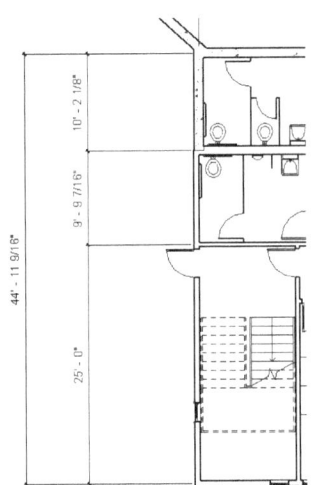

Note that the two locked dimensions did not change. Only the unlocked dimension updated.

25. Close without saving.

Dimensions and Constraints

Command Exercise

Exercise 5-2 – Modifying Dimension Text

Drawing Name: **dimtext.rvt**
Estimated Time to Completion: 15 Minutes

Scope

Replace Dimension Text
Restore Dimension Text
Modify Dimension Text

Solution

1. Activate the **Ground Floor Admin Wing** floor plan.

   Floor Plans
      Ground Floor
      Ground Floor Admin Wing
      Lower Roof
      Main Floor
      Main Floor Admin Wing
      Main Roof

2. 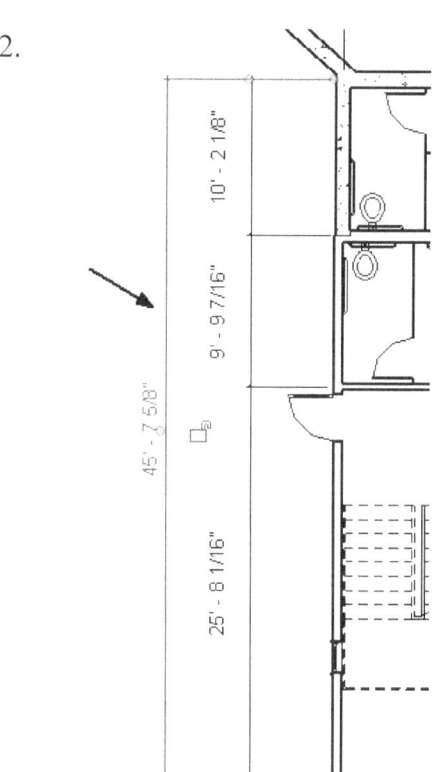 Select the dimension indicated.

   Double left click on the dimension text to bring up the dimension text dialog.

5-9

3. Enable **Replace with Text**.
   Enter **45′ 8″**.

   Press **OK**.

4. You will get an error message stating that you cannot change the numeric value without moving the element to correspond with that value.

   Press **Close**

5. Enable **Replace with Text**.
   Enter **OVERALL**.

   Press **OK**.

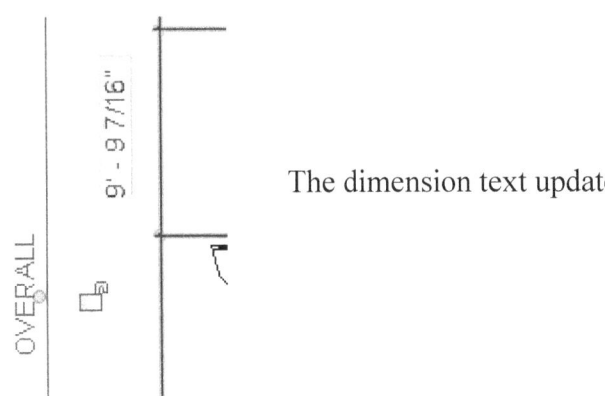

The dimension text updates.

6. Activate the **Manage** ribbon.

   Select **Project Units** on the Settings panel.

7. Select the **Length** button under the Format column.

8. Set the Rounding **To the nearest ½″**.

   Press **OK** until all dialogs are closed.

Dimensions and Constraints

Note that the dimensions update to the new rounding.

9. Double left click on the 10′ 2″ dimension to be edited.

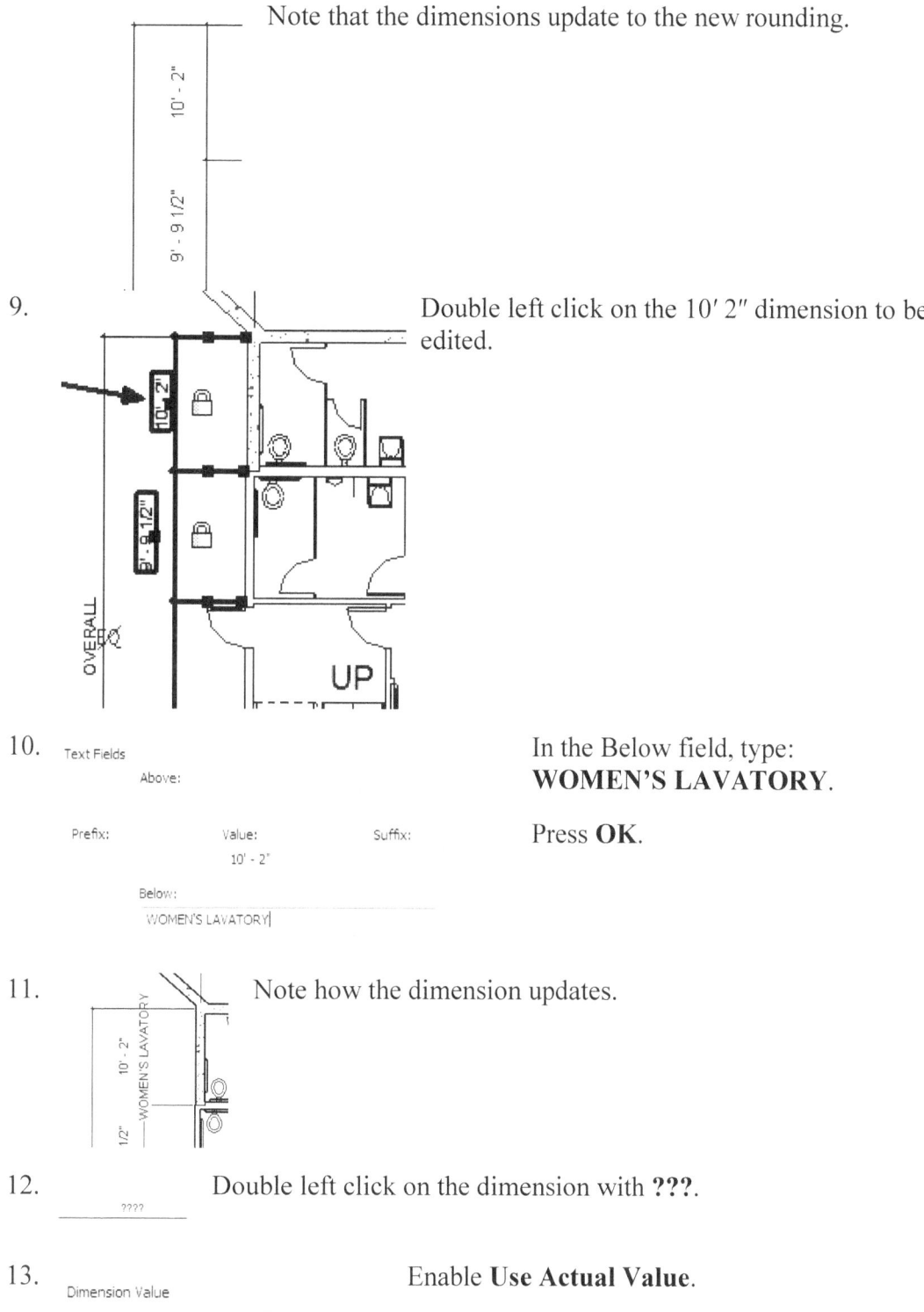

10. In the Below field, type: **WOMEN'S LAVATORY**.

    Press **OK**.

11. Note how the dimension updates.

12. Double left click on the dimension with **???**.

13. Enable **Use Actual Value**.

    Press **OK**.

14. Close without saving.

Command Exercise

Exercise 5-3 – Converting Temporary Dimensions to Permanent Dimensions

Drawing Name: **i_dimensions.rvt**
Estimated Time to Completion: 10 Minutes

Scope

Place dimensions using different options.
Convert temporary dimension to permanent

Solution

1. Activate the **Ground Floor Admin Wing** floor plan.

2. 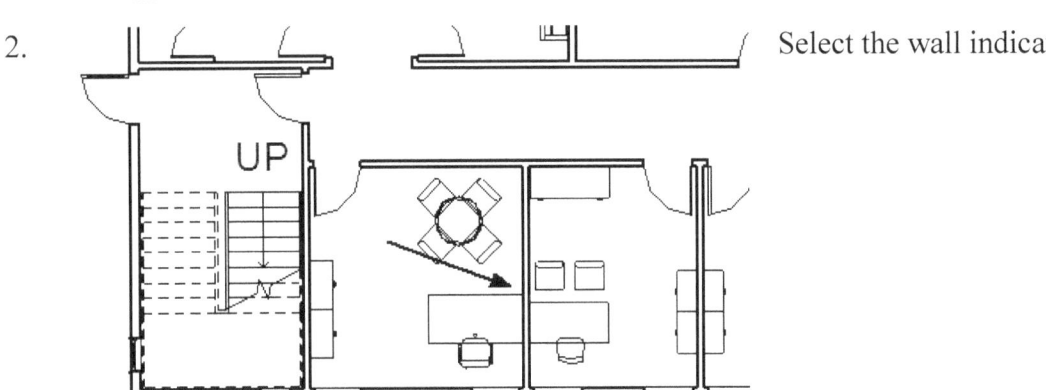 Select the wall indicated.

3. 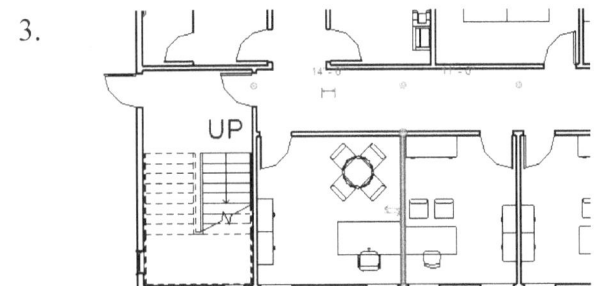 Two dimensions will appear.

   There is a small dimension icon visible.

   This icon converts a temporary dimension to a permanent dimension.

   Left click on this icon.

Dimensions and Constraints

4. 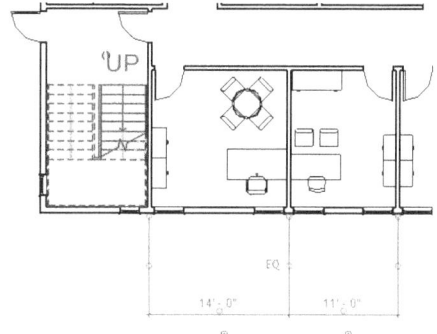 The dimensions are now converted to permanent dimensions.

   Drag the dimensions below the view.

5. Click on the dimensions so they highlight.

   Click on the witness line grip indicated.

   Notice how each time the grip is clicked on, the witness line shifts from the wall centerline to the face of the wall.

6. Select the grip at the endpoint of the witness line. Drag the endpoint to increase the gap between the wall and the witness line.

   Release the grip.

7. Select the witness line grip indicated.

   Drag the witness line to the outer wall. Note that if the witness line is set to the outer wall face it maintains that orientation.

   Release.

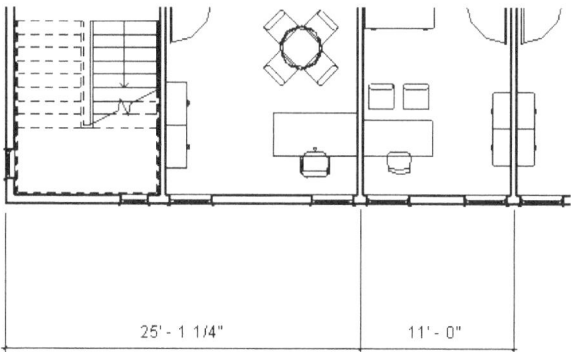 The dimension updates.

   Note the new dimension value.

8. Close without saving.

Command Exercise

Exercise 5-4 – Applying Constraints

Drawing Name: **i_constraints.rvt**
Estimated Time to Completion: 20 Minutes

Scope

Using the Align tool to constrain elements

Solution

1. Activate the **Main Floor** floor plan.

2. 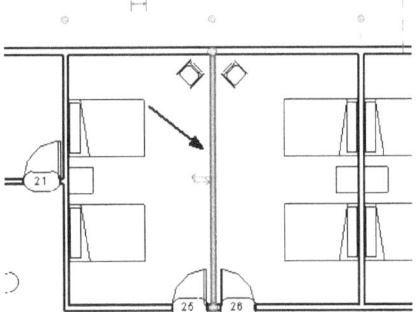 Select the interior wall located between Door 25 and Door 26 as indicated.

3. The associated temporary dimensions will become visible.

   Left click on the permanent dimension toggle to convert the dimensions to permanent dimensions.

Dimensions and Constraints

4. Left click anywhere in the display window.

    In the Properties pane:

    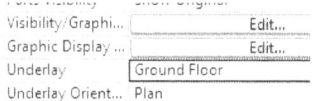
    Scroll down to the **Underlay** parameter.

    Set the Underlay to **Ground Floor**.

    This will make the ground floor visible in the graphics window. Elements on the ground floor will be displayed in a lighter shade.

5. 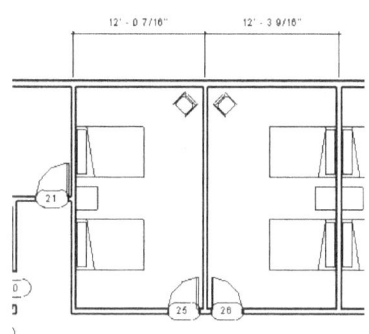 Note in the upper left corner the interior walls are not aligned.

6. Activate the **Modify** ribbon.

    Select the **Align** tool from the Modify panel.

7. Prefer: Wall faces   Set the preference to **Wall faces** on the Options bar.

8. 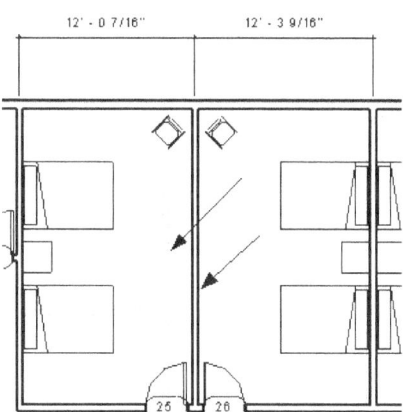 Select the right side of the ground floor interior wall as the source for the alignment.

5-15

9. 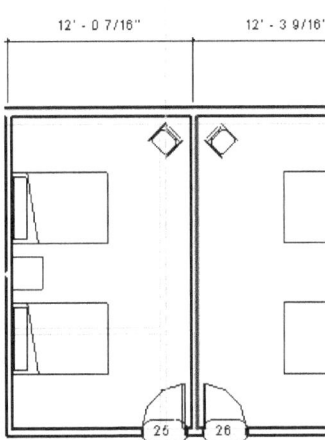 Select the right side of the main floor interior wall as the target for the alignment (this is the wall that will be shifted).

10. A warning dialog will appear due to the door location. Ignore the warning. ***Don't move the door or wall.***

11. 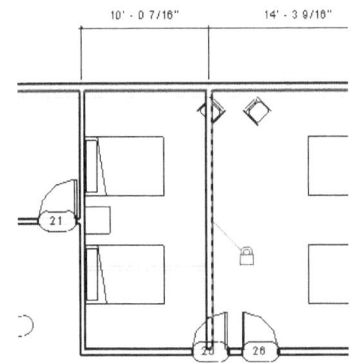 Lock the alignment in place so that the walls remain constrained together.

12. Right click and select Cancel to exit the Align mode.

13. Activate the **Ground Floor** floor plan.

Dimensions and Constraints

14. 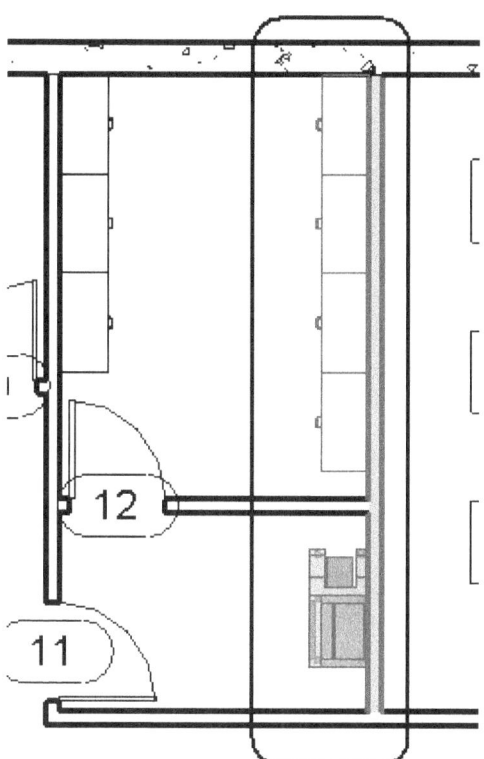 Select the wall as well as the cabinets and copier next to the wall.

15. Select the **Move** tool on the Modify ribbon.

16.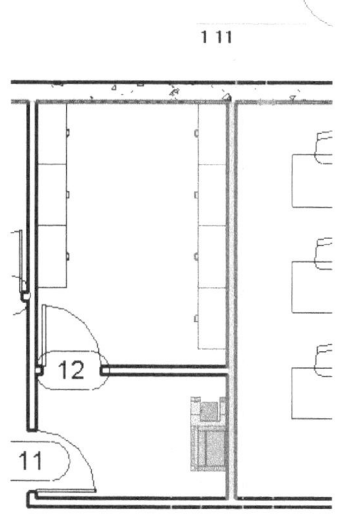

Select a base point.

Move the selected elements to the right **1' 11"**.

Left click in the window to release the selected elements.

17. Activate the **Main Floor** floor plan.

5-17

18. 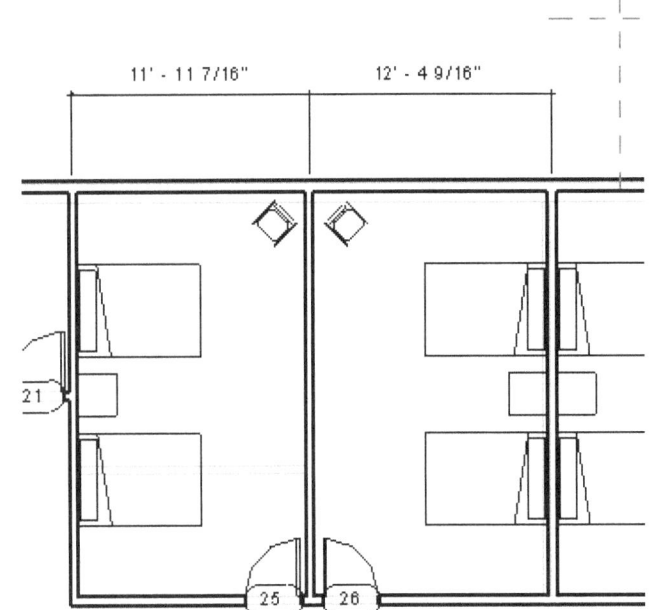 Note that the main floor interior wall has also shifted (the walls remained aligned because they are locked together) and the dimension has updated.

19. Close without saving.

Dimensions and Constraints

Command Exercise
Exercise 5-5 – Equality Formula

Drawing Name: **equality_formula.rvt**
Estimated Time to Completion: 10 Minutes

Scope

Using an Equality Formula

Solution

1. Activate the **Main Floor** floor plan.

2. Select the EQ dimensions.
   Right click and disable EQ display.
   Note the value is 16' 4".

3. With the dimension still selected, go to the Properties pane.

   Under Equality Display:

   Select **Equality Formula**.

4. Select **Edit Type**.

5. Scroll down.
   Select the button next to Equality Formula.

6. Add Number of Segments and Length of Segment to Label Parameters.

   Organize so that Number of Segments is on Row 1; Length of Segment is on Row 2, and Total Length is on Row 3.

5-19

7.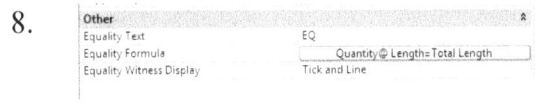
   Add a Suffix for Number of Segments - @.
   Add a Suffix for Length of Segment - =.
   Place a space in each Prefix field.

   This builds an equation that reads

   <Number of Segments> @ <Length of Segment> = <Total Length>

   Press **OK**.

8. The Equality Formula is updated.

   Press **OK**.

9. Left click in the graphics window to release the selection and update the dimension.

   If you don't see spaces, go back and add spaces to the prefix fields.

10. Close without saving.

Command Exercise

Exercise 5-6 – Alternate Dimensions

Drawing Name: **Alternate Dimensions.rvt**
Estimated Time to Completion: 10 Minutes

Scope

Apply Alternate Dimensions

Solution

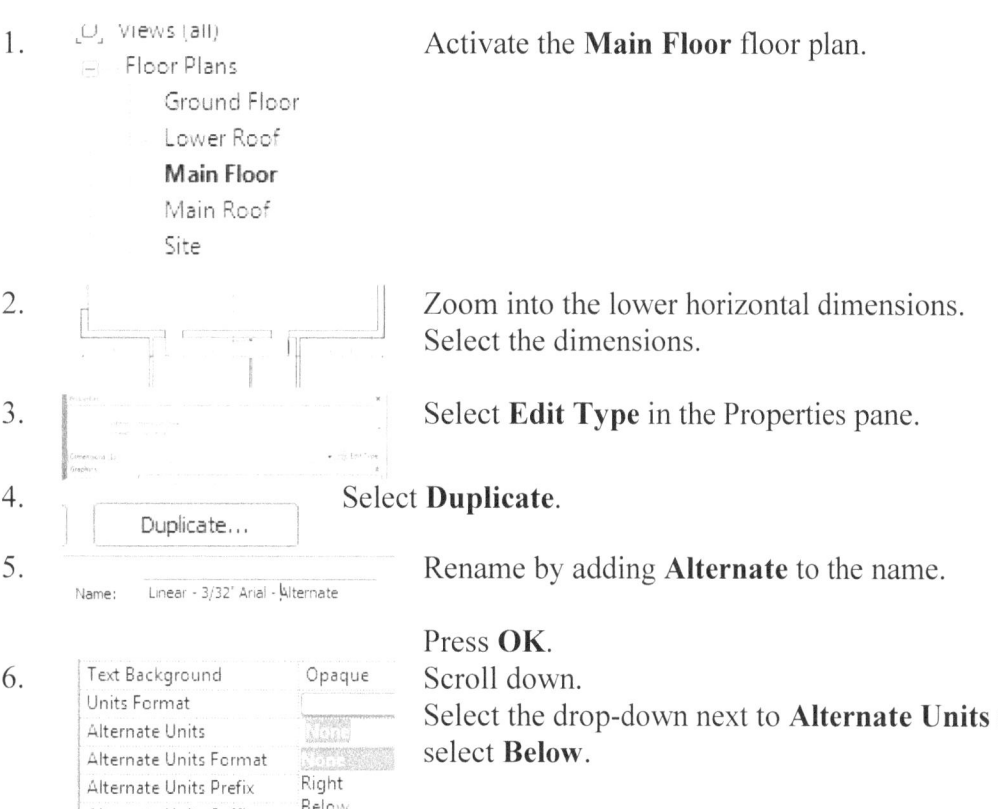

1. Activate the **Main Floor** floor plan.

2. Zoom into the lower horizontal dimensions. Select the dimensions.

3. Select **Edit Type** in the Properties pane.

4. Select **Duplicate**.

5. Rename by adding **Alternate** to the name.

6. Press **OK**.
   Scroll down.
   Select the drop-down next to **Alternate Units** and select **Below**.

7. Type **mm** in the Alternate Units Suffix field.
   Press **OK**.

8. *The dimensions now display with both Imperial and millimeter units.*

9. Close without saving.

Command Exercise

Exercise 5-7 – Multi-Segmented Dimensions

Drawing Name: **multisegment.rvt**
Estimated Time to Completion: 10 Minutes

Scope

Add and delete witness lines to a multi-segment dimension

Solution

1. Activate the **Main Floor** floor plan.

2. Select the lower dimension so it highlights.

3. On the ribbon, select **Edit Witness Lines**.

4. Select the center of each door indicated.

   Then, left pick below the building to accept the additions.

5. Drag the dimensions down so you can see them easily.

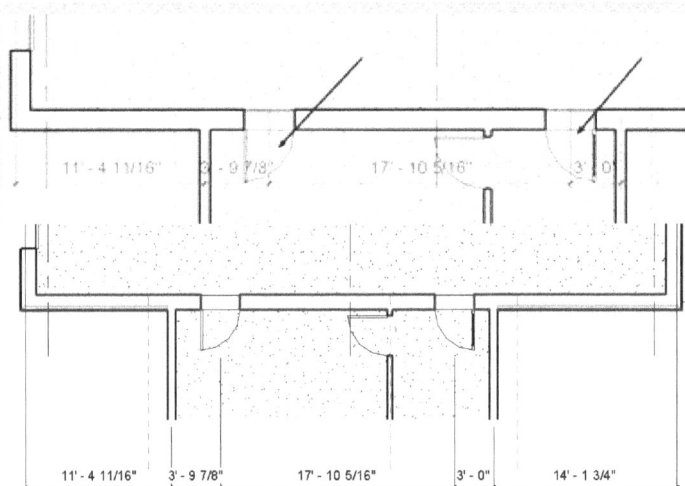

5-22

Dimensions and Constraints

6. Select the dimension.
   Select **Edit Witness Lines**.
   Add the witness line at the wall indicated.

7. Select the dimension.
   Select **Edit Witness Lines**.
   Add the witness line at the two reference planes indicated.

8.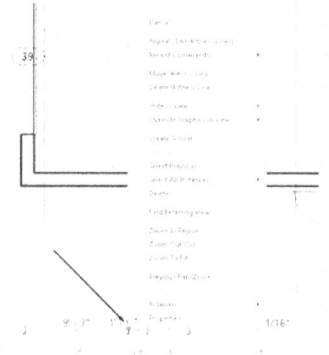

   Select the dimension.

   Select the grip on the witness line for the left reference plane.

   Right click and select **Delete Witness Line**.

9.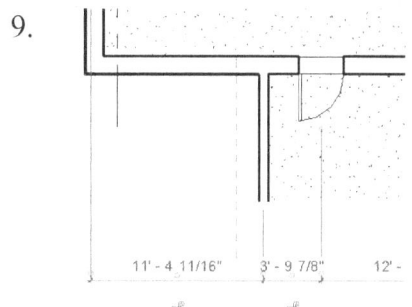

   The dimension should update with the witness line removed.

5-23

10. 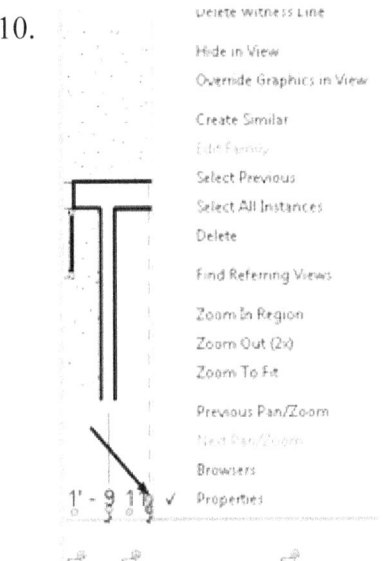 Repeat to delete the witness line on the right reference plane.

11. The dimension should update.

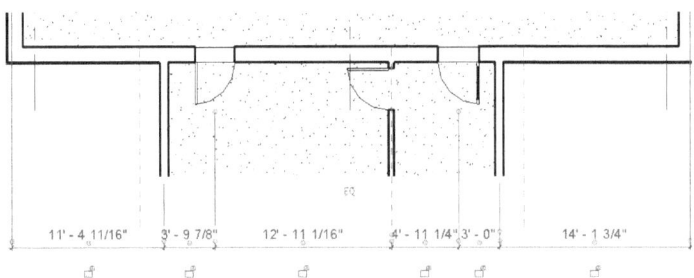

12. Close without saving.

Practice Exam

1. Constraints can be used to:
    A. Prevent other users from moving an element in a shared project.
    B. Lock two elements at a fixed distance.
    C. Keep two elements equally spaced.
    D. Alert the user when someone changes or moves an element.

2. To change a dimension's value:
    A. Use the Change Dimension tool.
    B. Click on Edit Witness Line.
    C. Select the element to be re-dimensioned.
    D. Click on the dimension.

3. True or False: You can override a dimension value with a new dimension value.

4. Select the TWO types of dimensions:
    A. PERMANENT
    B. TEMPORARY
    C. PERPETUAL
    D. INVISIBLE

5. To create a dimension using feet and inches, you can use this keystroke to separate the feet and inches values:
    A. COMMA
    B. SEMI-COLON
    C. SPACE
    D. COLON

6. A _____ allows you to locate elements equidistant from one another.
    A. multi-segmented dimension
    B. single-segmented dimension
    C. scope box
    D. reference plane

7. You are editing a dimension in Revit. You type 0 48. Revit will translate this to:
    A. 4' 0"
    B. 48' 0"
    C. 48"
    D. 4' 8"

8. You are placing a wall in Revit. You use the listening dimension to set the wall length. You enter '**5.5**'. Revit places the wall and it is _____ long.

> A. 5' 6"
> B. 5 1/2"
> C. 5' 0"
> D. 5' 5"

9. To change the position of an element, such as a door or wall"

> A. Place a permanent dimension, then modify the dimension.
> B. Select the element, then modify the temporary dimension
> C. Select the element and move it
> D. Use the Measure tool

10. The three types of dimensions used in Revit are:

> A. LINEAR
> B. ANGULAR
> C. TEMPORARY
> D. PERMANENT
> E. CONTINUOUS
> F. LISTENING

Answers:
1) B & C; 2) C; 3) False; 4) A & B; 5) C; 6) A; 7) A; 8) A; 9) B; 10) C, D, & F

Lesson Six

Developing the Building Model

This lesson addresses the following User and Professional exam questions:

- Floors
- Ceilings
- Ceiling Lighting Fixtures
- Stairs
- Railings
- Roofs

Floors are level-based system families. A floor is added on the active level. The top of the floor is aligned to the level with its thickness projected downwards. You can also offset the floor so it is above or below a level using Element Properties.

Floors can be sloped. Floors can also be created by creating a mass element and selecting a face on the mass element.

Openings can be added to floors either by modifying the floor sketch or adding an opening. Railings can be added to either sloped or flat floors.

Ceilings are level-based system families. Ceilings can be controlled in a similar manner as floors using element properties.

Both ceilings and floors are Model elements.

Roofs are also level-based system families.

Stairs and railings are system families. They use profiles, which can be loaded. When you click the starting point of the stairs in plan view, the number of treads is calculated based on the distance between the floors and the maximum riser height. You can adjust the maximum riser height in the stairs element properties.

Railings consist of rails and balusters.

Stairs are automatically created assuming that they are going up from the level where they are placed. You can change the direction of the stairs (flip from going up to going down) by using the control arrow.

Command Exercise

Exercise 6-1 – Modifying a Floor Perimeter

Drawing Name: **i_floors.rvt**
Estimated Time to Completion: 15 Minutes

Scope

Modify a floor.
Determine the floor's perimeter.

Solution

1. Activate **the Main Floor Admin Wing** floor plan.

   Floor Plans
   — Ground Floor
   — Lower Roof
   — Main Floor
   — **Main Floor Admin Wing**
   — Main Roof
   — Site

2. Window around the area indicated.

   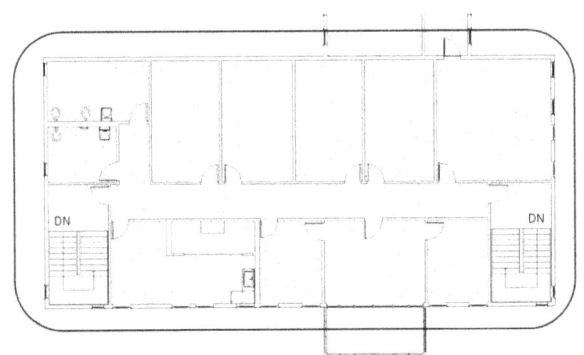

3. Select the **Filter** tool located in the lower right corner of the window.

4. Select **Check None** to disable all the checks.

   Then check only the **Floors**.

   Press **OK**.

Category:	Count:
<Room Separation>	1
Casework	5
Curtain Panels	12
Curtain Wall Grids	5
Curtain Wall Mullions	16
Doors	15
☑ Floors	1
Lines (Lines)	16
Plumbing Fixtures	7
Railings	4
Rooms	13
Stairs	2
Walls	27
Windows	18
Total Selected Items:	1

Developing the Building Model

5. | Dimensions | |
|---|---|
| Slope | |
| Perimeter | 254' 8" |
| Area | 3254.30 SF |
| Volume | 3457.69 CF |
| Thickness | 1' 0 3/4" |

In the Properties pane:

Note that the floor has a perimeter of 254' 8".

6. 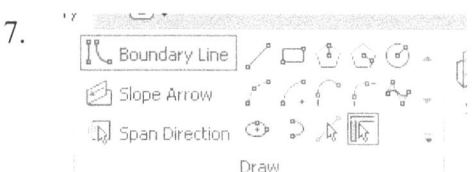   Select **Edit Boundary** under the Mode panel.

7. Select **Pick Walls** mode under the Draw panel.

8. 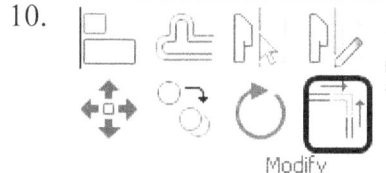   Uncheck the Extend into wall (to core) option.

9. Select the three walls indicated.

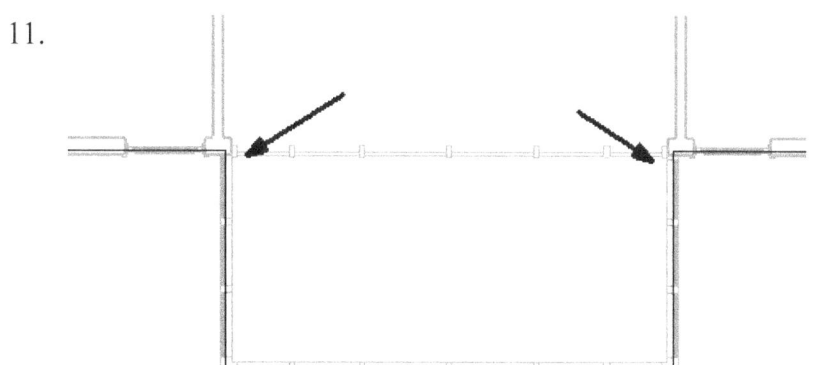

10. Select the **Trim** tool on the Modify panel from the ribbon.

11. Trim the two corners indicated so that there is no wall in the section between the two vertical walls on the upper ends.

6-3

12. Select the **Green Check** to **Finish Floor** from the ribbon.

13. Select **No**.

14. Note that the floor has a perimeter of 270' 8".

    Press **OK** to close the dialog.

15. Close without saving.

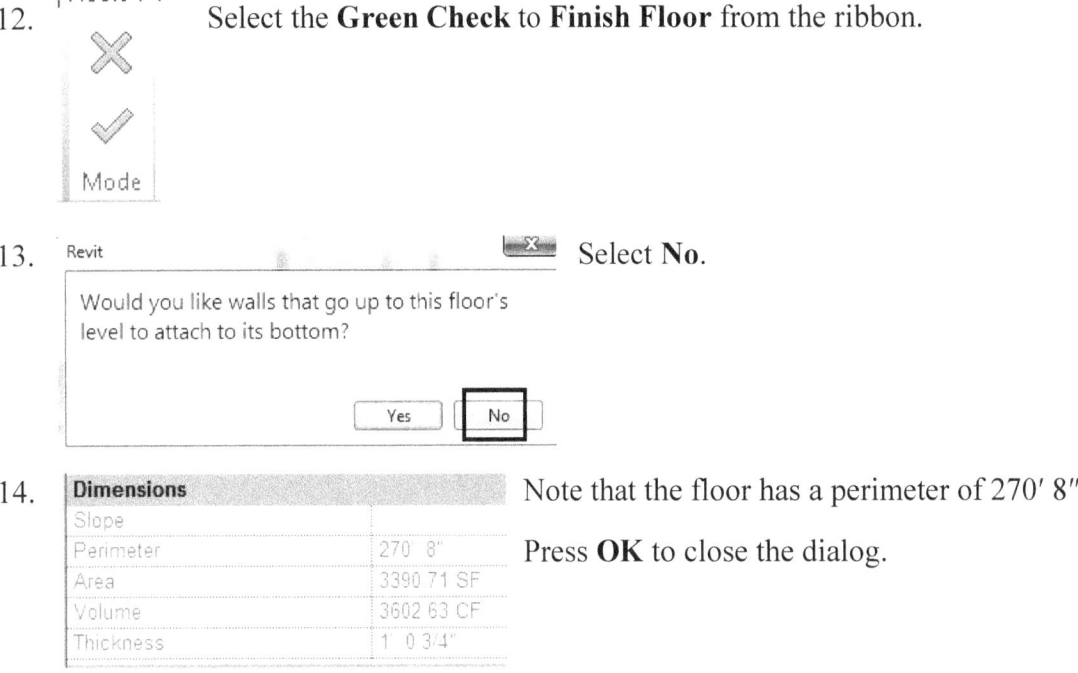

In the exam, you may be asked to modify a floor's sketch and then enter the resulting area, perimeter, or volume. Practice this exercise until you can get the correct perimeter value and feel comfortable using the tools.

Developing the Building Model

Command Exercise

Exercise 6-2 – Modifying a Ceiling

Drawing Name: **i_ceilings.rvt**
Estimated Time to Completion: 15 Minutes

Scope

Modify a ceiling.
Rotate a ceiling grid.
Determine the ceiling's elevation.

Solution

1. Ceiling Plans
   Ground Floor
   Ground Floor Admin Wing

   Activate the **Ground Floor Admin Wing** ceiling plan.

2. 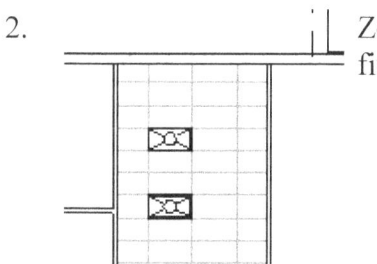 Zoom into the area where the ceiling grid and lighting fixtures are placed.

3. 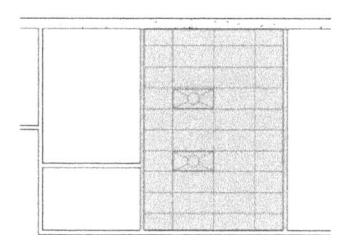 Window around the ceiling and lighting fixtures to select.

4.  Select the **Filter** tool.

5. Verify that only the ceiling and lighting fixtures are selected.

   Press **OK**.

6. 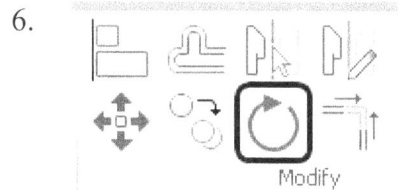 Select the **Rotate** tool from Modify panel on the ribbon.

6-5

7. Select a base point that is horizontal and to the right of the selected elements.

8. 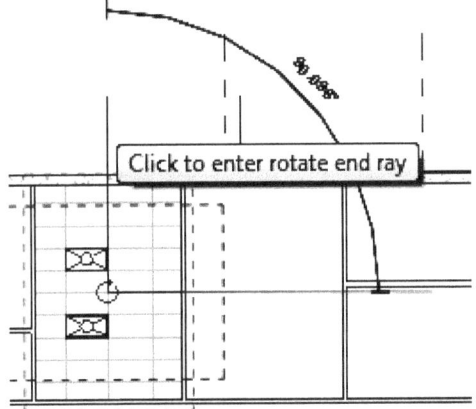 Move the cursor above the selected elements.

   Enter **90** to ensure a 90-degree rotation.

   Press **ESC** to release the selection.

9. 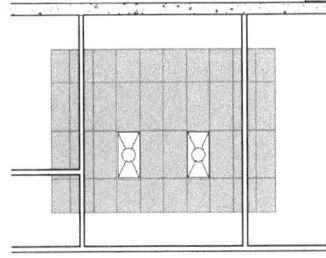 Select the ceiling so it highlights.

10.  Select **Edit Boundary** from the Mode panel.

11. 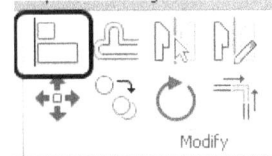 Select the **Align** tool from the Modify panel on the ribbon.

Developing the Building Model

12. 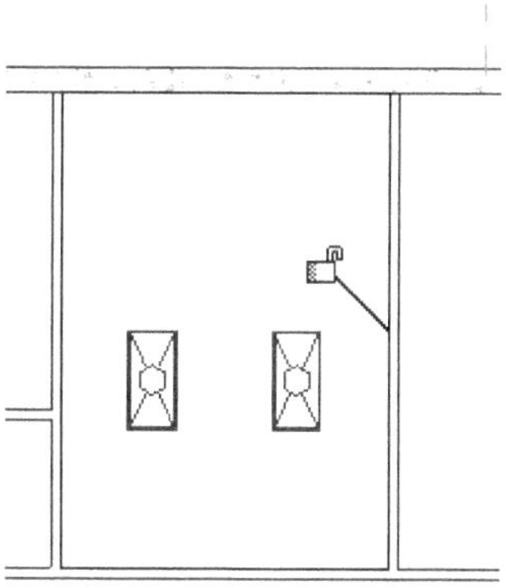 Align the ceiling boundary to the inner face of the walls.

13. 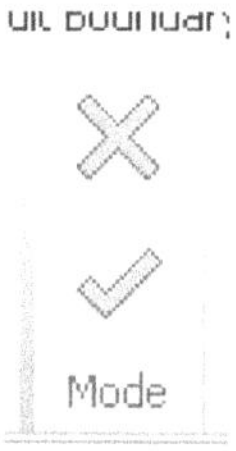 Select the **Green Check** to **Finish Ceiling**.

14. 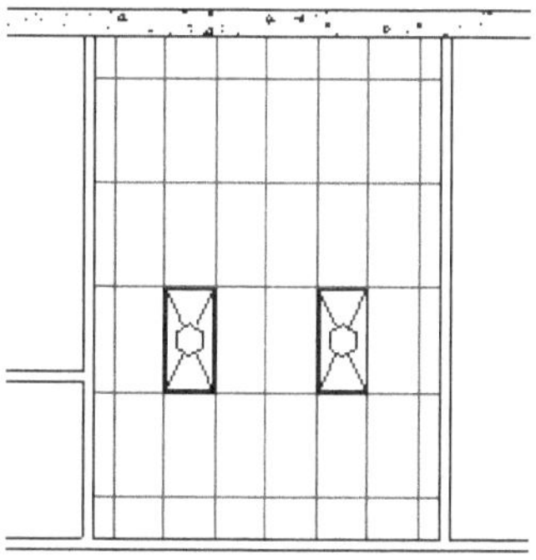 The grid updates.

    Left click in the window to release the selection.

15. 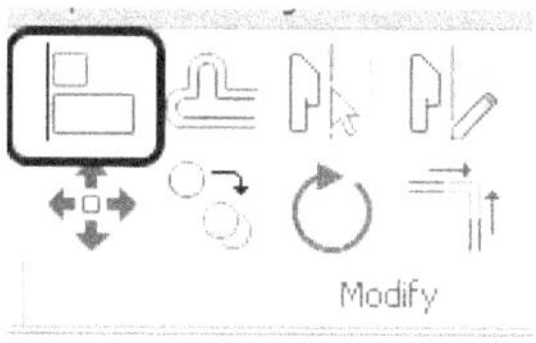 Select the **Align** tool from the Modify panel on the ribbon.

    R

16. 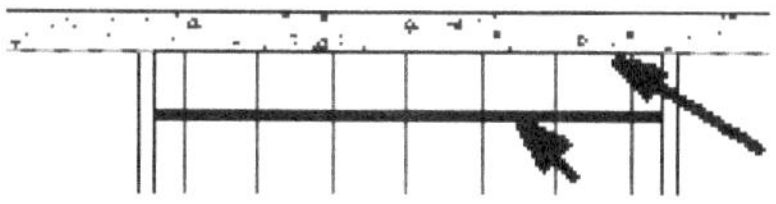 Select the inner face of the upper wall with the top horizontal gridline.

17. 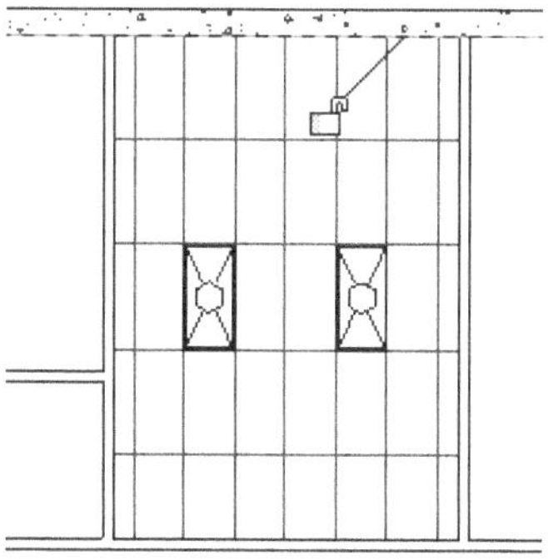 The ceiling grid updates so that all the grid tiles are whole in the vertical direction.

18. Select the ceiling ONLY using Filter or the TAB functions.

19. In the Properties pane:

    Note that the user can control the height offset from the level where the ceiling is placed.

    Note the perimeter, area, and volume values.

20. Close without saving.

The exam will ask the user to either place or modify a ceiling and then identify one of the element parameters, such as perimeter, area, or volume. Users should practice this exercise until they are comfortable with the tools.

Developing the Building Model

Command Exercise

Exercise 6-3 – Creating Stairs by Sketch

Drawing Name: **i_stairs.rvt**
Estimated Time to Completion: 20 Minutes

Scope

Place stairs using reference work planes.

Solution

1. Activate the **Ground Floor** floor plan.

2. 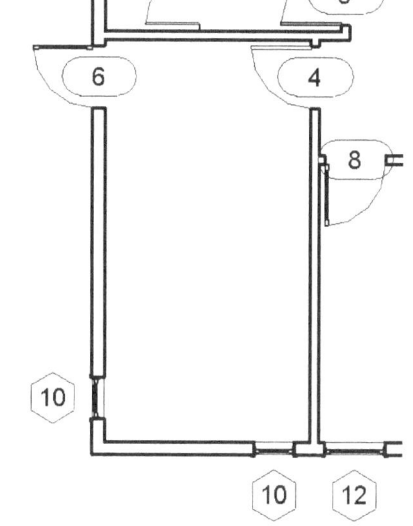 Zoom into the lower left corner of the building.

3. Select the **Reference Plane** tool from the Work Plane panel on the Architecture ribbon.

4. Offset: 2' 4" Set the Offset to **2' 4"** on the Options bar.

5. Select the **Pick Lines** tool from the Draw panel.

6-9

6. 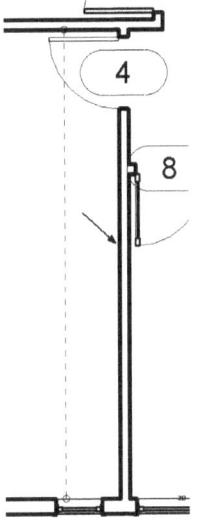 Place a vertical reference plane 2' 4" to the left of the wall where Door 4 is placed.

7. 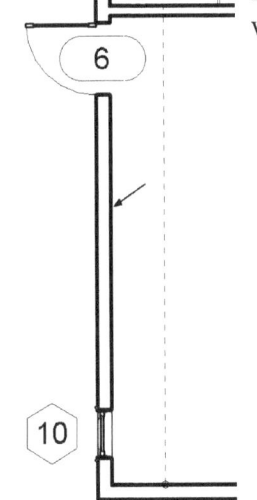 Place a vertical reference plane 2' 4" to the right of the wall where Door 6 is placed.

8. 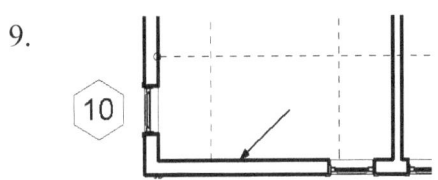 Set the Offset to **4' 4"** on the Options bar.

9. Place a horizontal reference plane 4' 4" above the wall where Window 10 is located.

10. Set the Offset to **7' 4"** on the Options bar.

Developing the Building Model

11. 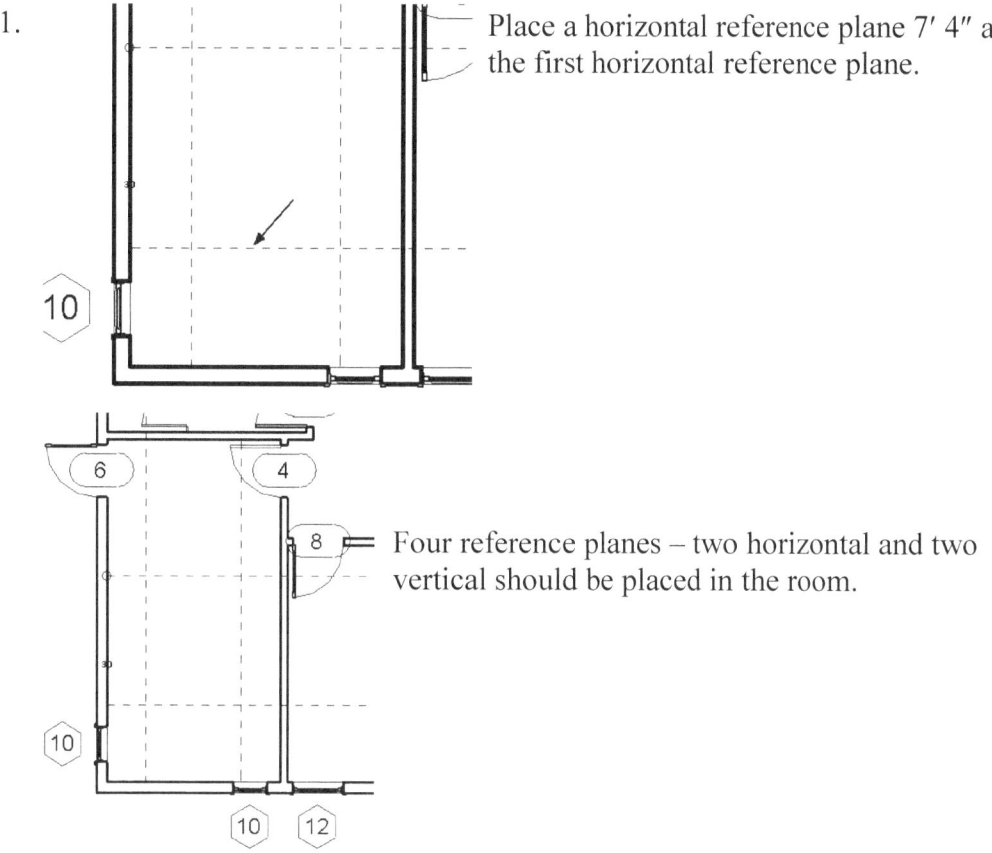 Place a horizontal reference plane 7' 4" above the first horizontal reference plane.

Four reference planes – two horizontal and two vertical should be placed in the room.

12. Activate the Architecture ribbon.

Select the **Stair by Component** tool from the Circulation panel.

13. Select **Edit Type** from the Properties pane.

14. Change the Maximum Riser Height to **8"**.
Set the stair width is set to **4' 0"**.

Press **OK**.

15. Set the desired number of risers to **16**.

6-11

16. Select the **Run** tool from the Draw panel on the ribbon.

17. Set the Location line to **Run: Center** on the Options bar.

18. Pick the first intersection point indicated for the start of the run.

    Pick the second point indicated.

19. Moving clockwise, select the lower intersection point #3 and then the upper intersection point #4.

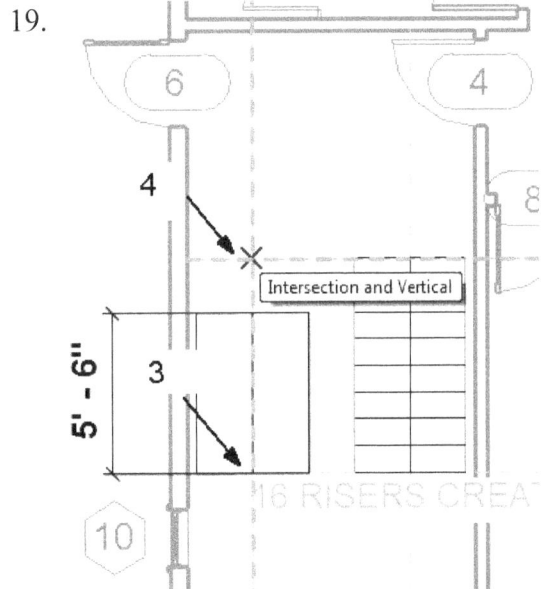

20. 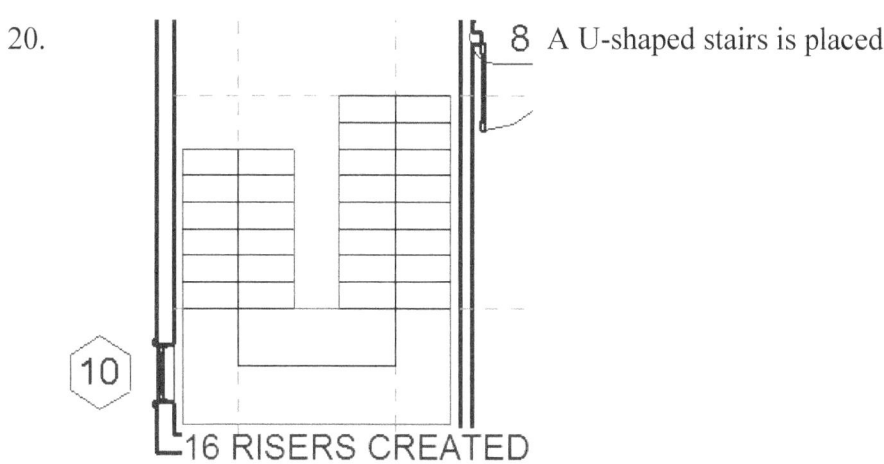 A U-shaped stairs is placed.

21.  Select the **Green Check** on the Mode panel to **Finish Stairs**.

22. Select the stairs.

    On the Properties pane:
    Locate the value for the Actual Tread Depth.

    Locate the value for the Actual Riser Height.

23. Close without saving.

Command Exercise

Exercise 6-4 – Modifying Assembled Stairs

Drawing Name: **modifying_stairs.rvt**
Estimated Time to Completion: 20 Minutes

Scope

Edit an Assembled Stair
Control the number of risers
Modify a landing

Solution

1. Activate the {3D} view.

2. 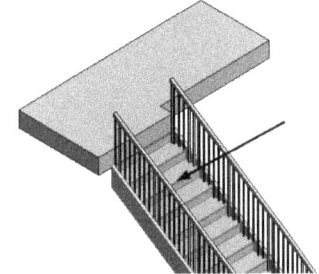 Select the Stairs.

   Be sure you select the stairs, not the railing to landing. You should see Assembled Stair on the Properties panel if it is selected properly.

   Assembled Stair
   MTL Pan (C 9x15)

   Click on **Edit Stairs** on the ribbon.

Developing the Building Model

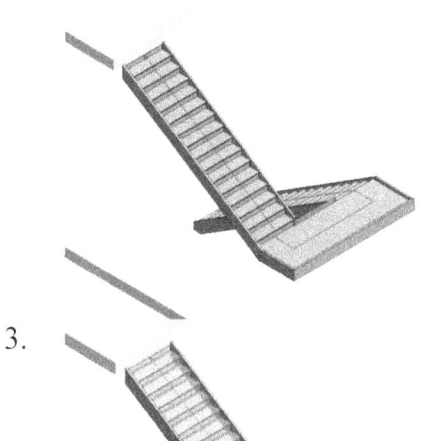

The view display changes.

Note that the railings and stringers visibility have been turned off so you can focus on the stairs.

3. Select the top run.

Note you can control the characteristics of each section of the stair (runs and landings).

4. In the Properties pane:

Uncheck **End with Riser**.

5.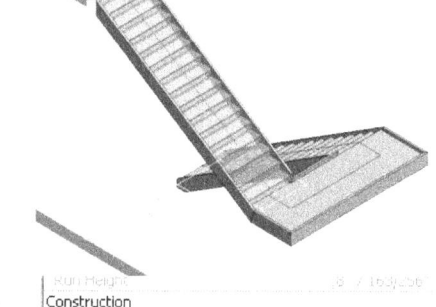

Note that the stair updates.

Left click in the display window to release the stair selection.

Note that we are still in Edit Stair mode - check the ribbon!

6. Activate **Level 1**.

7. Select the top run again.

Select the round bubble indicated not the arrow.

6-15

8. 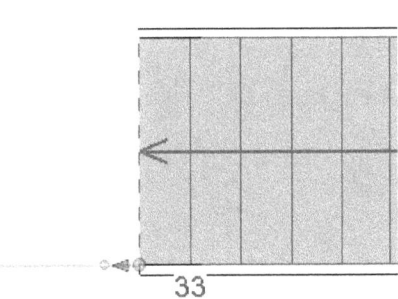 Drag the filled dot handle to align with the floor-edge.

   Note that the riser count updates.

9. 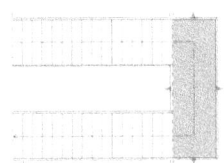 Select the Landing.

   This activates shape handles.

10. 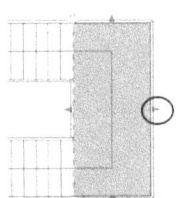 Using the left shape handle extend the landing so it is aligned with the end points of the arc on the floor.

11. 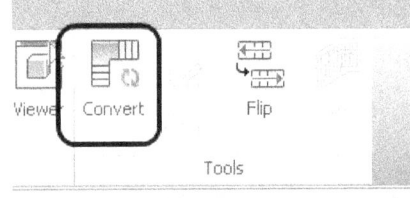 With the landing still selected/highlighted:

    Select the **Convert** button on the ribbon.

12. 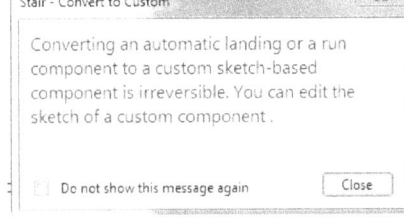 You will see a warning message that the landing will become a sketch-based component.

    Press **Close**.

13. 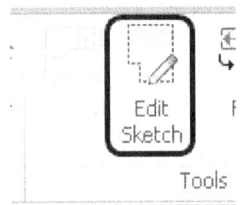 Click on **Edit Sketch** on the ribbon.

14.

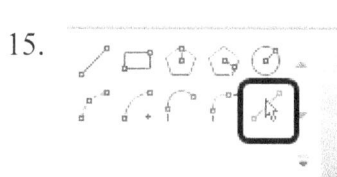

The landing is now in sketch mode.

15.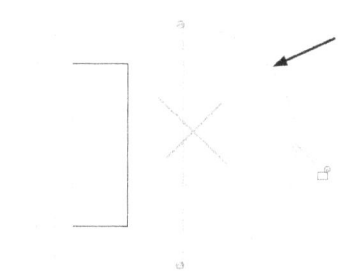

Select the **Pick** tool.

16.

Pick the arc which is part of the floor.

Delete the right vertical line.

Trim the boundary so the horizontal lines are tangent to the arc.

Your boundary should look like this.

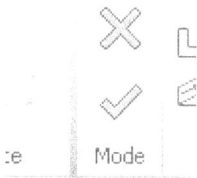

Select Green Check.

If you get an error message, use the TRIM command to clean up the points where the arc connects with the lines.

17.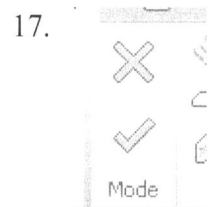

Select Green Check to finish editing the stairs.

18. Activate the **{3D} view**.

19. Inspect the modified stairs.

20. Save as *ex6-4.rvt*.

Developing the Building Model

Command Exercise

Exercise 6-5 – Creating a Stair by Component

Drawing Name: **component_stairs.rvt**
Estimated Time to Completion: 45 Minutes

Thanks to Robert Manna, Krista Manna, and David Light!

Scope

    Create a Stair by Component
    Modify the Width
    Modify a run sketch
    Re-position a run
    Re-position a landing
    Modify a floor's sketch

The walls in this exercise have been set to have transparency. This is done using Visibility/Graphics overrides. An exercise on how to do this is in Lesson 9- Exercise 9-7.

Solution

1. Activate the **Level 1** floor plan.

2. Activate the Architecture ribbon.

    Select the **Stair→Stair by Component** tool.

3. Select the **Railing** button on the ribbon.

4. 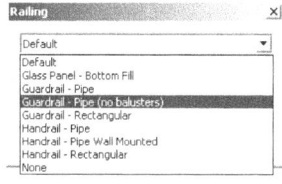 Select **Guardrail- Pipe (no balusters)** from the drop-down list.

    Press **OK**.

5. On the Options bar:
   Set the Location Line to **Run:Left**.
   Set the Offset to **0' 0"**.
   Enable **Automatic Landing**.

6.
7.  Activate the **Run** button.

8. Start the run at the intersection of the right reference plane and the inside finish face of the bottom horizontal wall (1).
   Then pick at the intersection of the left reference plane and the inside finish face of the bottom horizontal wall (2).

9. Continue the run by picking the intersection of the left reference plane and the inside finish face of the top horizontal wall (3).

   Next pick the intersection of the right reference plane and the inside finish face of the top horizontal wall (4).

10. Place a third run on top of the first run using points 1 and 2. (Repeat step 6)

    The third run will display as light grey lines.

11.  Click on the **Green Check** on the ribbon to complete the stairs.

12. 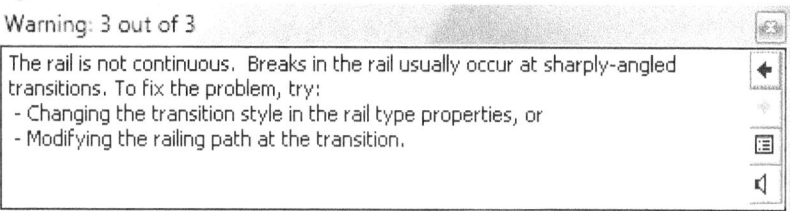 You may see some warnings about errors.
Review the errors – we will be fixing them.
Close the error dialog box by left clicking on the x in the upper right corner.

13. Activate a **{3D}** view.

14.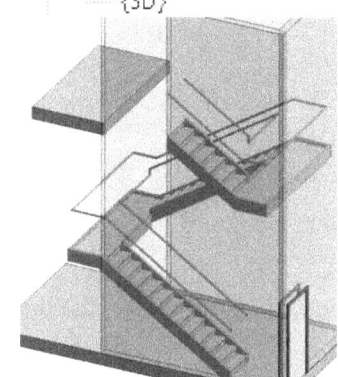

    You can see there are some necessary tweaks.

    Select the stairs so they are highlighted.

15. 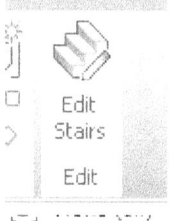 Select the stairs and select **Edit Stairs** on the ribbon.

16.  Activate **Level 1**.

17. Select the second run.

18. 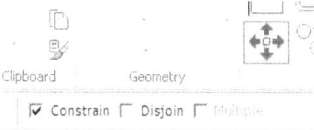 Select the **Move** tool.
Enable **Constrain** on the Options bar.

    This sets ORTHO on.

19.	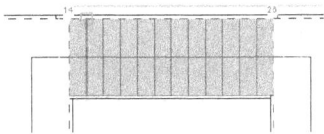	Move the run 11" to the right.
20.	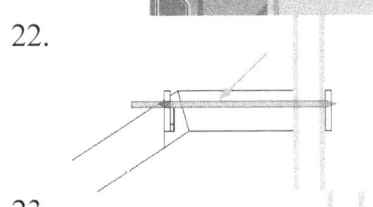	You will see an error message about the landing. Close the error dialog.
21.	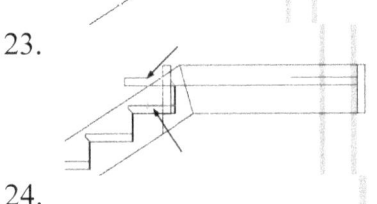	The landing now extends too far into the wall and needs to be adjusted. Activate **Section 1**.
22.		Select the landing indicated. Select the MOVE tool.
23.		Select the top of the landing as the start point for the move operation. Select the top of the tread as the end point for the move operation.
24.		*The landing position adjusts so it is flush with the inside face of the wall.*

25.	Floor Plans
	  Level 1
	  Level 2
	  Site
	Ceiling Plans
		Activate **Level 2**.

26.	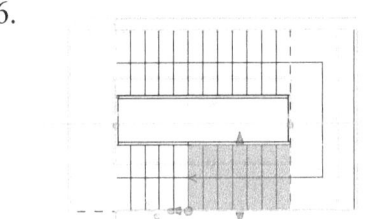	Select the third run which is on top of the first run.

Developing the Building Model

27. Select the MOVE tool.
Move the third run 11" to the left.

28. Select the third run of the stairs (the run which was just moved).

29. Select the **Convert** button on the ribbon.

30. *If you see a warning dialog, just close it.*

31. Select **Edit Sketch**.

32. Select the **Start-End-Radius arc** tool.

33. Draw an arc from the endpoint of the far right riser line to the finish face of the top horizontal wall endpoint.

34. Delete the boundary line indicated by the x-mark.

35. Select the **Extend Multiple** tool.

6-23

36.  Select the arc as the boundary for the extension.
Then select all the risers to extend them to meet the arc.

37. Select the **Trim** tool.

38. Select the arc and the far left riser line to trim.

39. Select **Green Check** to close the sketch edit.

40. Select **Green Check** to close the stairs edit.

41. Ignore the error warning for the rails.
We will fix that later.

42. Switch to a 3D view and orbit to inspect your modified stairs.

43. Save as *ex6-5.rvt*.

Developing the Building Model

Command Exercise
Exercise 6-6 – Railings

Drawing Name: **railings.rvt**
Estimated Time to Completion: 45 Minutes

Scope

Create a new railing type.
Change the properties to use extensions and terminations.
Modify a railing path to edit joins.
Modify the position of supports.

Solution

1. 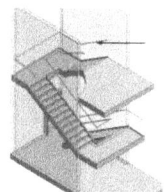 Activate the **3D view**.

   Select the inside railing.

2. 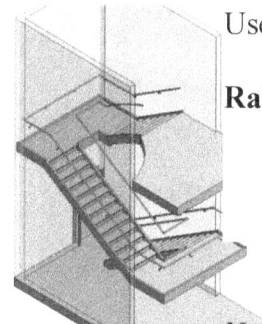 Use the Type Selector to change the railing type to:

   Railing – Wall Mounted.

   Railing
   Railing - Wall Mounted

   Note that the hand rail has changed to show side supports.

3. In the Project Browser, under Families:

   Locate the **Handrail Type** under Railings.

4. Right click on the **Pipe - Wall Mount**.

   Select **Type Properties**.

6-25

5. Change the Hand Clearance to **0' 3"**.
   Change the Height to **3' 6"**.

6. Under Supports:
   Change the Spacing to **3' 0"**.

7. Under Supports:
   Change the Family to **Metal - Circular- Pillar**.

8. Under Extension (Beginning/Bottom):
   Set the Extension Style to **Floor**.
   Set the Length to **1' 6"**.

9. Under Extension (Beginning/Bottom):
   Set the Extension Style to **Wall**.
   Set the Length to **1' 6"**.

10. Click in the Material column.

    Set the Material to **Aluminum 6061-AHC**.

11. Under Terminations:
    Set the Beginning/Bottom Termination to **Termination - Aluminum-Floorbase**.
    Set the End/Top Termination to **Termination-Pipe- 1 1/2"**.

    Press **OK** to close the dialog.

    Notice the new terminations at the top and the bottom of the railing.

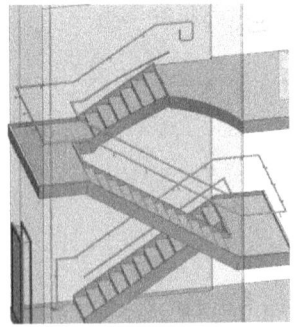

Developing the Building Model

12.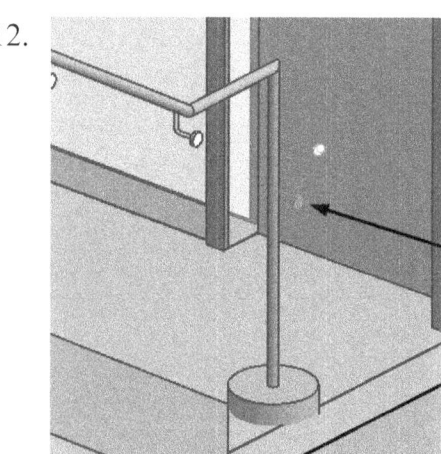

Using the TAB key select just the support located on the floor extension.

Note that there is a PIN fixing the support in place and preventing any edits.

Unpin the support.

Press DELETE.

This will remove the support.

13.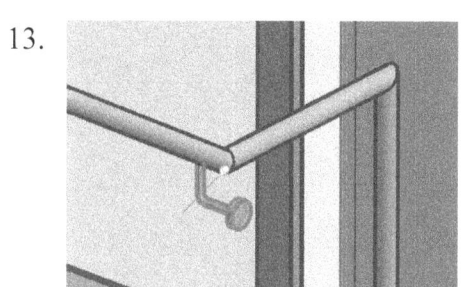

Notice that all the other supports are still in place. Using the TAB key select just the support located right before the turn on the floor extension.

Note that there is a PIN fixing the support in place and preventing any edits.

Unpin the support.

14.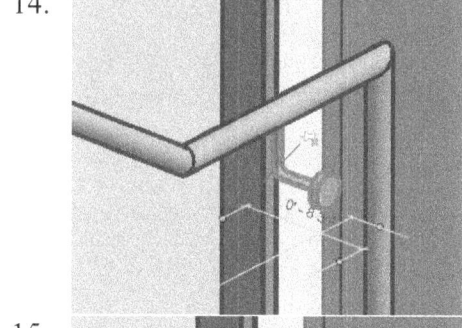

Move the support so it is centered on the horizontal part of the extension.

If you re-pin the support, it will move back to it's original position, so leave it unpinned.

Left click anywhere in the display window or press ESC to release the selection.

15.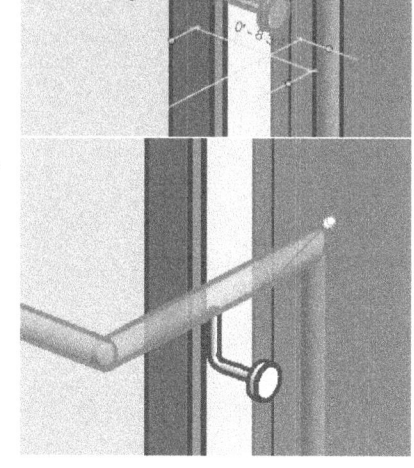

Use the TAB key so just the rail where the support was moved is selected and not the supports.

Supports can also be copied to a new location using the COPY tool.

16.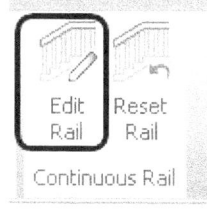

Select the outside railing.
Select **Edit Rail** on the ribbon.

6-27

17. Select **Edit Path** on the ribbon.

18. Select the **Fillet** tool on the Draw panel.

19. On the Options bar:
Enable **Radius**.
Set the value to **2"**.

20. 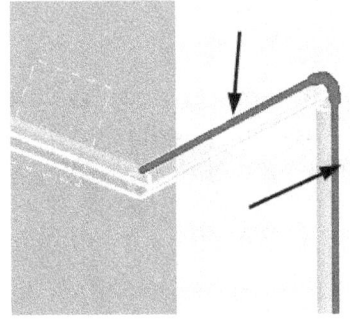 Select the two lines indicated to form the fillet.

21. Select **Edit Rail Joins**.

22. 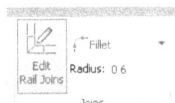 Click on the intersection above the new fillet.

23. 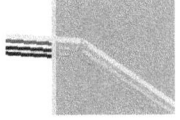 On the ribbon:
Set the join type to **Fillet**.
Set the Radius to **6"**.

24.  Pan up towards the top of the railing to next rail join.
Select the intersection so you see an x.

25. On the ribbon:
Set the join type to **Fillet**.
Set the Radius to **6"**.

26.  Pan to the next rail join.
Select the intersection so you see an x.

Developing the Building Model

27.  On the ribbon:
Set the join type to **Fillet**.
Set the Radius to **6"**.

28. 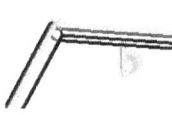 Pan to the next rail join.
Select the intersection so you see an x.

29.  On the ribbon:
Set the join type to **Fillet**.
Set the Radius to **6"**.

30.  Pan to the next rail join.
Select the intersection so you see an x.

31. 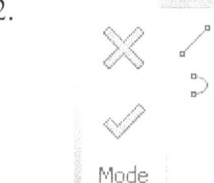 On the ribbon:
Set the join type to **Fillet**.
Set the Radius to **6"**.

Keep moving along the rail until all remaining rail joins are set to a fillet of 6".

32. 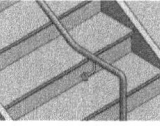 Select **Green Check** twice to complete the Edit.

33.  Note how the railing updated.

34. 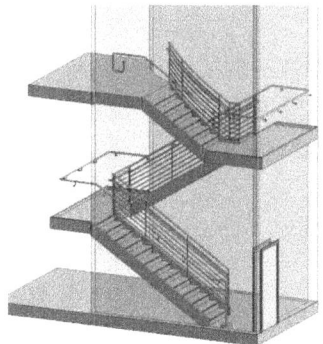 Select the outside railing and set it to **Guardrail-Pipe**.

35. Save as *ex6-6.rvt*.

6-29

Command Exercise

Exercise 6-7 – Creating a Roof by Footprint

Drawing Name: **i_roofs.rvt**
Estimated Time to Completion: 15 Minutes

Scope

Create a roof.
Determine the roof's volume.

Solution

1. Activate the **T.O. Parapet** floor plan.

2. 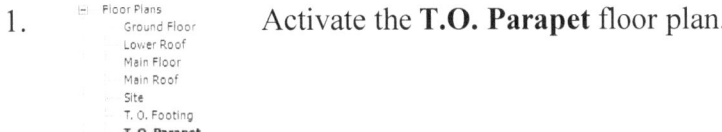 Select **Roof by Footprint** from the ribbon.

3.  Select **Steel Truss - Insulation on Metal Deck - EPDM** using the Type Selector on the Properties panel.

4. 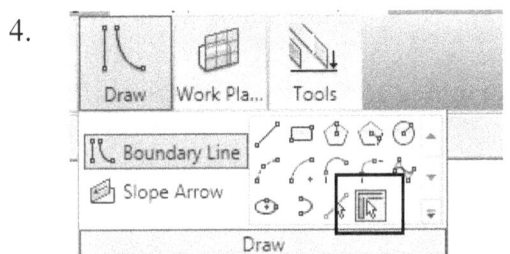 Select **Pick Walls** mode.

5.  On the Options bar:

   Uncheck **Defines slope**.
   Set the Overhang to **0'-0"**.
   Uncheck **Extend to Wall Core**.

6. Select all the exterior walls.

7. 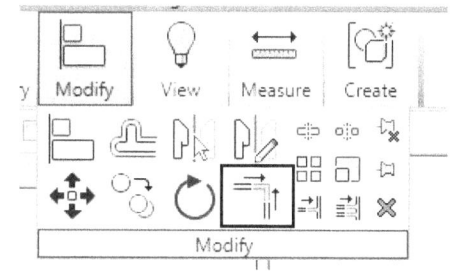 Use the **Trim** tool from the Modify panel to create a closed boundary, if necessary.

8. Align the boundary lines with the exterior face of each wall.

9. 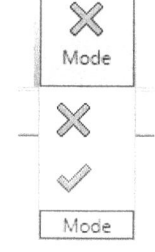 Select the **Green Check** on the Model panel to **Finish Roof**.

10.  Select the roof.

    In the Properties panel:

    Note the volume of the roof.

    The volume should be 6620.22 CF.

    If your volume is not the same, repeat the exercise.

11. Close without saving.

Command Exercise

Exercise 6-8 – Creating a Roof by Extrusion

Drawing Name: **i_roofs_extrusion.rvt**
Estimated Time to Completion: 30 Minutes

Scope

Create a roof by extrusion.
Modify a roof.

Solution

1. Activate the **3D view.**

2. 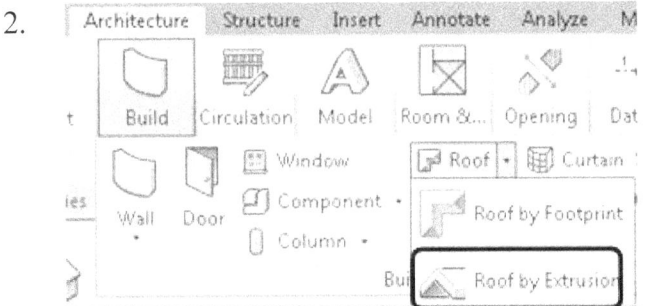 Activate the Architecture ribbon.

   Select **Roof by Extrusion** under the Build panel.

3. 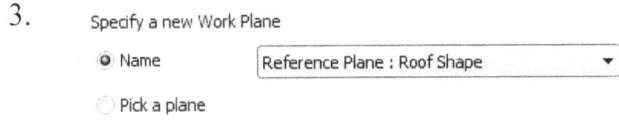 Enable **Name**.
   Select **Roof Shape** from the list of reference planes.
   Press **OK**.

4. 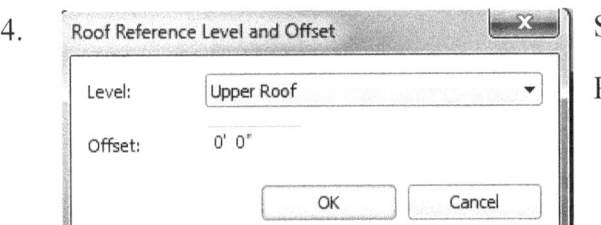 Select **Upper Roof** from the list.

   Press **OK**.

5. 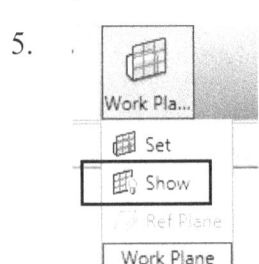 Select the **Show** tool from the Work Plane panel.

   This will display the active work plane.

Developing the Building Model

6. Right click on the ViewCube's ring.

   Select **Orient to a Plane**.

7. Enable **Name**.

   Select **Roof Shape** from the list of reference planes.

   Press **OK**.

8. Select the **Start-End-Radius Arc** tool from the Draw panel.

9.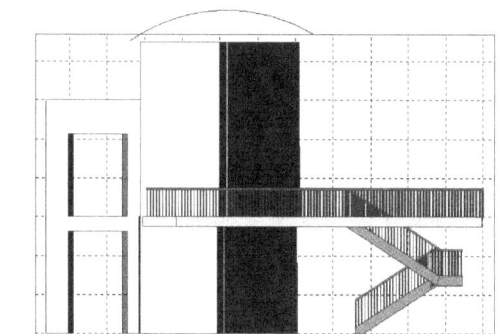

   Draw an arc over the building as shown.

10. Select the **Green Check** under the Mode panel to finish the roof.

11. 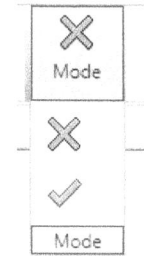 Switch to a 3D view.

12. 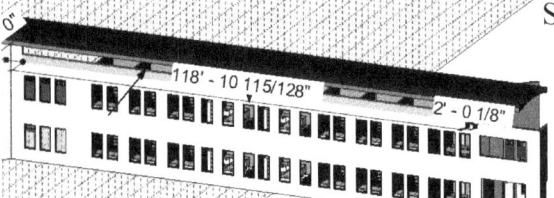 Select the wall below the roof.

13. 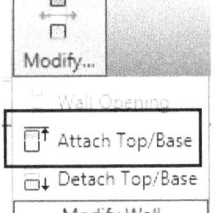 Select **Attach Top/Base** from the Modify Wall panel.

    Then select the roof.

    The wall will adjust to meet the roof.

    Go around the building and attach the appropriate remaining walls.

14. 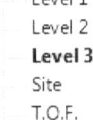 Activate the **Level 3** floor plan.

15. Activate the Architecture ribbon.

    Select the **Roof By Footprint** tool from the Build panel.

16. Select the **Pick Walls** tool from the Draw panel.

17. On the Options bar:

    Enable **Defines slope**.
    Set the Overhang to **2′ 0″**.
    Enable **Extend to wall core**.

18. 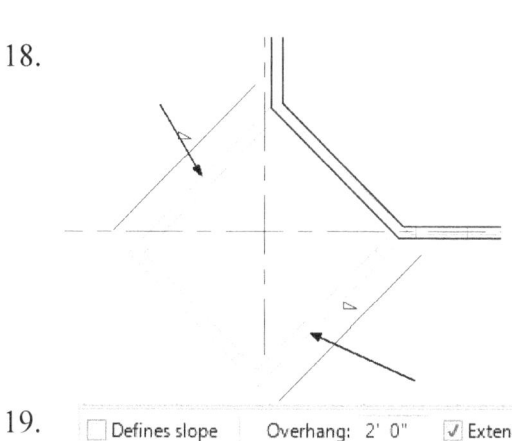 Select the outside of edge of the two walls indicated.

19. ☐ Defines slope   Overhang: 2' 0"   ☑ Extend to wall core   Disable **Defines slope**.

20. 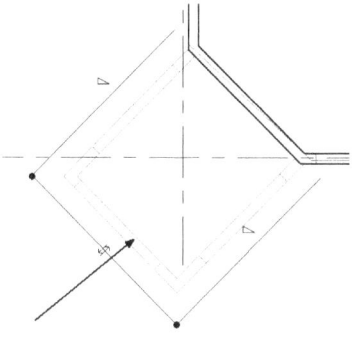 Select the outside edge of the walls indicated.

21. ☐ Defines slope   Overhang: 0' 0"   ☑ Extend to wall core   Set the Overhang to **0' 0"**.

22. Select the **Pick Line** tool from the Draw panel.

23. 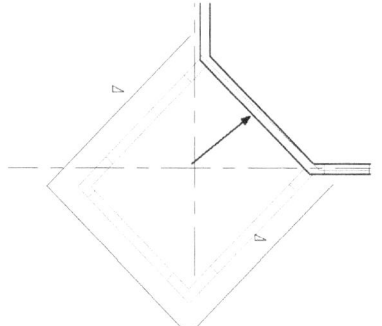 Pick the exterior side of the wall indicated.

24. Select the **Trim** tool from the Modify panel.

25. 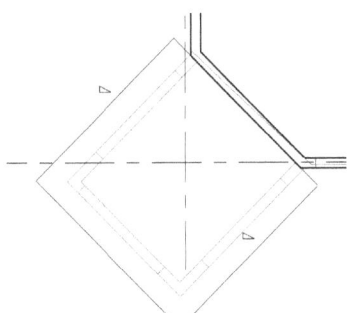 Trim the roof boundaries so they form a closed polygon.

26. Select **Edit Type** in the Properties pane.

27. Type: Generic - 9" Set the Type to **Generic- 9"**. Press **OK**.

28. Dimensions Slope 6"/12" Set the Slope to **6"/12"**.

29. Select the back and front boundary and lines and uncheck **Defines Slope** on the Options bar.

    Defines Slope

30. Once the sketch is closed and trimmed properly, select the **Green Check** on the Mode panel.

31. 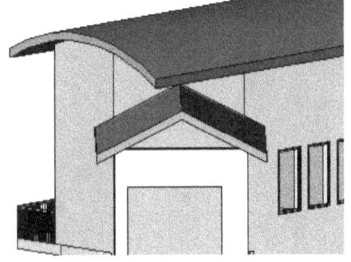 Return to a 3D view.

32.  Select the front wall.

33. 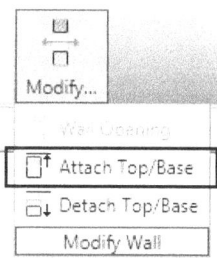 Select Attach Top/Base from the Modify Wall panel.

34.  Select the roof to attach the wall.

35. Close without saving.

Command Exercise

Exercise 6-9 – Modifying a Roof Join

Drawing Name: **roof-join.rvt**
Estimated Time to Completion: 30 Minutes

Thanks to Solomon Smith and Wil Wiens!

Scope

    Adding a roof join
    Modifying a roof slope
    Modifying a roof footprint
    Modifying a wall profile

Solution

1.   ⊟—3D Views
          {3D}

    Activate the **{3D}** view.

2.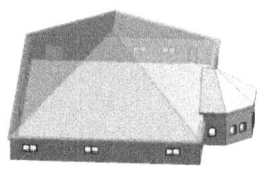

    Select the big roof.

3. In the Properties panel:

    Note the slope is set to **9/12**.

4.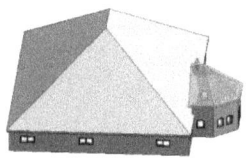

    Select the little roof.

5. In the Properties panel:

    Note the slope is set to **9/12**.

6. Activate the **Modify** Ribbon.

    Select **Join/Unjoin Roof** in the Geometry panel.

7. Select the exposed edge of the smaller roof.

8. Select the adjoining face of the large roof.

9. The small roof extends to join the large roof.

10. Select the small roof.

11. Change the slope to **6"/12"**.

12. Select the large roof.

13. Select **Edit Footprint** from the ribbon.

14. Activate **Level 4.**

15. Select the **Pick Line** tool.

16. Select the overlapping edges of the bay roof.

    The two vertical lines should be offset in 1'-0".

17. Use the TRIM and DIVIDE tools to clean up the roof outline.

18. Select **Green Check** to finish the roof.

19. Select the **West** elevation.

20. Select the wall indicated.

21. Select **Edit Profile** from the ribbon.

22. Select the **Pick Line** tool.

23. Select the roof edge.

Developing the Building Model

24.  Use the TRIM tool to eliminate the overlapping edges.

25. Select the **Green Check**.

26.  Switch to a 3D view to inspect the west side of the wall.

27. Select the **East** elevation.

28. Select the wall indicated.

29. Select **Edit Profile** from the ribbon.

30. 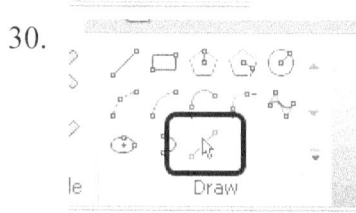 Select the **Pick Line** tool.

31. Select the roof edge.

32. Use the TRIM tool to eliminate the overlapping edges.

33. 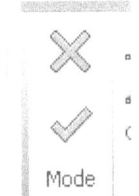 Select the **Green Check**.

34.  Switch to a 3D view and inspect the walls and roofs.

35. 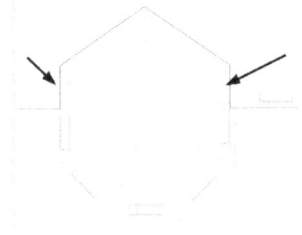 To eliminate any gaps:
Switch to a wire frame display.
Switch to a top view.
Align the boundary of the large roof with the exterior face of the walls indicated.

36. Save as ex6-9.rvt.

Developing the Building Model

Command Exercise

Exercise 6-10 – Creating a Sloped Ceiling

Drawing Name: **i_ceilings.rvt**
Estimated Time to Completion: 30 Minutes

Scope

Create a sloped ceiling.

Solution

1. Activate the **Ground Floor Admin Wing** ceiling plan.

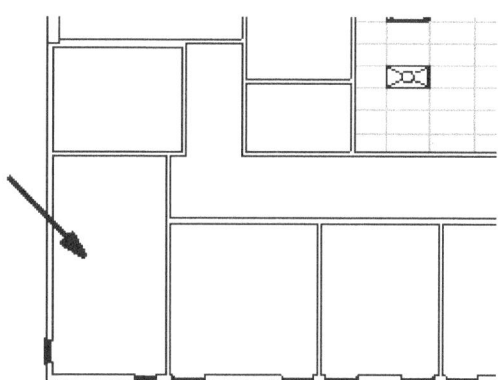

A sloped ceiling will be placed in the room indicated.

2. Select the **Ceiling** tool from the Build panel on the Architecture ribbon.

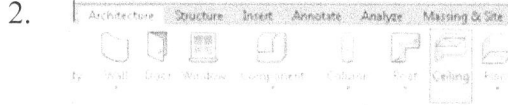

3. Select **Sketch Ceiling** from the Ceiling panel.

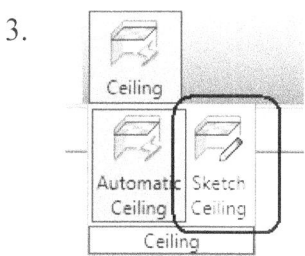

4. 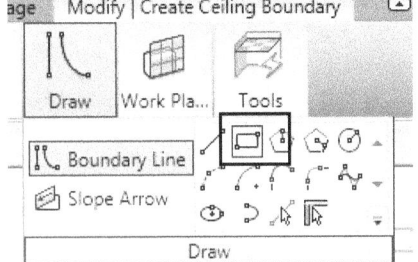 Select the **Rectangle** tool from the Draw panel.

5. 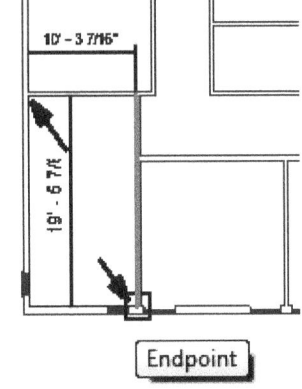 Draw a rectangle inside the room by selecting opposing corners.

6. Select the **Slope Arrow** tool from the Draw panel to place a slope.

7. Select the Midpoint at the top of the rectangle.

8. Bring the arrow down to indicate the direction of the slope.

   Pick to place the endpoint so it is coincident with the lower wall.

9. On the Properties pane:

   Under Specify: Select **Slope**.

Developing the Building Model

10. 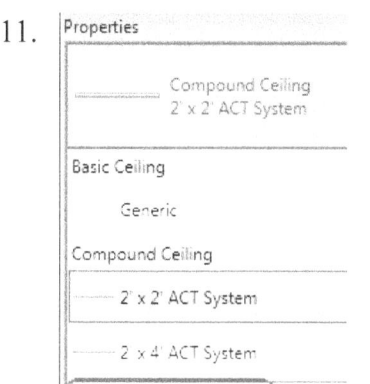 Set the Height Offset to **6"**.

Set the Slope to **2"/12"**.

Left click in the drawing window to release the selection.

11. 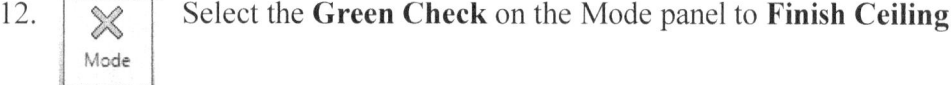 In the Properties pane:

Use the Type Selector to set the ceiling type to **GWB on Mtl. Stud**.

12. 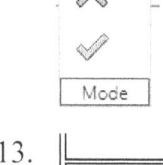 Select the **Green Check** on the Mode panel to **Finish Ceiling**.

13. The ceiling does not display a hatch pattern by default.

6-45

14.  Window around the room where you just placed the ceiling.

15.  Select the **Filter** tool.

16. 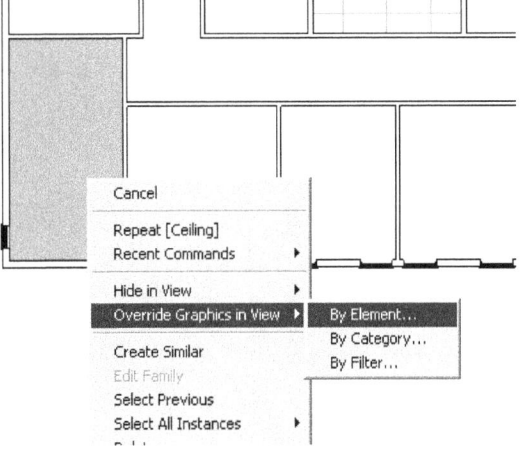 Enable only the Ceiling.

    Press **OK**.

17. Right click and select **Override Graphics in View→By Element**.

18. 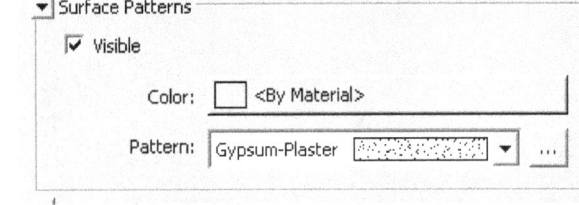 Expand the Surface Patterns section.

    Set the Pattern to **Gypsum-Plaster**.

    Press **OK**.

Developing the Building Model

19. 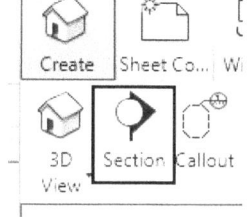 Activate the **Ground Floor** floor plan.

20. Select the **Section** tool from the Create panel on the View menu.

21. 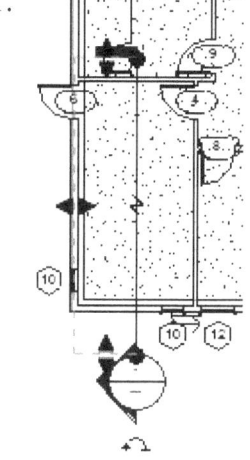 Place the section so it intersects the room with the sloped ceiling.

    Double click on the section head bubble to activate the section view.

22. 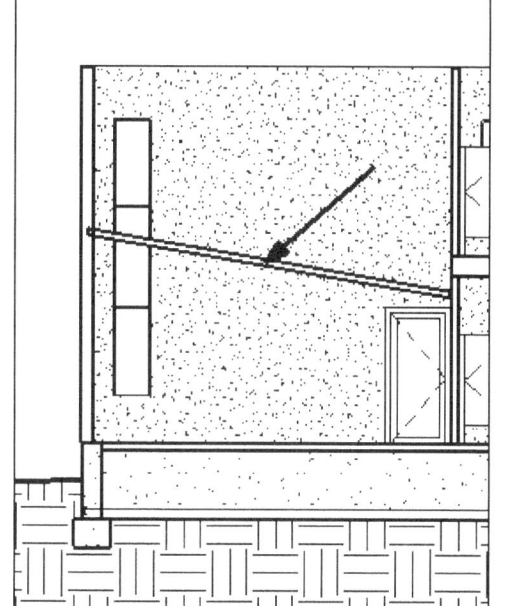 The sloped ceiling is shown.

    Select the ceiling.

6-47

23.

Constraints	
Level	Ground Floor
Height Offset From Level	7' 0"
Room Bounding	✓
Dimensions	
Slope	1/4" / 12"
Perimeter	59' 6 5/8"
Area	200.52 SF
Volume	71.02 CF
Identity Data	
Comments	
Mark	
Phasing	
Phase Created	New Construction
Phase Demolished	None

In the Properties pane:

Set the Height Offset to **7' 0"**.

Set the Slope to **1/4"/12"**.

Press **Apply**.

24. 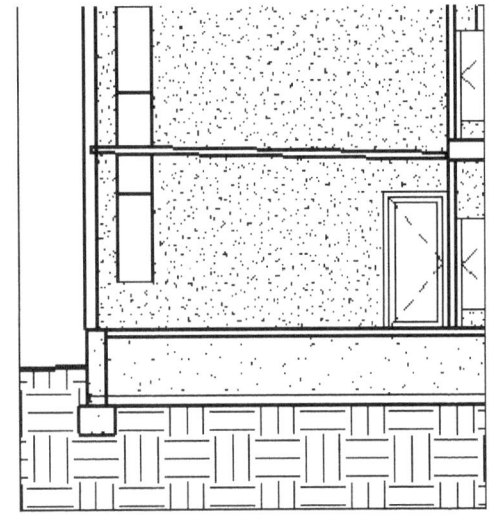 The ceiling updates with the new values.

25. Close without saving.

Practice User Exam

1. To create a partial ceiling:

    A. Use a sketch
    B. Create an area first
    C. Use the partial ceiling tool
    D. Create modified ceiling

2. The boundary of a floor is defined by:

    A. Model Lines
    B. Reference Lines
    C. Slab Edges
    D. Sketch Lines
    E. Drafting Lines

3. For the stairs shown, the vertical green lines represent:

    A. Risers
    B. Boundary Lines
    C. Stringers
    D. Run
    E. Railings

    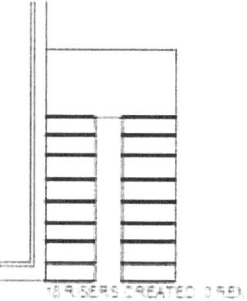

4. True or False: Stairs can be placed on multiple levels at a time.

5. True or False: You can change the direction of stairs using the flip arrows.

6. Identify the type of roof:

    A. Gable
    B. Hip
    C. Tar and Gravel
    D. Flat
    E. Shingled

    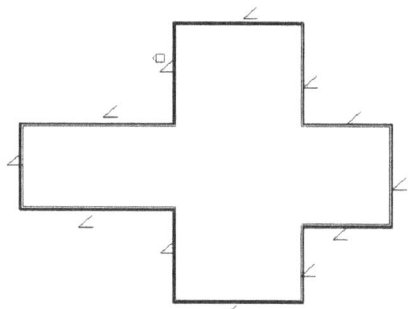

7. Name the three methods used to place a roof:

    A. By face
    B. By profile
    C. By footprint
    D. By extrusion
    E. By sketch

8. True or False: The extrusion of a roof can extend only in a positive direction from the selected work plane.

9. True or False: When attaching a wall to a roof, the profile of the wall does not change.

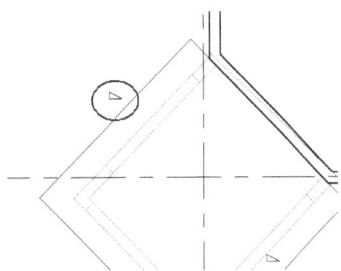

10. The symbol circled indicates:
    A. Slope
    B. Top Constraint
    C. Roof Boundary
    D. Gutter

11. To place a railing:
    A. Draw a path
    B. Select a host element
    C. Click to place
    D. Work in section view

12. You can change a Generic 12" Floor to a Wood Joist 10" Floor using the:
    A. Properties filter
    B. Options Bar
    C. Type Selector
    D. Modify ribbon

13. To create a stair landing, use the _____ tool.
    A. Run
    B. Boundary
    C. Riser
    D. Landing

14. To attach the top of a wall to a roof:
    A. Drag the top of the wall by grips
    B. Select the wall, check on Attach Top/Base and select roof
    C. Define wall attachments in instance properties
    D. Use the ALIGN tool

Answers:
1) A; 2) D; 3) B; 4) True; 5) True; 6) B; 7) A, C & D; 8) False; 9) False; 10) A) Slope; 11) B; 12) C; 13) B; 14) B

Lesson Seven

Detailing and Drafting

This lesson addresses the following User and Professional exam questions:

- Callouts
- Tags
- Detail Views
- Text
- Callout view of a Section
- Drafting View
- Revision Clouds
- Revision Schedules
- Grid Guides

A callout is a view that is placed in a plan, section, detail or elevation view to create a more detailed view of part of the building model. The area that the callout defines in the view is called the callout bubble. The callout bubble has a leader or connecting line to a callout head which displays the detail number and sheet number. All these parts – callout bubble, callout head, and callout leader – are referred to as a callout tag.

When you create a callout, a new view is automatically created. There are two types of callout views: Callout in Parent View and Callout in Detail View.

A detail view is a view of a specific area of a plan, elevation or section view. This view provides a greater level of detail at a larger scale than the parent view. Detail views can be linked to callouts.

Drafting views are 2D views that depict a small area of the building model. These views do not contain any model elements at all. Users can create a library of drafting views to depict foundation work, stairs and railings, wall framing, etc., and use them across multiple projects.

Revision clouds and revision schedules are used to track changes to a project. Users can enter the changes into a project and then issue the changes to a sheet once the changes are made.

Command Exercise

Exercise 7-1 – Creating Drafting Views

Drawing Name: **c_libraries_and_details.rvt**
Estimated Time to Completion: 30 Minutes

Scope

Use an existing AutoCAD detail drawing and convert it into a Revit drafting view.
Add a callout to a section view.
Add a section view with callout to a sheet.
Add a drafting view to a sheet.

Solution

1. Activate the **View** ribbon.

   Select **Drafting View** from the Create panel.

2.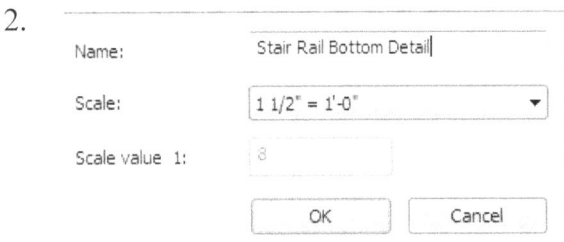

   Name the new Drafting View: **Stair Rail Bottom Detail**.
   Set the scale to 1 ½" = 1'-0"

   Press **OK**.

3.

   In the Project Browser, note that the Drafting Views category has been added and the new view is listed.
   The window will open to the new view which is empty.

4. Activate the **Insert** ribbon.

   Select **Import CAD** from the Import panel.

DETAILING AND DRAFTING

5.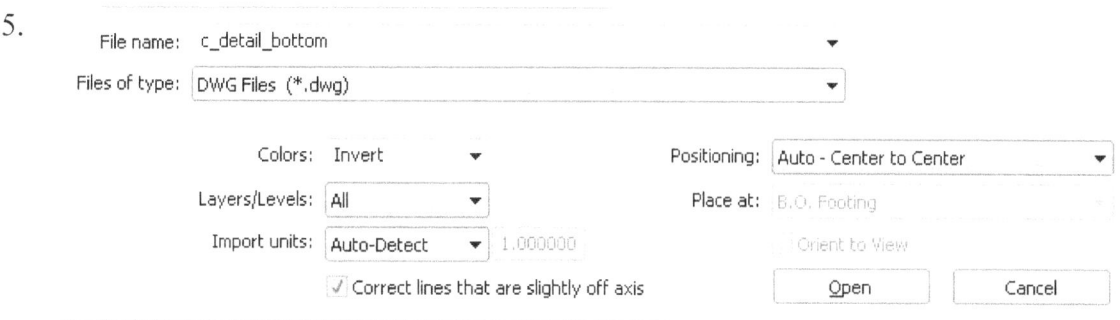

   Locate the *c_detail_bottom.dwg* file.
   Set Colors to **Invert**.
   Set positioning to **Auto- Center to Center**.
   Press **Open**.

6.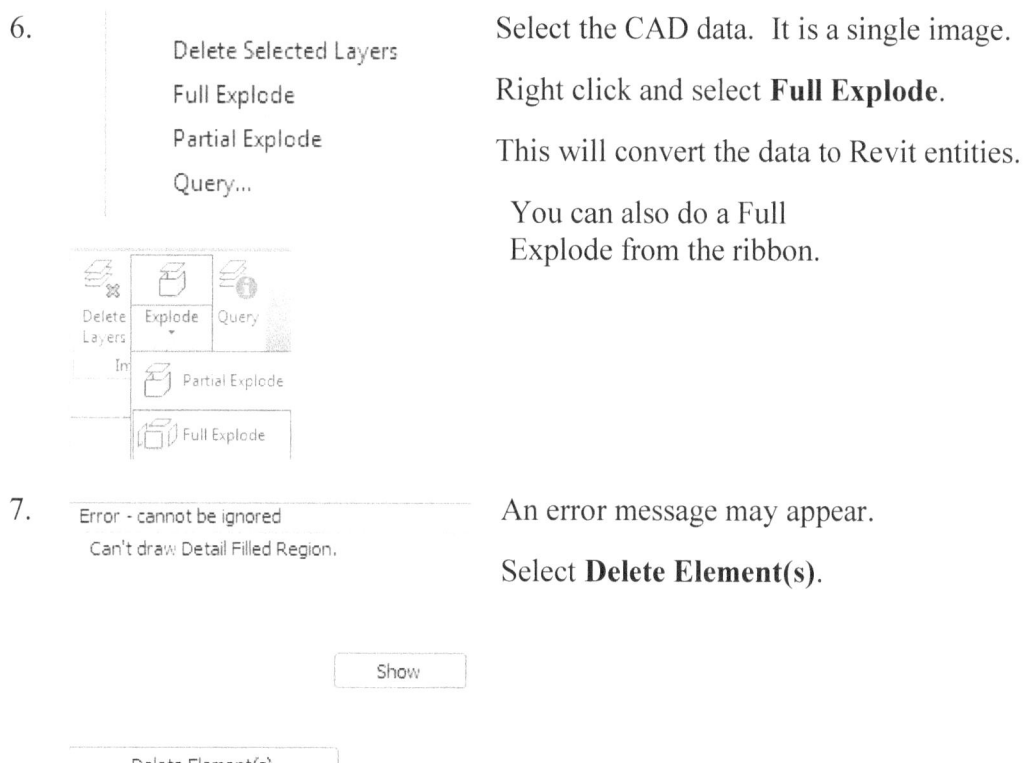

   Select the CAD data. It is a single image.

   Right click and select **Full Explode**.

   This will convert the data to Revit entities.

   You can also do a Full Explode from the ribbon.

7.
   Error - cannot be ignored
   Can't draw Detail Filled Region.

   An error message may appear.

   Select **Delete Element(s)**.

   Show

   Delete Element(s)

   Revit was unable to convert the detail filled region, but you can add that later.

8.
   West
   Sections (Building Section)
   Stair Section

   Next we add a callout for the drafting view.
   Activate the **Stair Section** under Sections.

7-3

9.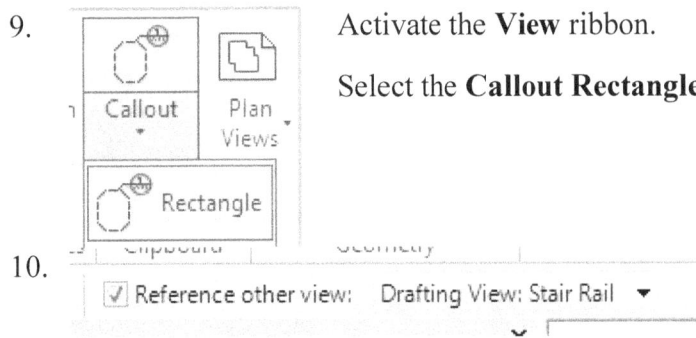

    Activate the **View** ribbon.

    Select the **Callout Rectangle** tool from the Create panel.

10. Place a check on the **Reference other view** box.

    In the drop-down list, select the **Stair Rail Bottom Detail** for the view to be referenced.

11.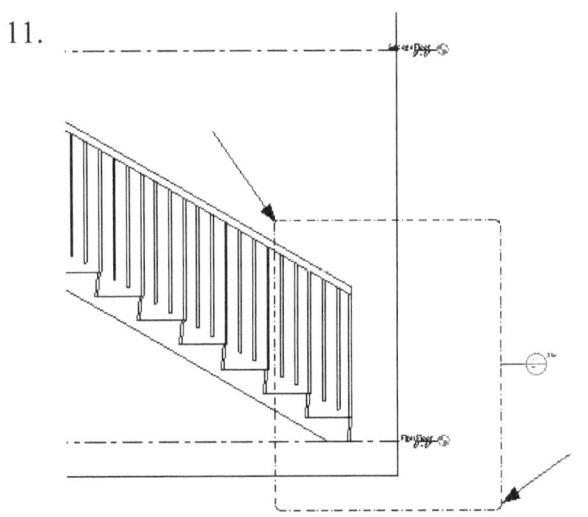

    Place the callout outline by picking in the two locations indicated.

12.

    Some users don't like to see the viewport outline. This is the rectangle around the view.

    Use the **Hide Crop Region** toggle on the View Control bar to hide the viewport outline.

13.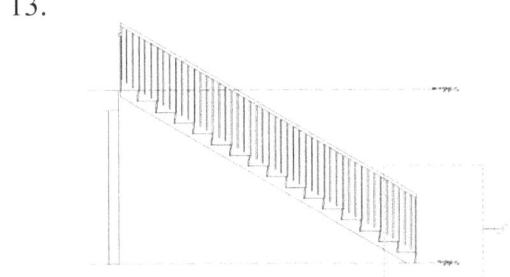

    The rectangular viewport is no longer visible.

Detailing and Drafting

14. Zoom into the callout bubble.
    The bubble data is blank until the drafting view is added to a sheet.

15. Activate the **Stair Section** sheet.

16. 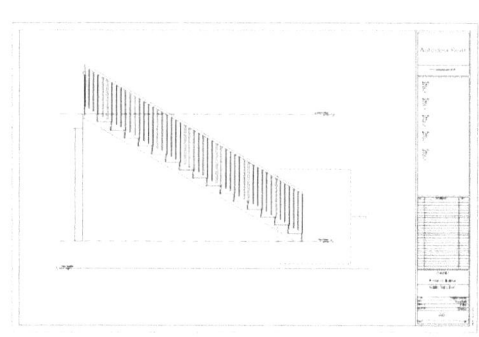 Drag and drop the Stair Section from the browser onto the sheet.

17. Activate the **Stair Details** sheet.

18. 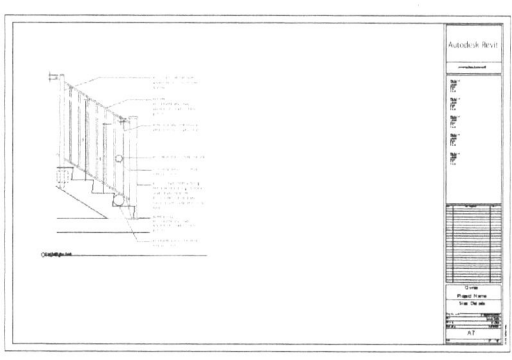 Drag and drop the Drafting View for the bottom of the stairs onto the sheet.

19. Activate the **Stair Section** view under Sections.

20. 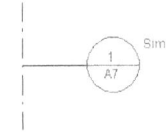 Zoom into the callout bubble.

    Note that the callout is now filled in with the sheet information.

    Double left click on the bubble.

21. This navigates the user to the drafting view.

22. Activate the **Stair Section** view under Sections.

23. Select the callout handle.

24. Select **Edit Type** in the Properties pane.

25. 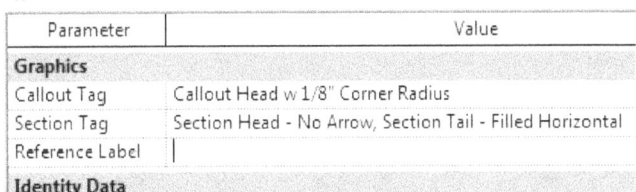 Delete the word **Sim** in the Reference Label field.

    Press **OK**.

26. 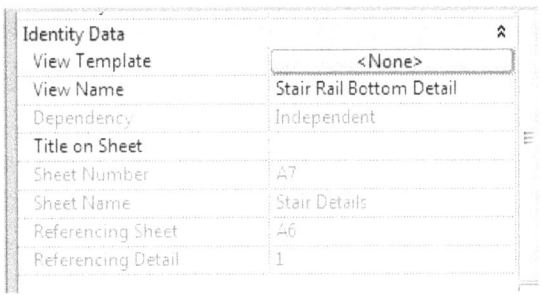 Activate the **Stair Rail Bottom Detail** view.

    In the Properties pane:

    Note that the Referencing Sheet is listed under the Identity Data

    None of the data is editable in this dialog.

27. Save as *ex7-1.rvt*.

Extra Challenge

Create a callout for the top of the stairs using c_detail_top.dwg as the drafting view.

Add the new drafting view to the Stair Details sheet.

1. Create a new drafting view.
2. Import the CAD file.
3. Add the callout to the top of the stairs on the section view.
4. Drag the new drafting view to the Stair Details sheet.

Command Exercise

Exercise 7-2 – Reassociate a Callout

Drawing Name: **callout_redefine.rvt**
Estimated Time to Completion: 10 Minutes

Scope

Change the drafting view for a callout

Solution

1. Activate the **North Elevation** view.

2.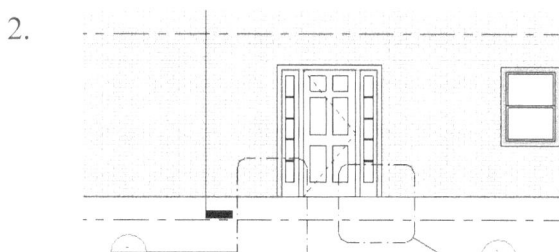

   There are two call-outs on the exterior door.

   The callout on the right references a sheet with a drafting view.

   The callout on the left has no sheet or view references.

3.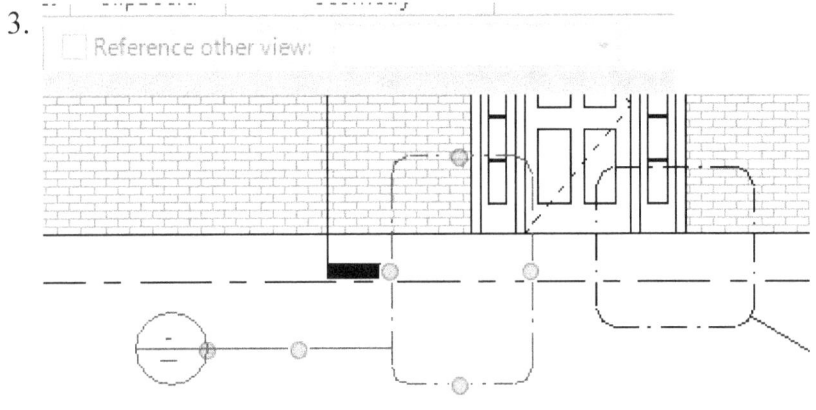

Select the left callout with the blank bubble.

Note that the Reference other view is grayed out.

If you do not assign a reference view when you place a callout, you can't create the link later.

4.  Open Sheet **A7 - Stair Details**..
This is the sheet referenced by the callout.

5. Open Sheet **A8- Door Details**.
The callout should be pointing to the Exterior Door Threshold Detail.

6.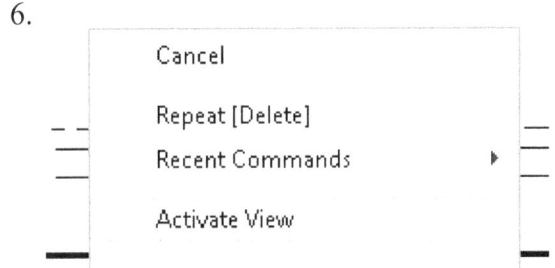

Right click on the elevation view on the sheet.

Select **Activate View**.

Detailing and Drafting

7. 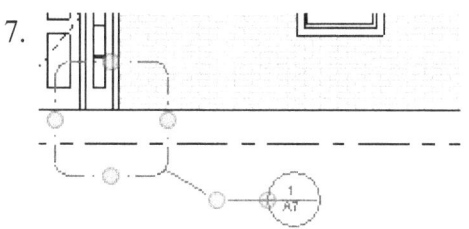 Select the right callout.

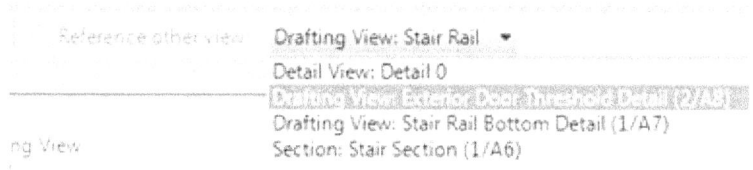

8. On the Options bar:
   Select the **Exterior Door Threshold Detail** from the drop-down list.

9. Right click and **Deactivate View.**

10. 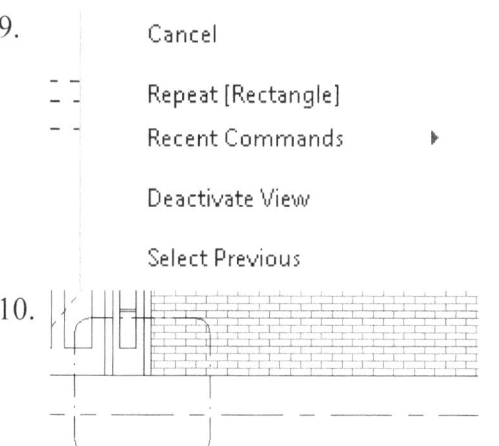 Note how the callout updates with the correct sheet and view number.

11. Close without saving.

Command Exercise
Exercise 7-3 – Filled Regions

Drawing Name: **ex7-1.rvt**
Estimated Time to Completion: 25 Minutes

Scope

Use Filled Regions to modify elements in a drafting view

Solution

12. Activate the **Stair Rail Bottom Detail** view.

13. Select all the text.

    Use the Type Selector in the Properties panel to change the text to **1/4" City Blueprint**.

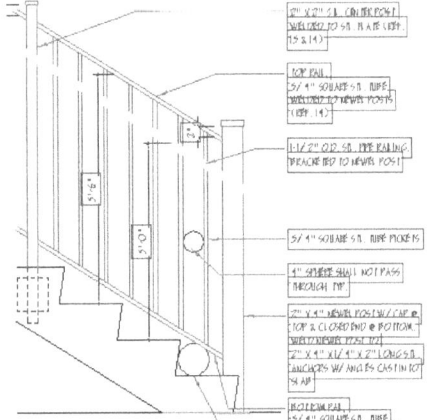

3. Activate the Annotate ribbon.

   Select **Detail Component**.

4. Select **Break Line** on the Type Selector.

5. 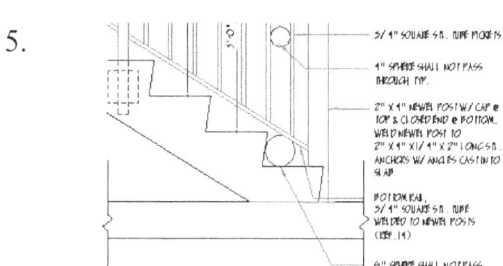 Use the SPACE bar to rotate and place a break line at each side of the floor.

6. 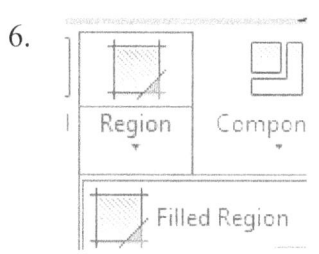 Select the **Filled Region** tool.

7. Use the **Rectangle** tool to trace the floor area.

8. Select the **Floor Filled Region** from the Type Selector.

9. 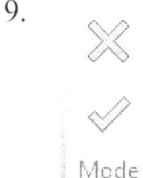 **Green Check** to finish.

10. 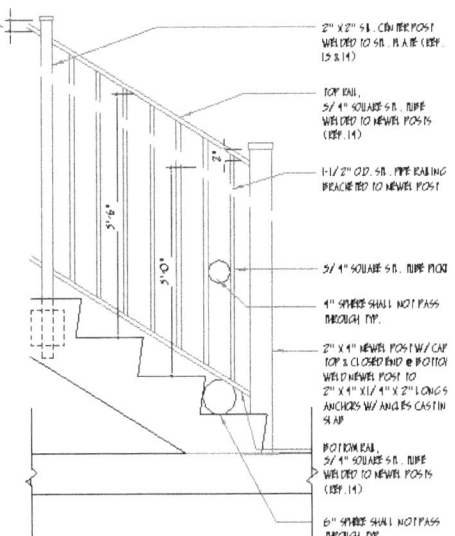 The filled region is displayed.

11. 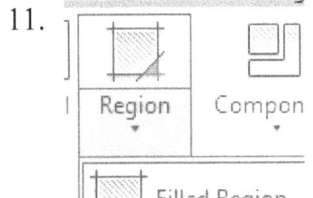 Select the **Filled Region** tool.

12. 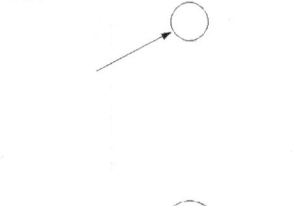 Use the **Pick Line** tool to trace the two circles.

13. Select the two circles.

14.  Select the **DO NOT PASS SPHERE** filled region from the Type Selector.

Detailing and Drafting

15.  **Green Check** to finish.

16. 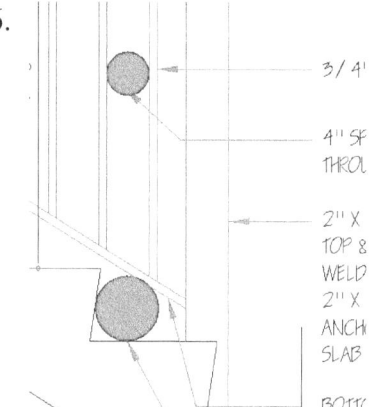 The filled region is displayed.

    Left click to release the selections.

17. 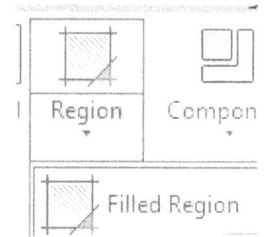 Select the **Filled Region** tool.

18. 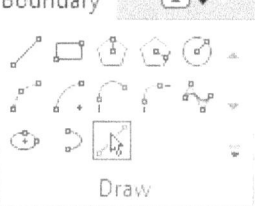 Use the **Pick Line** tool to trace the two circles.

19. 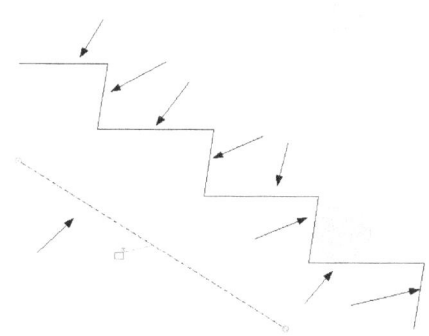 Select the riser, stringer, and tread lines.

7-13

20.  Use the LINE tool to close the filled region.

21. Select the **Treads** Filled Region from the Type Selector.

22.  **Green Check** to finish.

23. 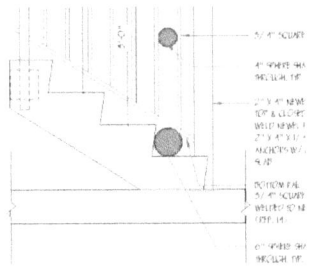 The filled region is displayed.

24. Save as *ex7-3.rvt*.

Command Exercise

Exercise 7-4 – Callout from Sketch

Drawing Name: **callout_sketch.rvt**
Estimated Time to Completion: 25 Minutes

Scope

Object Settings
Callout by Sketch
Drafting View
Duplicate view as Dependent
Visibility/Graphics
Masking Region
Convert Dependent view to Independent

Solution

1. Activate the Manage ribbon.

   Select **Object Styles**.

2. Select the Annotation tab.

   Set the color of the Callout Boundary to **Blue**.

   Set the Line Pattern to **Dash dot dot**.

   Press **OK**.

3. Activate **Level 1**.

4. Highlight Level 1.
   Right click and select **Duplicate View→Duplicate as Dependent**.

5. Rename the view **Level 1 – Stairs Detail**.

   Press **OK**.
6. The view should be listed under the Level 1 view.

7. Use the Crop Region to crop the view to show only the stairs area.

8. Select one of the furniture elements. Right click and select **Hide in View→Category**.

   Repeat to hide the section line.

9. Activate Level 1.

   Note the furniture is not visible in Level 1.

10. Type **VV** to bring up the Visibility/Graphics dialog.

    Enable **Furniture**.

    The furniture is visible again.

DETAILING and Drafting

11. Activate **Level 1 – Stairs Detail**.

12. Activate the Annotate ribbon.

    Select **Masking Region**.

13. Draw a rectangle around the furniture.

14. Select Green Check.

    The furniture is no longer visible.

15. Activate Level 1.

    Note the furniture is not visible in Level 1.

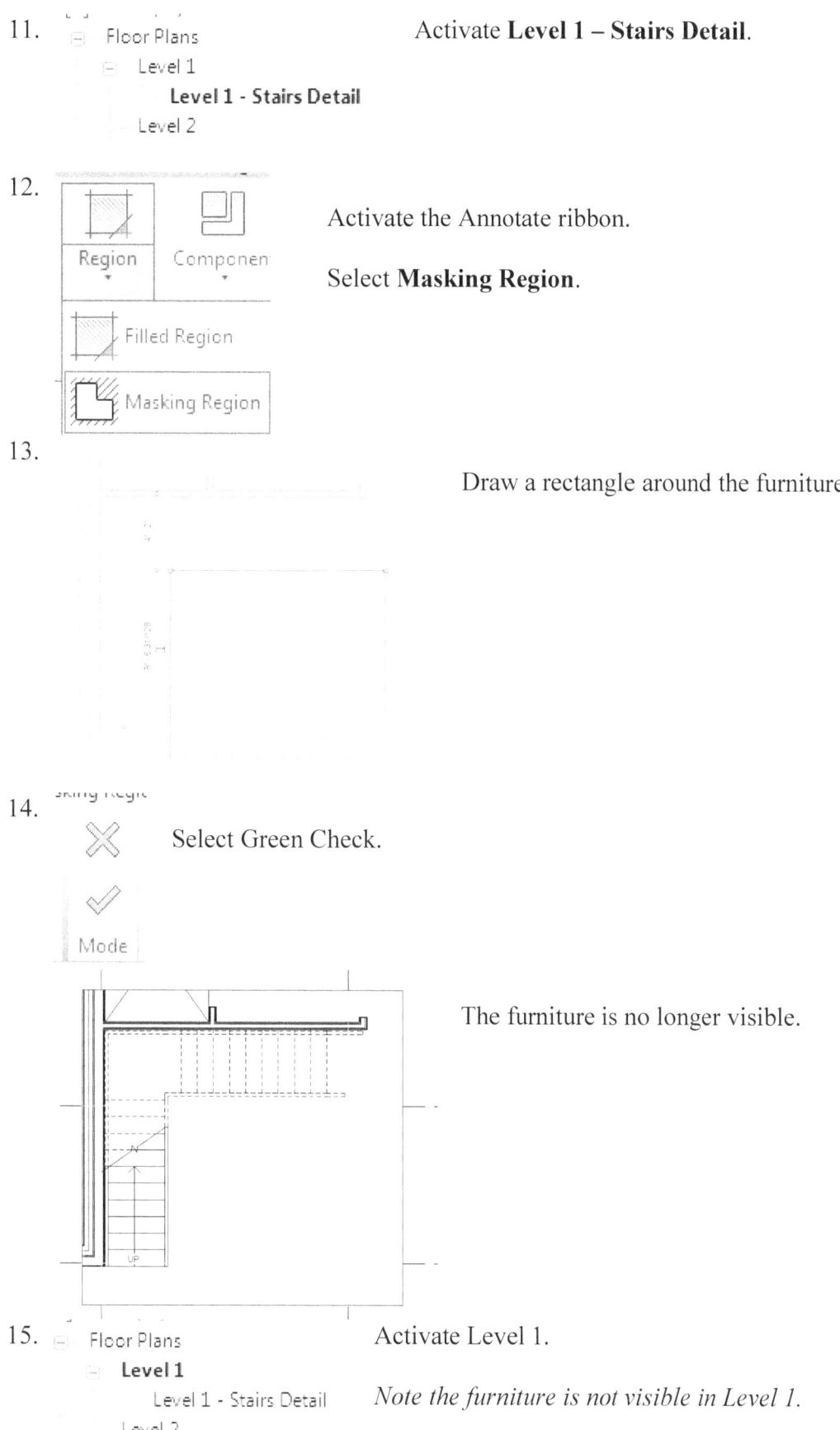

7-17

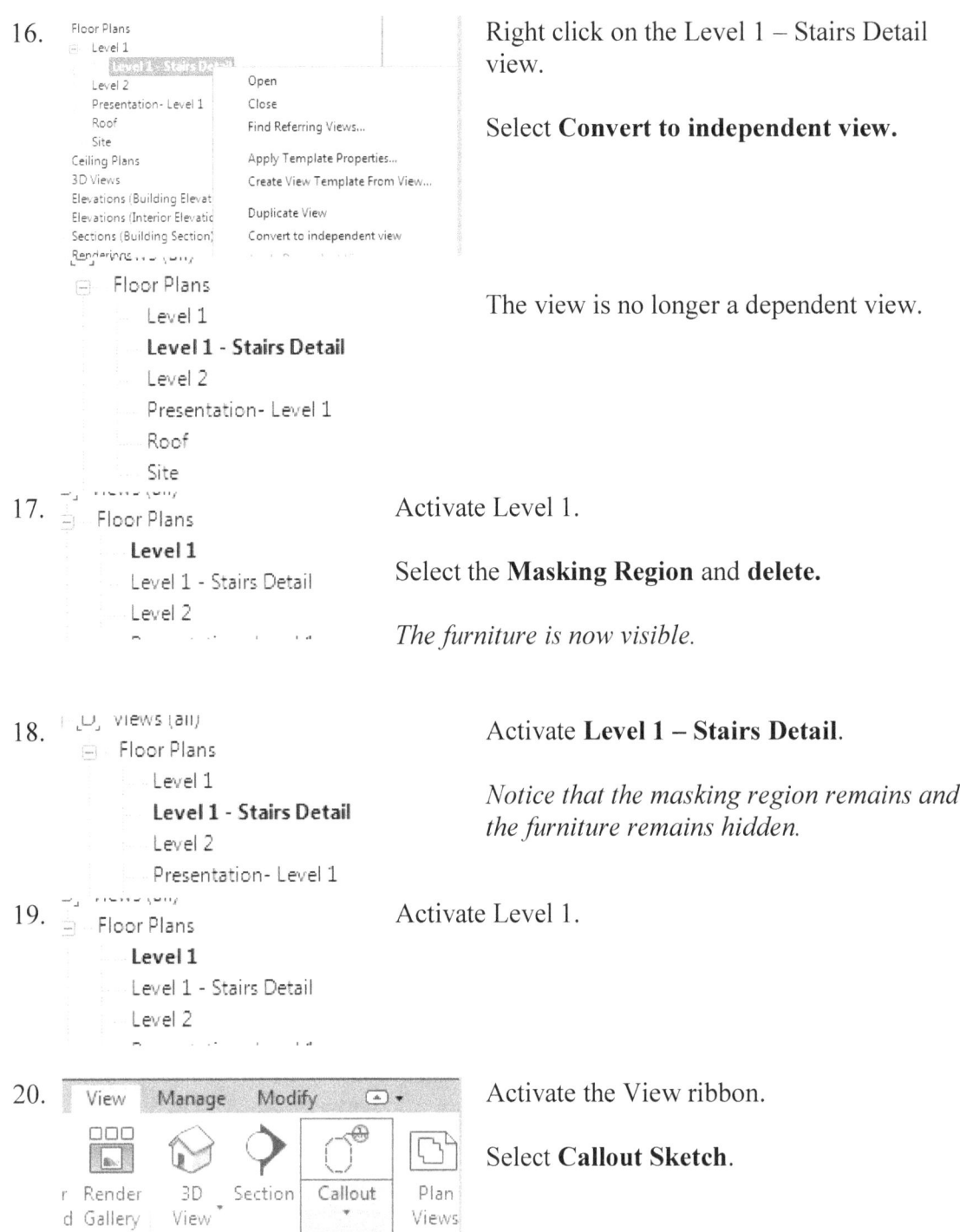

16. Right click on the Level 1 – Stairs Detail view.

    Select **Convert to independent view.**

    The view is no longer a dependent view.

17. Activate Level 1.

    Select the **Masking Region** and **delete.**

    The furniture is now visible.

18. Activate **Level 1 – Stairs Detail**.

    Notice that the masking region remains and the furniture remains hidden.

19. Activate Level 1.

20. Activate the View ribbon.

    Select **Callout Sketch**.

7-18

21.   On the Options bar:

    Enable **Reference other view**.
    Select **Level 1 – Stairs Detail,**

22. Use the LINE tool to draw an outline around the stairs.

23. Select **Green Check**.

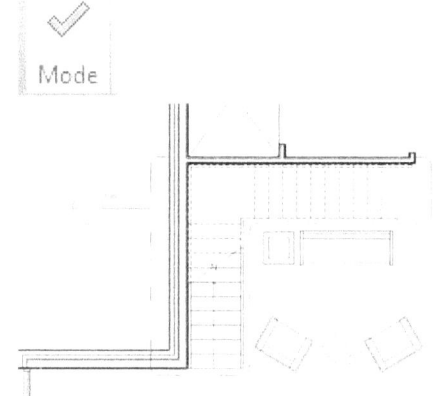

A Callout section is created.

Notice it uses the line type and color you selected in Object Styles.

Double left click on the callout bubble.

24. **Level 1 - Stairs Detail**

    The Level 1 –Stairs Detail view is activated.

25. Save as *ex7-3.rvt*.

Command Exercise

Exercise 7-5 – Save and Re-Use a Drafting View

Drawing Name: **c_ details.rvt**
Estimated Time to Completion: 10 Minutes

Scope

Save a drafting view for re-use.
Insert the saved file into a project.

Solution

1.  Highlight the **Stair Rail Bottom Detail** in the Drafting Views (Detail) category in the Project Browser.

2. Right click and select **Save to New File**.

3. Browse to your work folder and select **Save**.

4. Start a new project.

5. Press **OK**.

6. Activate the **Insert** ribbon.

   Select **Insert from File→Insert Views from File** from the Import panel.

7. Locate the file you just saved.

   Press **Open**.

8.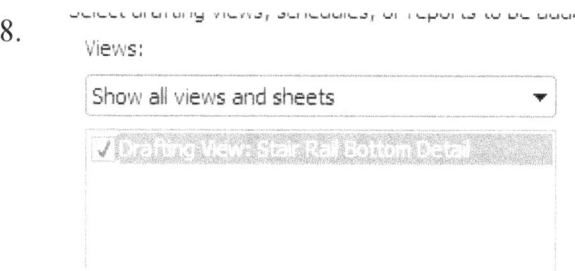

All views are listed.

Only one view is available for import – the view you saved.

Press **OK**.

9. 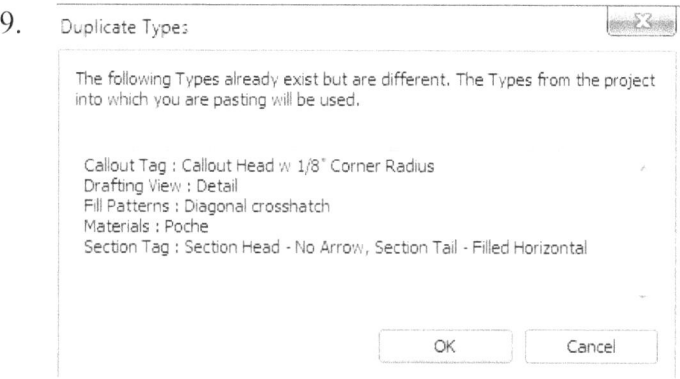 You may see this dialog having to do with types which are different.

Press **OK**.

10. The view is listed in the browser and is activated.

11. You can ignore any warnings and just close this dialog box.

12. Close the files without saving.

Command Exercise

Exercise 7-6 – Adding Tags

Drawing Name: **i_dimensions.rvt**
Estimated Time to Completion: 15 Minutes

Scope

Add room tags to a view.
Duplicate view with Detailing
Duplicate view as dependent
Change the room tags assigned type.

Solution

1. Activate the **Ground Floor Admin Wing**.

2. Mouse over the different rooms and note that room elements have been placed in each room.

3. Activate the **Annotate** ribbon.
   Select the **Tag All** tool from the Tag panel.

4. Select the **Room Tag**.

   Press **OK**.

5. Press **Yes** if you see this dialog.

7-22

DETAILING and Drafting

6. 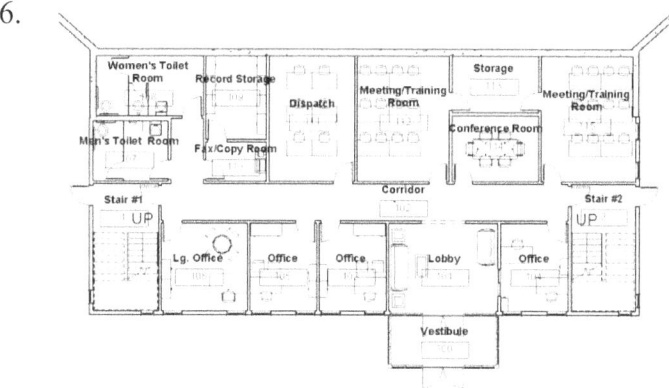 Room tags are placed in all rooms.

7.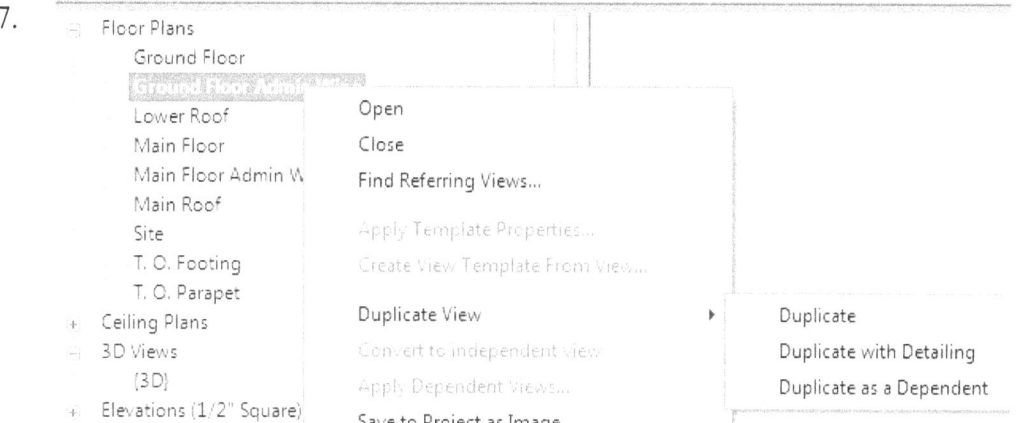

Highlight the view in the browser.

Right click and select **Duplicate View→Duplicate**.

8. 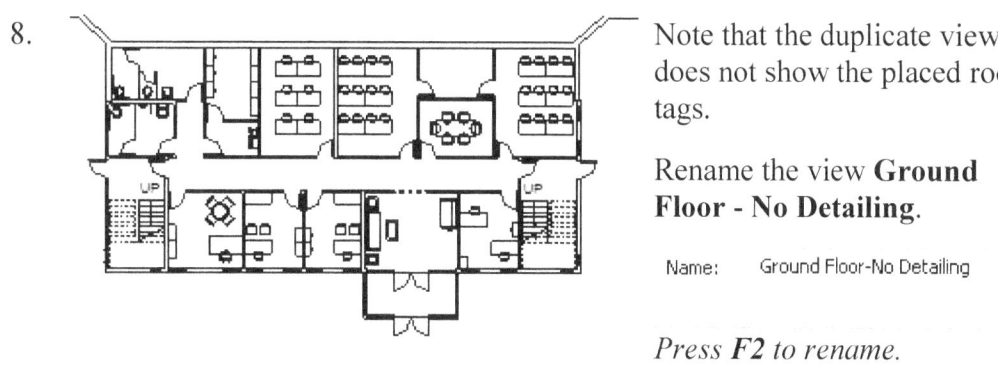 Note that the duplicate view does not show the placed room tags.

Rename the view **Ground Floor - No Detailing**.

Name: Ground Floor-No Detailing

*Press **F2** to rename.*

Press **OK.**

9.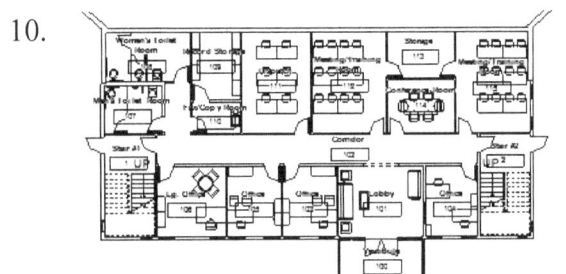

Highlight the Ground Floor Admin Wing floor plan view in the browser.

Right click and select **Duplicate View→Duplicate with Detailing**.

10.

Note that the new view includes the room tags.

Rename the view - **Ground Floor – Detailing**

Press **OK.**

11.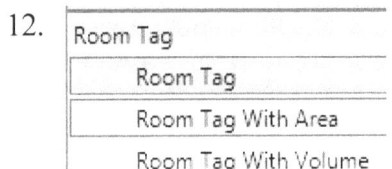

Select one of the room tags.

Right click and select **Select All Instances→Visible in View**.

12.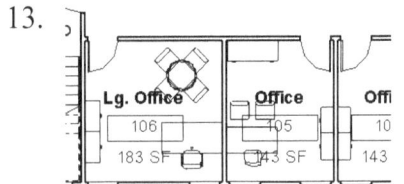

In the Type drop-down on the Properties pane, select **Room Tag with Area**.

13. The room tags update.

14.

Switch back to the original Ground Floor Admin Wing.

Note that the room tags for that view did not update.

15. Close without saving.

Detailing and Drafting

Command Exercise

Exercise 7-7 – Creating a Detail View

Drawing Name: **i_detail.rvt**
Estimated Time to Completion: 45 Minutes

Scope

Add detail components to a detail view.
Set a detail view to independent.
Change the clip offset of a detail view.
Use Repeating Details.
Use detail lines.

Solution

1. Activate the **Main Roof** floor plan.

2. 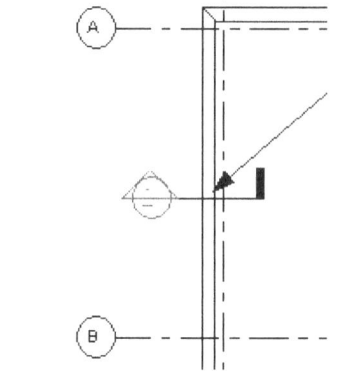 Double click on the section head located between A-B grids to activate that section view.

3. Note the name of the active section view in the browser.

4. 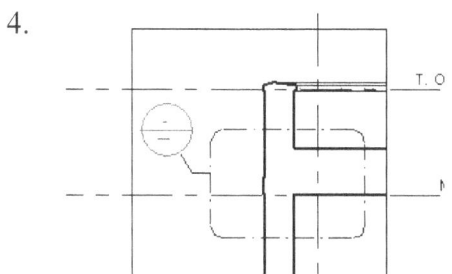 Double click on the callout bubble to activate the callout view.

7-25

5. Sections (Detail)
   Callout (2) of Section 1
   Callout of Section 1
   Callout of Section 2

   Note the name of the active callout section view in the browser.

6.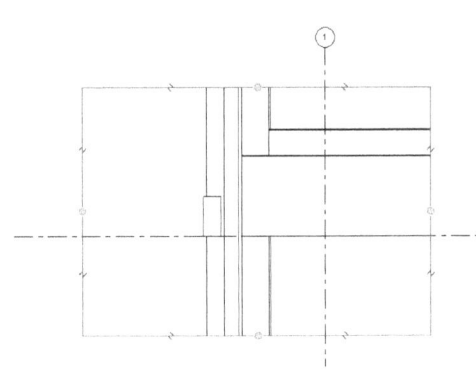

   Click on the crop window to activate the grips.

   The grips allow you to change the extents of the view.

7.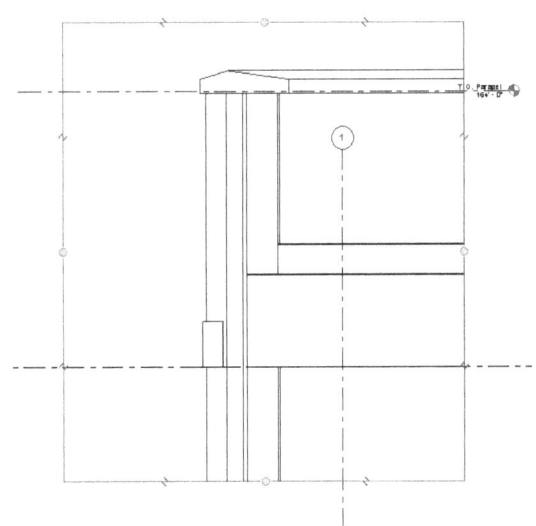

   Drag the crop region up using the grip bubble so that the roof parapet is shown.

8. Highlight the active Callout view in the browser.

   Right click and select **Rename**.

9. Change the name to **Roof Detail**.

   Press **OK**.

Detailing and Drafting

10.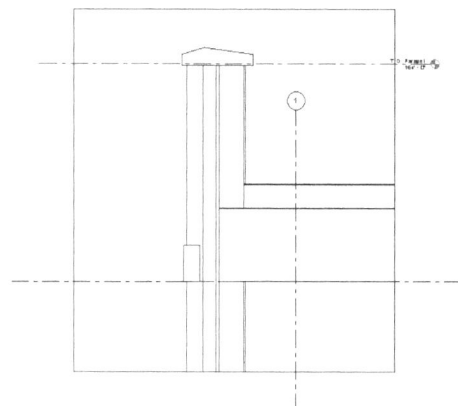

    In the Properties pane:

    Under Extents:
    Set the Far Clip Settings to **Independent**.

    Change the Far Clip Offset to **5' 0"**.

    Note that the view updates and no longer displays the far North wall.

11.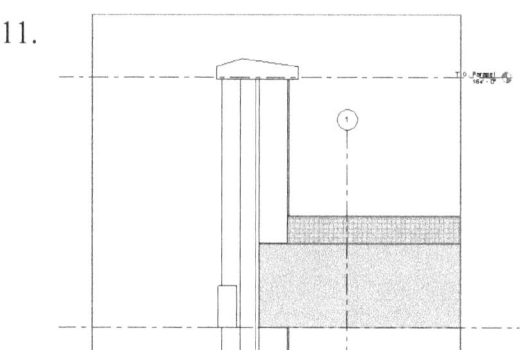

    Select the roof.

    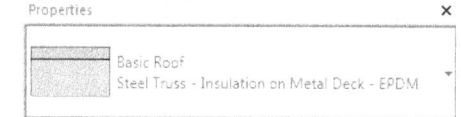

    If the roof is selected, it should be listed in the Properties pane.

12.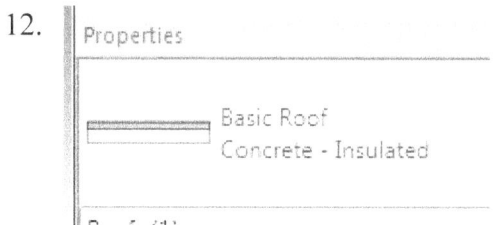

    Change the Roof type to **Concrete - Insulated** on the Properties pane.

    If the roof is changed in the section view, is it changed in the model?

13.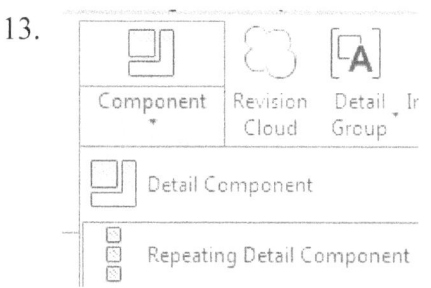

    Select the **Repeating Detail** tool from the Detail panel on the Annotate ribbon.

7-27

14. Set the repeating detail as **Brick** using the Type Selector.

15. 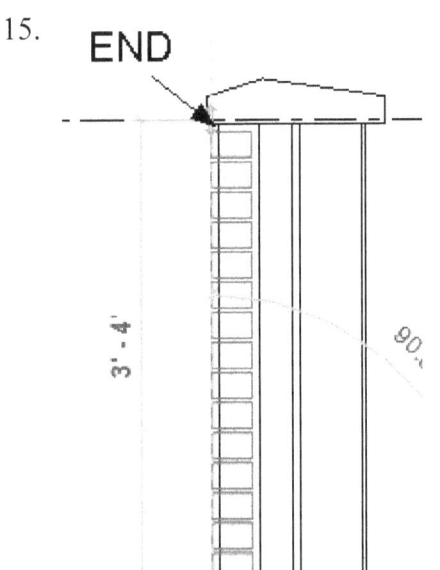 Pick the two points indicated to place the brick detail.

    Start at the bottom and end at the top where the T.O. Parapet level is.

16. Select the brick repeating detail just placed.

    Select the **Edit Type** on the Properties pane.

17. Note that the detail was placed using a **Fixed Distance** layout.

    Note that the detail can be rotated.

    Note the spacing is set to **2 5/8"**.

18. Select **Duplicate**.

19. Type **Mortar**.

    Press **OK**.

Detailing and Drafting

20. | Pattern | |
    |---|---|
    | Detail | Mortar Joint : Brick Joint |
    | Layout | Fixed Distance |
    | Inside | ✓ |
    | Spacing | 0' 2 5/8" |
    | Detail Rotation | None |

    For the Detail field, select **Mortar Joint: Brick Joint**.

    Press **OK**.

21. Select the **Repeating Detail** tool from the Detail panel on the Annotate ribbon.

22. Set the repeating detail as **Mortar** using the Type Selector.

23. 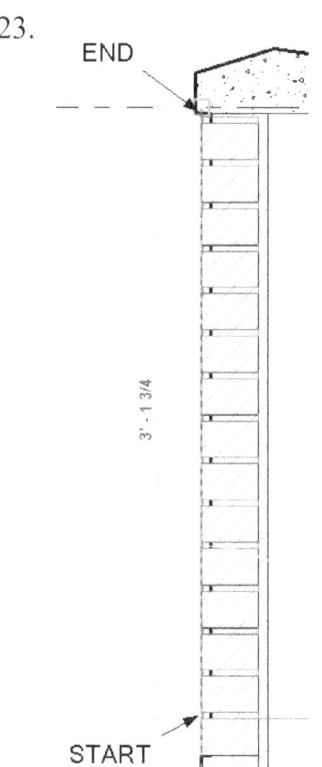 Start on top of the first brick placed and end at the TO Parapet level.

24. 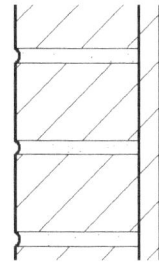 Zoom in to see the mortar joints.

7-29

25. Select the **Detail Component** tool from the Annotate ribbon.

26. Select the **3″ x 3″ Cant Strip** from the drop-down list on the Type Selector.

27. 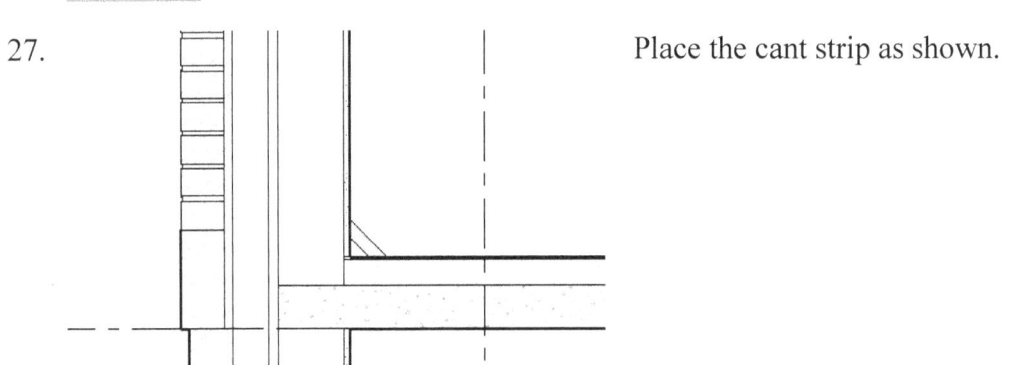 Place the cant strip as shown.

28. Select the **Detail Line** tool from the Annotate ribbon.

29. Select **Medium Lines** from the Line Style options drop-down list.

30. Enable **Chain**.
    Enter **1/4″** in the Offset box.

31. Select **Pick Lines** mode from the Draw panel.

32. 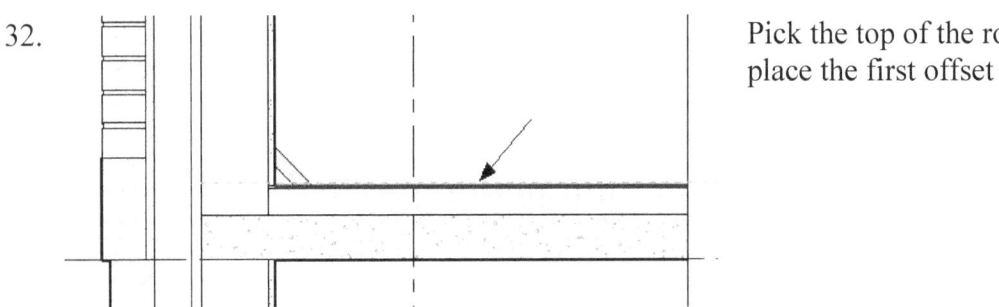 Pick the top of the roof to place the first offset line.

7-30

Detailing and Drafting

33. 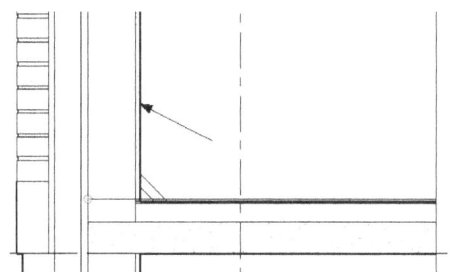 Pick the vertical wall.

34. Switch to **Draw Line** mode.

35. 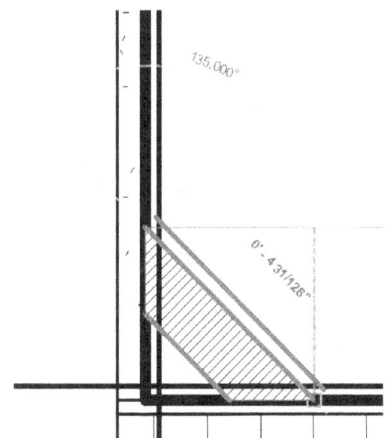 Place a line parallel to the Cant Strip.

    Set the offset to 1/4" and start the line at the top of the cant strip. This will orient the line above the can strip.

36. Use the **Trim** tool on the Modify Ribbon to clean up the detail lines.

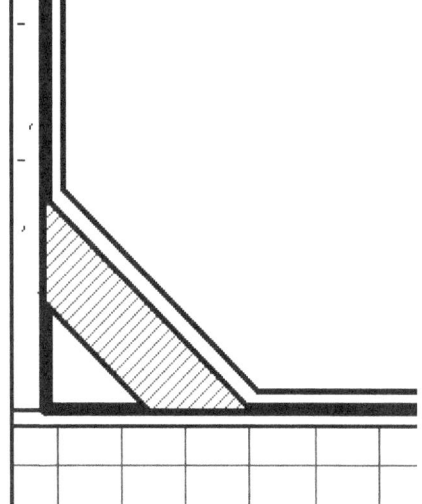

37.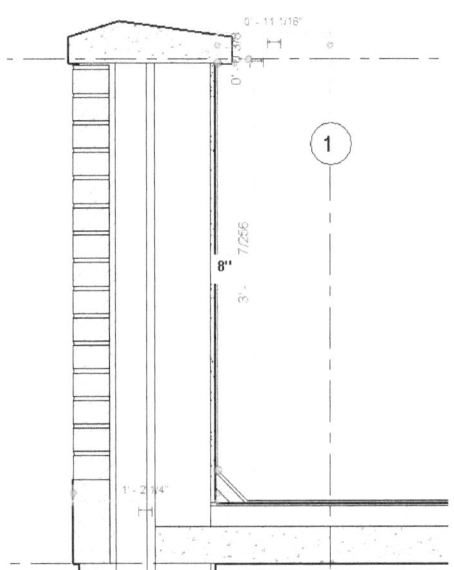

    Select the vertical line to enable the temporary dimension.

    Select the vertical line's vertical dimension.

    Change the value to 8".

38. 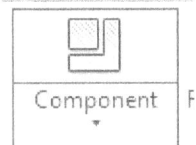 Select **Component→Detail Component** on the Annotate ribbon.

39. 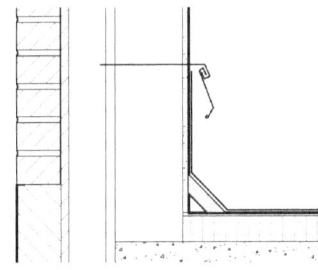 In the Properties pane:

    Select **Flashing - Cap** using the Type Selector.

40. 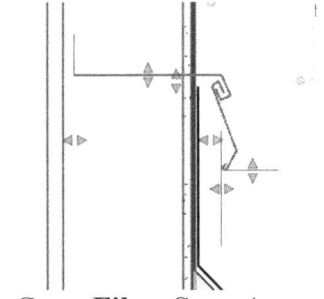 Place the Flashing Cap above the vertical line.

41. Select the flashing cap and use the grips to adjust the size.

42. Go to **File→Save As**.
    Save the file as *ex7-7.rvt*.

Detailing and Drafting

Command Exercise

Exercise 7-8– Creating a Detail Group

Drawing Name: **i_detail_groups.rvt**
Estimated Time to Completion: 20 Minutes

Scope

Add hidden lines.
Add a steel plate using Detail Component.
Add bolts to a plate.
Create a detail group.
Mirror a detail group.

Solution

1. Activate the **Beam-Column Connection** section view.

2. Activate the Annotate ribbon.

   Select the **Detail Component** tool from the Detail panel.

3. Select **Load Family** from the Mode panel.

4. Browse to the *Metal Fastenings* folder under *Detail Items\ Div 05 - Metals\ 050500 - Common Work Results for Metals*.

5. Open the *Steel Shear Plate - Face* family.

7-33

6. 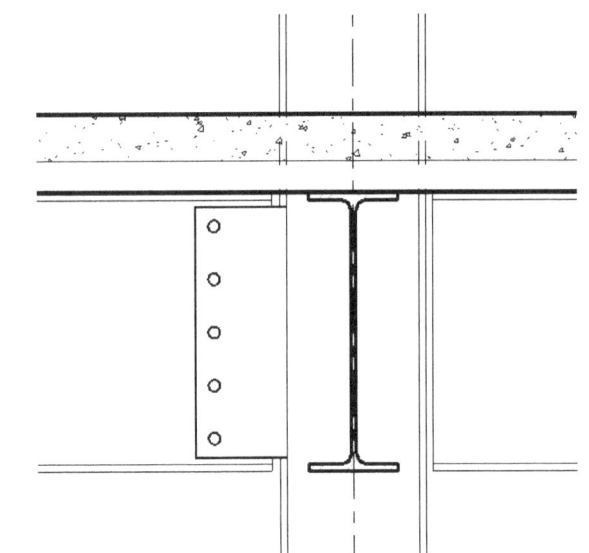 Press the SPACE bar to rotate the component prior to placing.

   Place on the left side of the column.

7. Activate the Annotate ribbon.

   Select the **Repeating Detail Component** tool from the Detail panel.

8. Select the **A325 - 5/8"** Plan detail on the Properties pane.

9. Select **Edit Type** in the Properties pane.

10. Set the Layout to **Fixed Number**.

    Press **OK**.

11. In the Properties pane:

    Change the number to **5**.

    Press **Apply**.

12. Place a bolt in the bottom hole.

    Then pick the top hole.

    Right click and select Cancel to exit the command.

    If the number reverts back to 3, place the detail. Release it. Re-select it and then change it again in the Properties pane and press Apply.

Detailing and Drafting

13.  Window around the steel plate and the placed bolts to select them.

    Select **Detail Group→Create Group** from the Detail panel on the Annotate ribbon.

14. 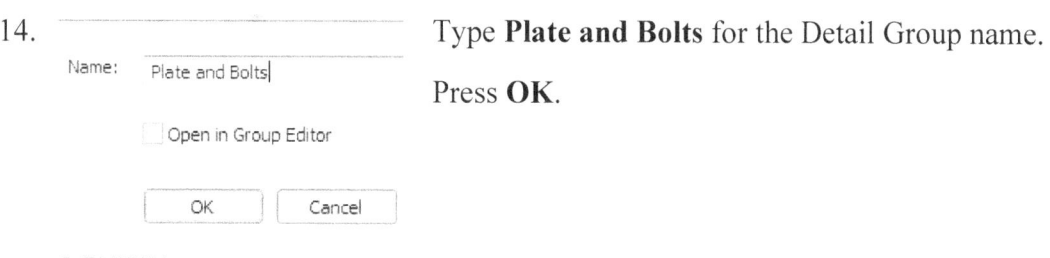 Type **Plate and Bolts** for the Detail Group name.

    Press **OK**.

15. 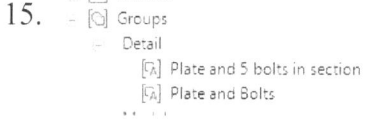 In the Browser, locate the detail group created under **Groups→Detail**.

16. Select the steel plate and bolts detail group in the display window. The group should highlight.

    Select **Mirror→Pick Axis** from the Modify panel.

17. Select the grid centered on the column as the axis.

    Click anywhere in the window to release the selection and complete the mirror command.

    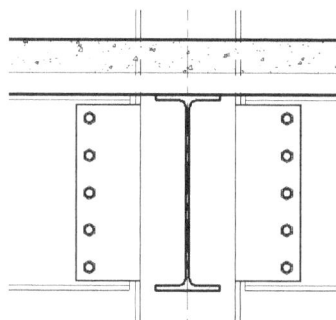

18. 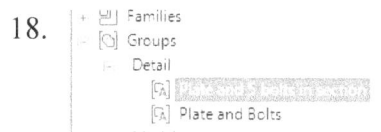 Locate the **Plate and 5 bolts in section** in the Project Browser.

7-35

19. 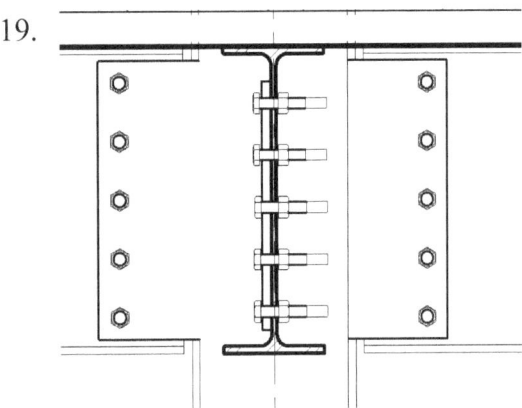 Drag and drop them into the location shown.

    Left click to place.

20. Close without saving.

detailing and Drafting

Command Exercise
Exercise 7-9 – Creating a Drafting View

Drawing Name: **i_drview.rvt**
Estimated Time to Completion: 50 Minutes

Scope

Create a drafting view.

Solution

1. Activate the **View** ribbon.

   Select the **Drafting View** tool from the Create panel.

2. Enter **Roof & Overflow Drain** for the Name.
   Set the Scale to **1 ½" = 1'0"**.

   Press **OK**.

3. A blank view opens.
   Note in the browser that we are in an active drafting view.

4. Select the **Reference Plane** tool from the Work Plane panel on the Architecture ribbon.

5. Place a vertical reference plane in the middle of the blank view.

6.  Select the **Detail Component** tool from the Detail panel on the Annotate ribbon.

7. Roof Drain Select **Roof Drain** from the Type Selector list on the Properties pane.

8. Place the roof drain **1′-3″** to the left of the reference plane.

   Right click and select Cancel to exit the command.

9. Select the **Filled Region** tool from the Detail panel on the Annotate ribbon.

10. 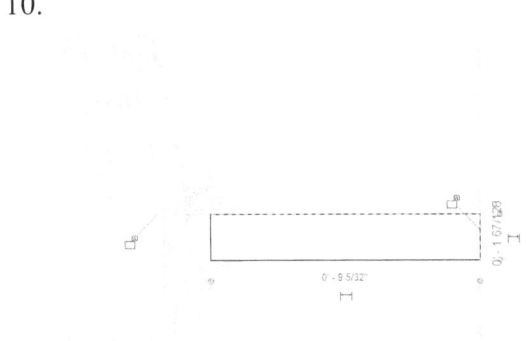 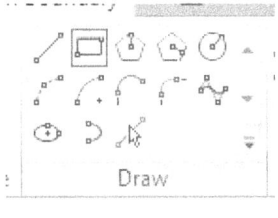 Use the **Rectangle** tool to draw a rectangle as shown.

11. Filled region Ortho Crosshatch - Small
    In the Properties pane:
    Select the Filled Region type to **Ortho Crosshatch – Small**.

12. 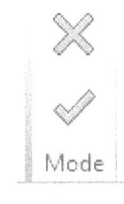 Select the **Green Check** under Mode to **Finish Region**.

13. Select the **Filled Region** tool from the Detail panel on the Annotate ribbon.

14. Use the **Rectangle** tool to draw a rectangle as shown.

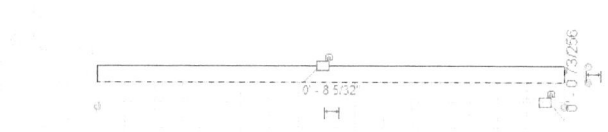
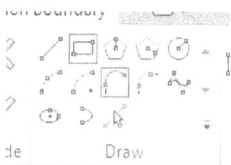

15.  In the Properties pane:

    Set the Filled Region to **Sand - Very Dense**.

16. Select the **Green Check** under Mode to **Finish Region**.

17. Select the **Detail Line** tool from the Detail panel.

18. Select **Thin Lines** on the Line Style.

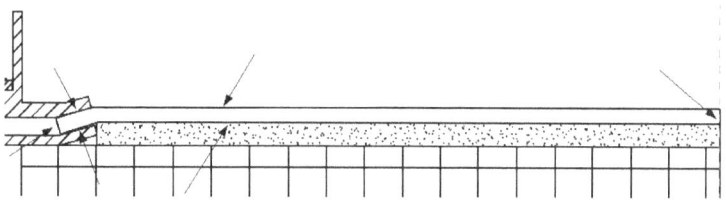

19. Draw the outline indicated with arrows.

20.  Window around the elements to select everything except the reference plane.

21. Select the **Mirror** tool using the **Mirror Pick Axis** option.
Select the reference plane to use as the axis.

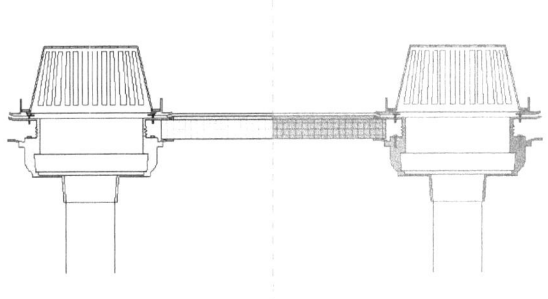

22. Window around the filled regions and detail lines to select.

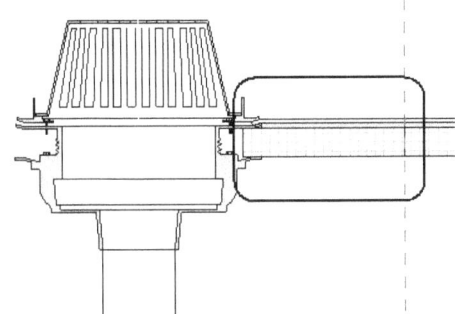

23. Select the **Mirror** tool using the **Draw Mirror Axis** option.

24. Select the midpoint of the roof drain to draw the axis.

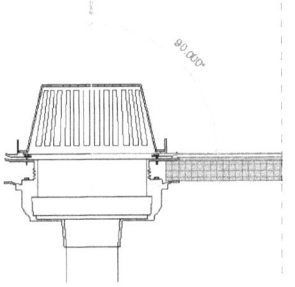

25. The elements are mirrored to the other side.

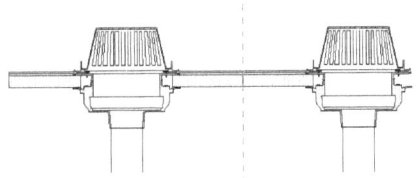

26. Select the mirrored elements.

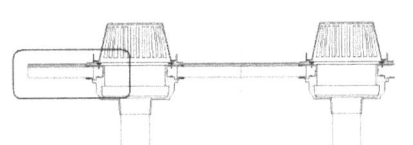

Detailing and Drafting

27. 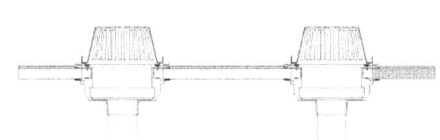 Select **Mirror → Pick Mirror Axis** from the ribbon.

28.  Select the reference plane to use as the mirror axis.

    The elements are mirrored.

29. Select the **Detail Component** tool from the Annotate ribbon.

30. Select **Break Line** from the Type Selector drop-down list on the Properties pane.

31. Position the cursor to the left of the detail view.

    Press the SPACEBAR to rotate the break line.

32. 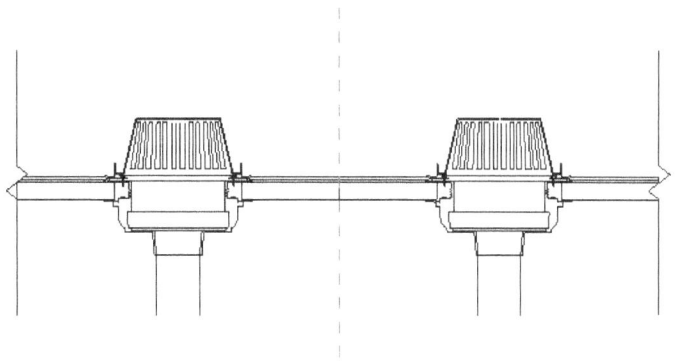 Place the break line so it is coincident with the vertical line at the end of the filled region element.

    Repeat for the right side.

7-41

33. Select the left roof drain.
34. On the Properties Pane:  Change the Ring Height to **2"**.

35. Select the **Aligned** dimension tool on the Annotate ribbon.

36. Place a dimension from center to center distance between the roof drains.

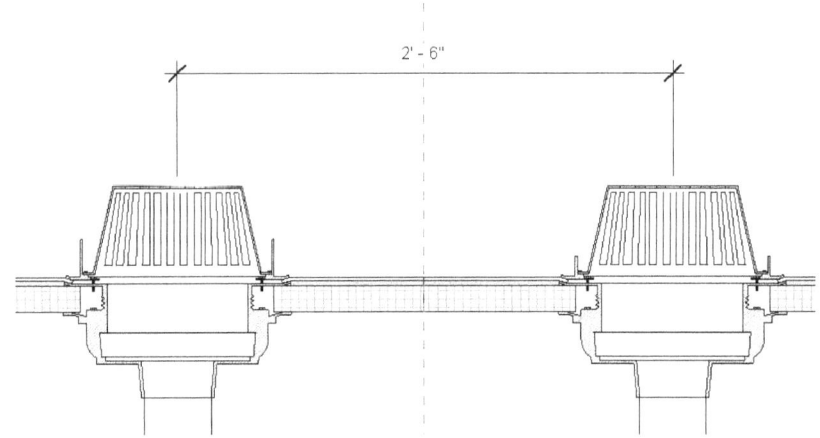

37. With the dimension selected, select **Linear w Center - 3/32" Arial** from the Type Selector drop-down list on the Properties pane.

38. The dimension changes to display centerline symbols.

Detailing and Drafting

39. 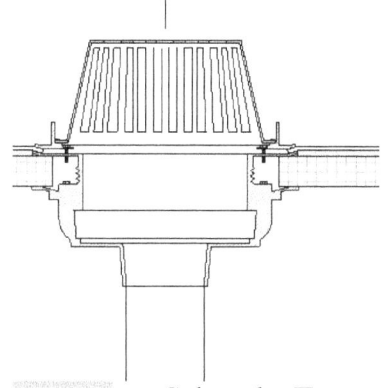 Zoom into the right roof drain.

40. Select the **Text** tool from the Annotate ribbon.

41. 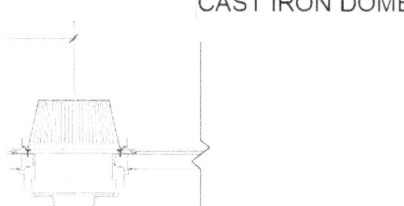 Enable the **Two Segments** leader option on the Format panel.

42. Place the text as shown.

43. 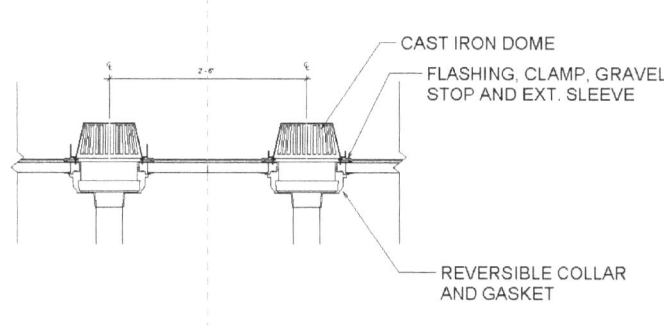 Add the text shown.

44. Close without saving.

7-43

Command Exercise

Exercise 7-10 – Revision Control

Drawing Name: **i_Revisions.rvt**
Estimated Time to Completion: 15 Minutes

Scope

Add a sheet.
Add a view to a sheet.
Setting up Revision Control in a project.

Solution

1. Activate the **View** ribbon.

   Select the **New Sheet** tool on the Sheet Composition panel.

2. Select the **Load** button.

3. Locate the *Titleblocks* folder.

   Locate the **D 22 x 34 Horizontal** titleblock.

   Press **Open**.

4. Highlight the **D 22 x 34 Horizontal** titleblock.

   Press **OK**.

Detailing and Drafting

5. 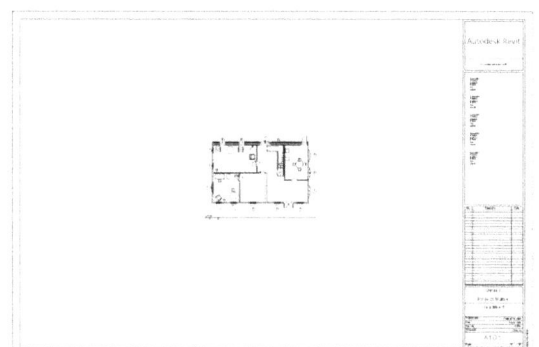 Drag and drop the **Level 1** floor plan onto the sheet.

6. Select the view.

   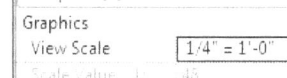 In the Properties pane:

   Set the View Scale to **1/4″ = 1′-0″**.

7. 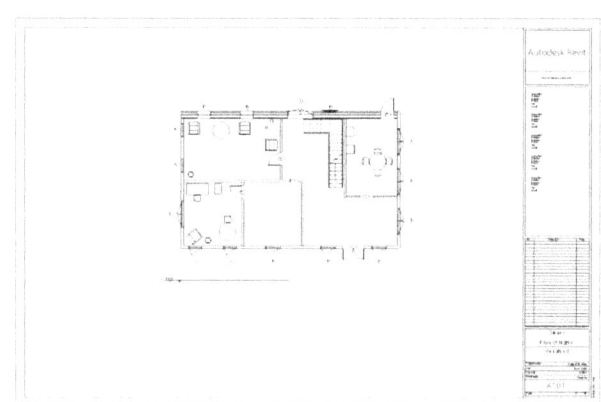 Position the view on the sheet.

8. To adjust the view title bar, select the view and the grips will become activated.

9. 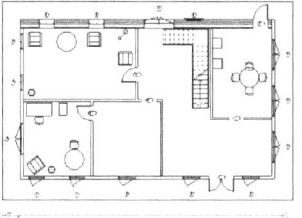 Activate the **View** ribbon.

   Select **Revisions** on the **Sheet Composition** panel.

10. This dialog manages revision control settings and history.

    Numbering can be controlled per project or per sheet. The setting used depends on your company's standards.

    Enable **Per Project**.

    One Revision is available by default. Additional revisions are added using the **Add** button.

11. 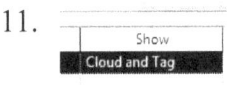 The visibility of revisions can be set to **None**, **Tag** or **Cloud and Tag**. Use the **None** setting for older revisions which are no longer applicable so as not to confuse the contractors.

    Set the revision to show **Cloud and Tag**.

12.  Select the **Options** button.

13. Change the sequence to remove the letters **I** and **O**.

    Press **OK**.

    Revit only allows numeric or alphabetic revisions – no combinations.

14. 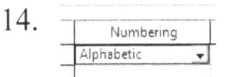 Set the Numbering to **Alphabetic**.

15. Enter the three revision changes shown.

    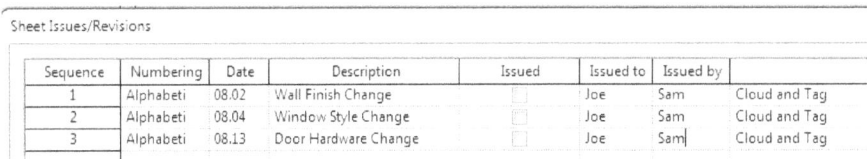

16. Press **OK** to close the dialog. *You can delete revisions if you make a mistake. Just highlight the row and press **Delete**.*

17. Save project as *ex7-10.rvt*.

Command Exercise

Exercise 7-11 – Modify a Revision Schedule

Drawing Name: **ex7-10.rvt**
Estimated Time to Completion: 20 Minutes

Scope

Modify a revision schedule in a title block.

Solution

1. Zoom in to the **Revision Block** area on the sheet.

   Note that the title block includes a revision schedule by default.

2.  Select the title block.

   Right click and select **Edit Family**.

3. Select the **Revision Schedule** in the Project Browser.

4. Select **Edit** next to Formatting in the Properties pane.

5. Change the First Column Header to **Rev**.

6. Select the Fields tab.

   Add the **Issued By** field.

   Do **NOT** remove the Revision Sequence field. This is a hidden field.

   Note that the Hidden field control is located on the Formatting tab. This is a possible question on the exam.

7. Order the fields as shown.

   Press **OK**.

8. The Revision Schedule updates.

Detailing and Drafting

9. To return to the title block sheet view, double left click on the – below Sheets (all) in the browser.

10. Adjust the column width of the schedule so it fits properly in the title block.

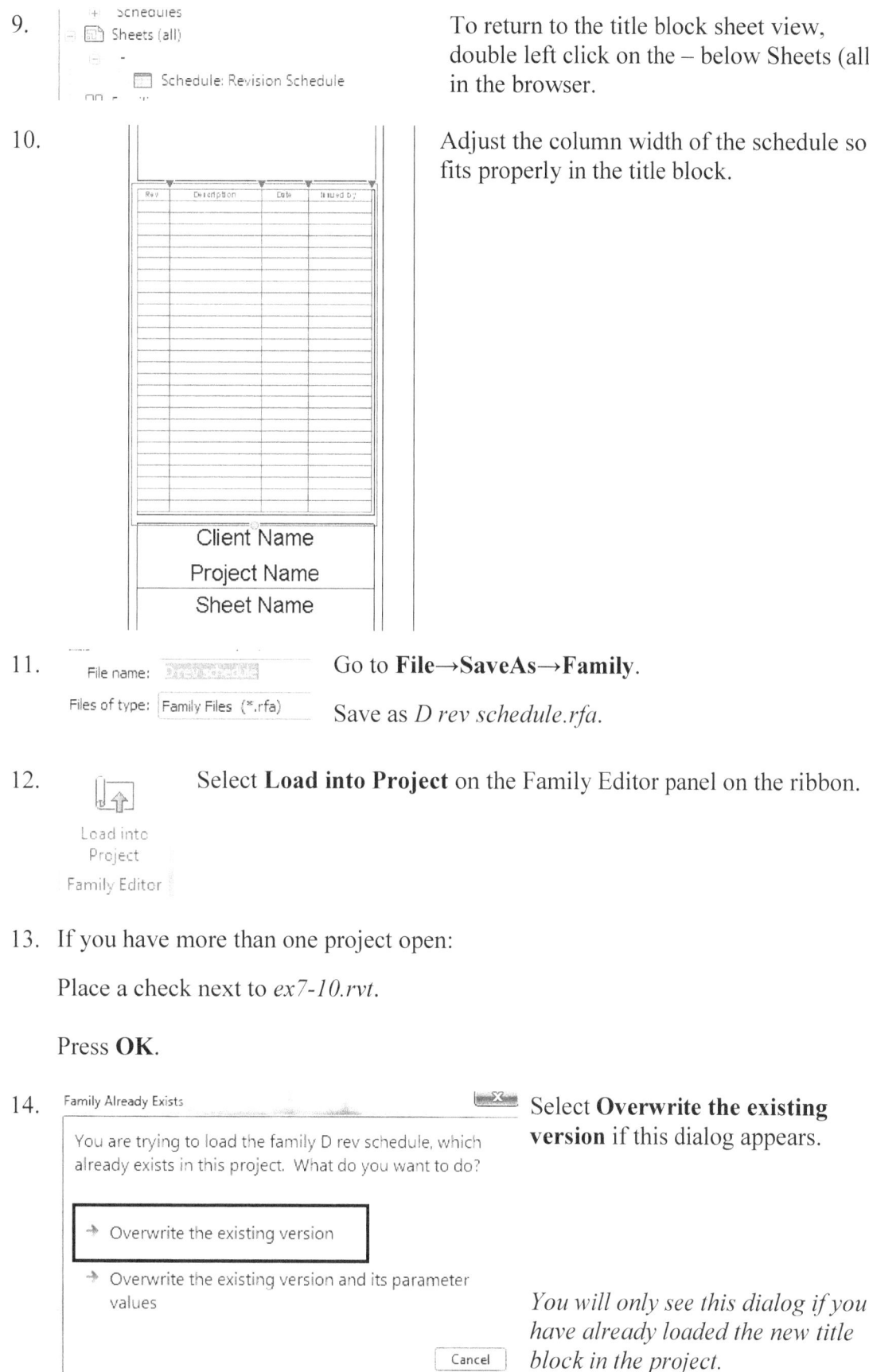

11. Go to **File→SaveAs→Family**.

    Save as *D rev schedule.rfa*.

12. Select **Load into Project** on the Family Editor panel on the ribbon.

13. If you have more than one project open:

    Place a check next to *ex7-10.rvt*.

    Press **OK**.

14. Select **Overwrite the existing version** if this dialog appears.

    You will only see this dialog if you have already loaded the new title block in the project.

7-49

15. If you see this dialog, just close it.

    This dialog will appear in a floor plan view.

16. Activate the Unnamed Sheet.

17. Select the title block so it is highlighted.

    Locate the **D rev schedule** title block in the pull-down list on the Properties pane and select.

18. Note the title block updates with the new revision schedule format.

19. Save as *ex7-11.rvt*.

Command Exercise

Exercise 7-12 – Add Revision Clouds

Drawing Name: **ex7-11.rvt**
Estimated Time to Completion: 30 Minutes

Scope

Add revision clouds to a view.
Tag revision clouds.

Solution

1. Activate the Unnamed sheet with the **Level 1** floor plan.

2. Activate the **Annotate** ribbon.

   Select **Revision Cloud** from the Detail panel.

3. On the Properties pane:

   Select the Revision that is tied to the revision cloud – **Wall Finish Change**.

4. Under Mark: Enter **A1**.

   Under Comments: Enter **Finish Changed to SW Gold Sunset**.

5.  Draw the Revision Cloud on the wall indicated.

   Note that clouds are drawn clockwise.

6. 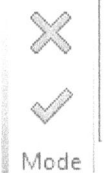 Select the **Green Check** on the Mode panel to **Finish Cloud**.

7.  Activate the **Annotate** ribbon. Select **Revision Cloud**.

8.  Select the Revision that is tied to the revision cloud – **Window Style Change**.

   Under Mark: Enter **B1**.

   Under Comments: Enter **Use Assy Code 301**.

9. Draw the Revision Cloud on the window indicated.

detailing and Drafting

10. Select the **Green Check** on the Mode panel to **Finish Cloud**.

11.  If you mouse over the revision cloud, you will see a tooltip to indicate what the revision is.

12. Activate the **Annotate** ribbon.

    Select **Revision Cloud** from the Detail panel.

13. On the Properties pane:

    Select the Revision that is tied to the revision cloud – **Door Hardware Change**.

    Under Mark: Enter **C1**.

    Under Comments: Enter **Use Hardware Set #230**.

14.  Draw the Revision Cloud on the door indicated.

15. Select the **Green Check** on the Mode panel to **Finish Cloud**.

7-53

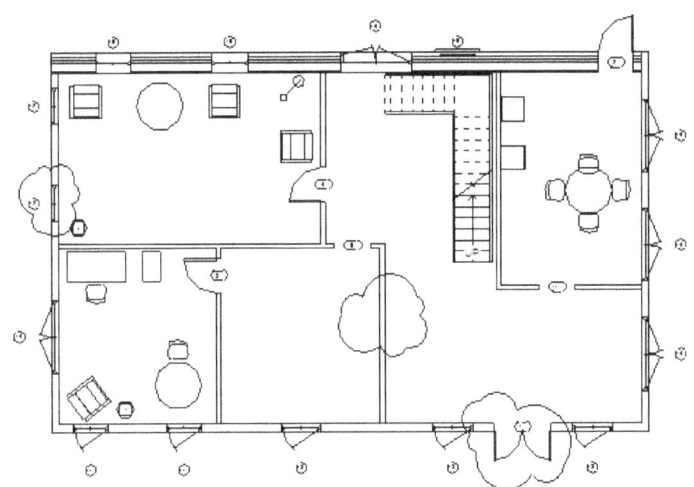

Three revision clouds have been placed in the view.

16. Zoom into the title block and note that the revision block has updated with the revisions that have been issued.

Rev	Description	Date	Issued by
A	Wall Finish Change	08.02	Sam
B	Window Style Change	08.04	Sam
C	Door Hardware Change	08.13	Sam

17. Select the **Tag by Category** tool from the Tag panel on the Annotate ribbon.

18.  Pick one of the revision clouds to identify the category to be tagged.

19. If you see this message: Press **Yes**.

   There is no tag loaded for Revision Clouds. Do you want to load one now?

20. Select the *Annotations* folder.

    Locate the **Revision Tag**.

    Press **Open**.

21. Select the revision cloud to add the tag.

22. 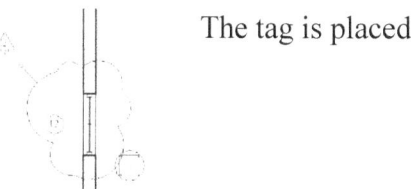 The tag is placed.

23. 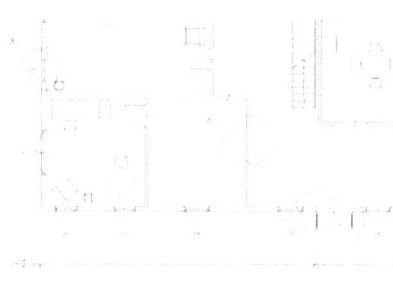 Repeat to add tags to the other two revision clouds.

24. Save as *ex7-12.rvt*.

Command Exercise

Exercise 7-13 – Aligning Views between Sheets

Drawing Name: aligning_views.rvt
Estimated Time to Completion: 30 Minutes

Scope

Using Grid Guides

Solution

1. Activate the sheet with **Level 1** floor plan.

2. In the Properties pane:

   Scroll down and note that Grid Guide is set to <None>.

3. Activate the **View** ribbon.

   Select **Guide Grid** on the Sheet Composition panel.

4. Press **OK**.

5. Select the Guide Grid.

   In the Properties pane:
   Set the Guide Spacing to **2"**.
   Press **Apply**.

Detailing and Drafting

6. 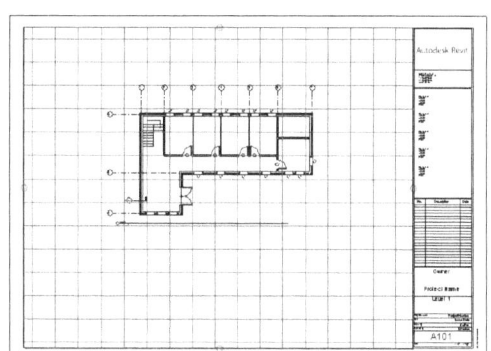 The grid updates.

   Use the blue grips to adjust the grid so it lies entirely inside the title block.

7. Select the viewport.

   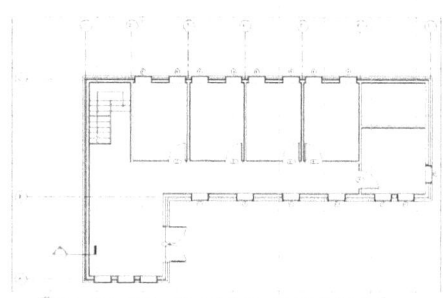

   Select the **Move** tool from the Modify Panel.

8. 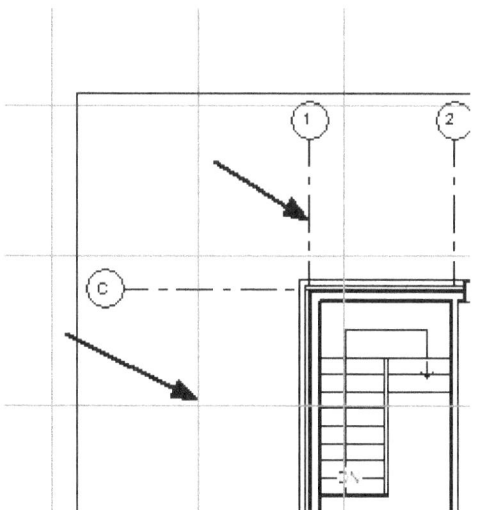 Select on the grid line 1 in the view.

   Then select the guide grid.

9. 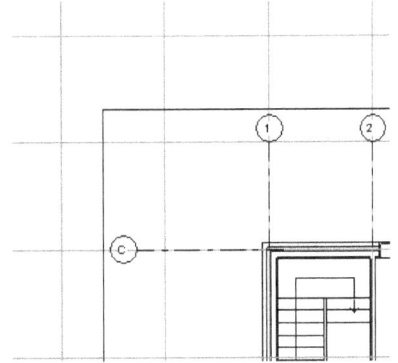 The view snaps to align with the grid.

   Repeat to shift grid line C into alignment with the grid guide.

10. 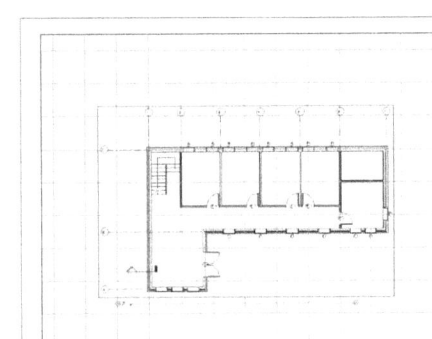 Zoom out and note which grid the view is aligned to.

11. Activate the sheet named **High Roof**.

12. In the Properties pane:

    Set the Guide Grid to **Guide Grid 1**.

13. Select the viewport.

    Select the **Move** tool from the Modify Panel.

14. 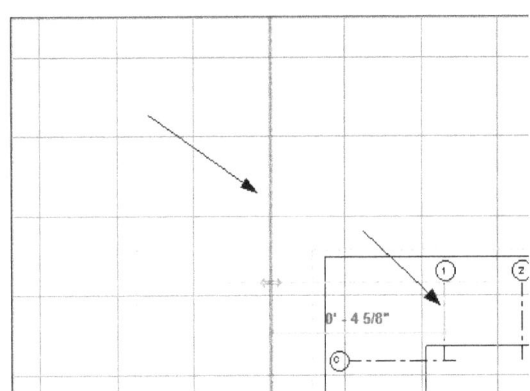 Select on the grid line 1 in the view.

    Then select the guide grid.

7-58

Detailing and Drafting

15. Repeat to shift grid line C into alignment with the grid guide.

16. Verify that the view is aligned to the same guide grid cells as sheet A101.

17. 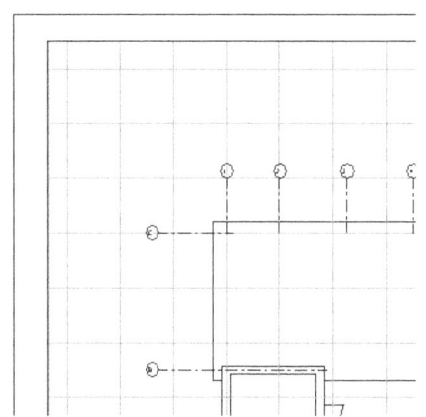 Type **VV**.
Select the **Annotation Categories** tab.
Disable visibility of the **Guide Grid**.
Press **OK**.

18. Select the viewport.
Select **Pin** from the Modify panel.

    This will lock the view to its current position.

19.  Activate the **Low Roof** sheet.

20. Select **Guide Grid** from the View ribbon.

21. Note you can select the existing guide grid.

    Press **OK**.

22. Note in the Properties pane the Guide Grid 1 is listed.

23. Select the view.
    On the Options bar:

    Set the Rotation on Sheet to **90° Clockwise**.

24. The view rotates.

25. Close the file without saving.

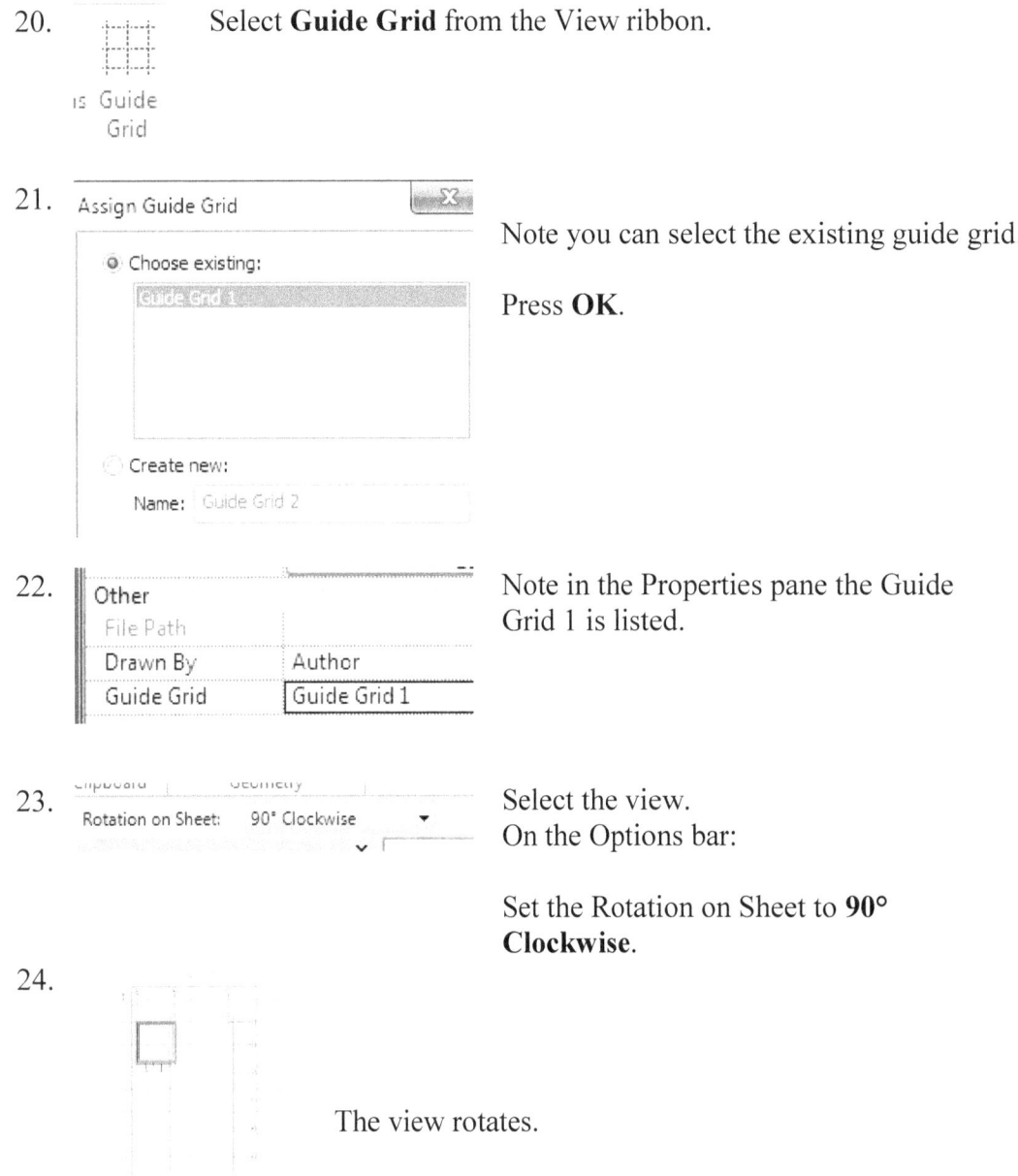

Practice Exam

1. Select the element which is NOT considered an annotation:

    A. Revision Cloud
    B. Elevation Marker
    C. Aligned Dimension
    D. Door Tag
    E. Text

2. Tags:

    A. Are View-specific
    B. Cannot be placed in an elevation view
    C. Are added or modified using the Text tool
    D. Display Parametric Information
    E. Scale with a view

3. True – False: After you place a revision cloud, you can add a tag to it.

4. True – False: A revision schedule displays information derived from revision clouds.

5. A callout:

    A. Crops a 3D view
    B. Tags Elements in a building model
    C. Shows an enlarged version of part of the parent view
    D. Generates separate views of elements of existing views

6. A detail view can be placed in which of the following:

    A. 3D View
    B. Plan View
    C. Elevation View
    D. Section View
    E. Camera View

7. Drafting Views can contain (select all that apply):

    A. Model elements
    B. Detail Lines
    C. Detail Components
    D. Filled Region
    E. Model Lines

8. A view title on a sheet can be moved independently from the view portion of the viewport, if you:

    A. Select only the view title.
    B. Enable Move View Title Independently on the Properties pane.
    C. Disable Move Title with View on the Properties pane.
    D. Modify the Viewport's Type parameters.

9. To rotate a view on a sheet:
    A. Select the view and then use the Rotate tool on the Modify panel.
    B. Set the view rotation in the Properties pane.
    C. Select the view and then set the view rotation on the Options bar.
    D. Change the Project North.

10. You can place Detail Components in all views EXCEPT:
    A. Plan
    B. Sections
    C. Sheets
    D. Legends

11. A Filled Region allows you to define its:
    A. Function
    B. Level Offset
    C. Boundary
    D. Phase

12. _____ can be used to align elements within and between sheets.
    A. Guide Grid
    B. Align
    C. Underlay
    D. Aligned Dimension

13. To lock a viewport's position:
    A. Right click and select Lock Position.
    B. Select the viewport, then Enable Lock Position on the Properties pane.
    C. Change the viewport's properties to lock position.
    D. Select the viewport, then use Pin from the Modify panel to lock position.

Answers
1) B; 2) A; 3) True; 4) False; 5) C; 6) B, C & D; 7) B, C & D; 8) A; 9) C; 10) C; 11) C; 12) A; 13) D

Lesson Eight

Construction Documentation

This lesson addresses the following User and Professional exam questions:

- Schedules
- Legends
- Rooms and Areas

A schedule is a type of view. It is a tabular display of information using element properties. Each element property is represented as a column field in the schedule.

The Schedules and Legends tools are located on the View ribbon.

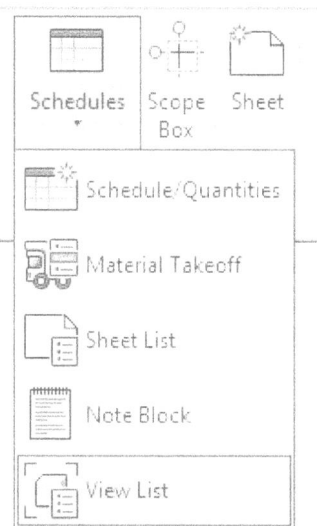

The **Schedule/Quantities** tool is used to create component schedules. Components are elements such as doors, windows, and rooms.

The **Material Takeoff** schedule lists the materials of any family used in the project.

A **Sheet List** is usually placed on the first sheet of the documentation set. It is a schedule of all sheets used in the project.

Note Blocks are useful for listing notes applied to elements in a project. For example, a user may have attached a note to several walls, which might have a description of the finish applied to each wall.

A **View List** is used to sort and itemize the views available in a project.

Command Exercise

Exercise 8-1 – Creating a Door Schedule

Drawing Name: **i_schedules.rvt**
Estimated Time to Completion: 5 Minutes

Scope

*Create a door schedule.
Use the sort and group feature to determine how many doors of a specific type are on a level.*

Solution

1. Activate the **View** tab on the ribbon.

   Select **Schedule/Quantities** from the Create panel.

2.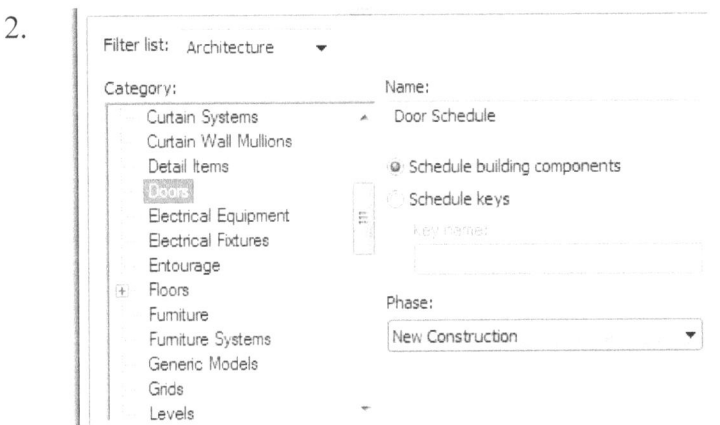

   Highlight the **Doors** category.

   Enable **Schedule building components**.

   Set the Phase to **New Construction**.

   Press **OK**.

Construction Documentation

3.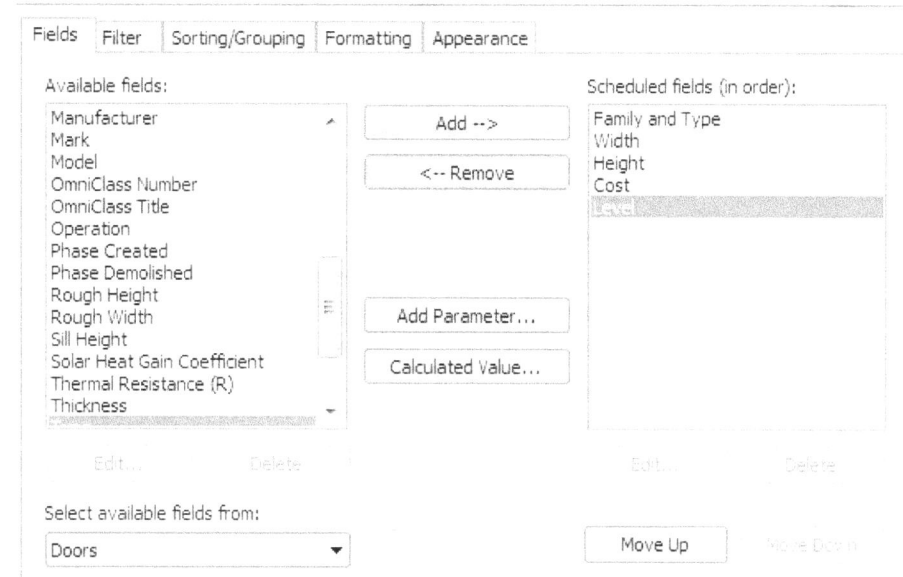

   Add the following fields:

   Family and Type
   Width
   Height
   Cost
   Level

4.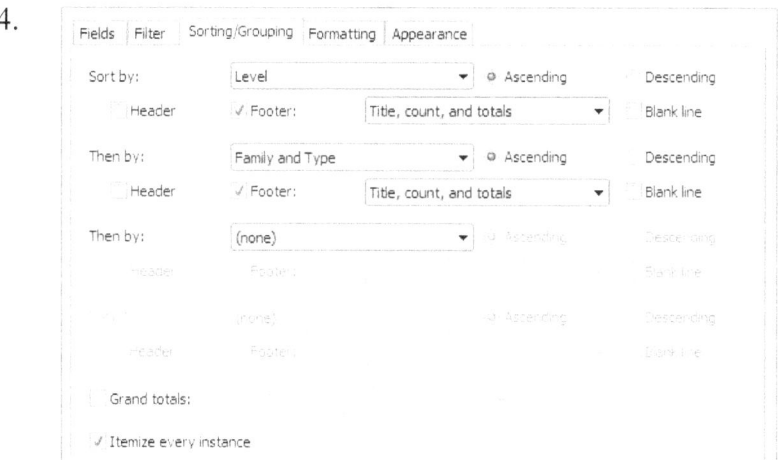

   Select the **Sorting/ Grouping** tab.

   Set Sort by: **Level**

   Enable **Footer**: **Title, count, and totals**

   Then by: **Family and Type**

   Enable **Footer**: **Title, count, and totals**

5. Select the **Formatting** tab.

   Highlight **Cost**. Disable **Show conditional format on sheets**. Enable **Calculate totals**.

6. Select the **Appearance** tab.

   Clear the **Blank row before data** checkbox.

7. Press **OK**.

8. Locate how many **Single-Flush Vision: 36″ x 84″** doors are placed on the Main Floor level.

   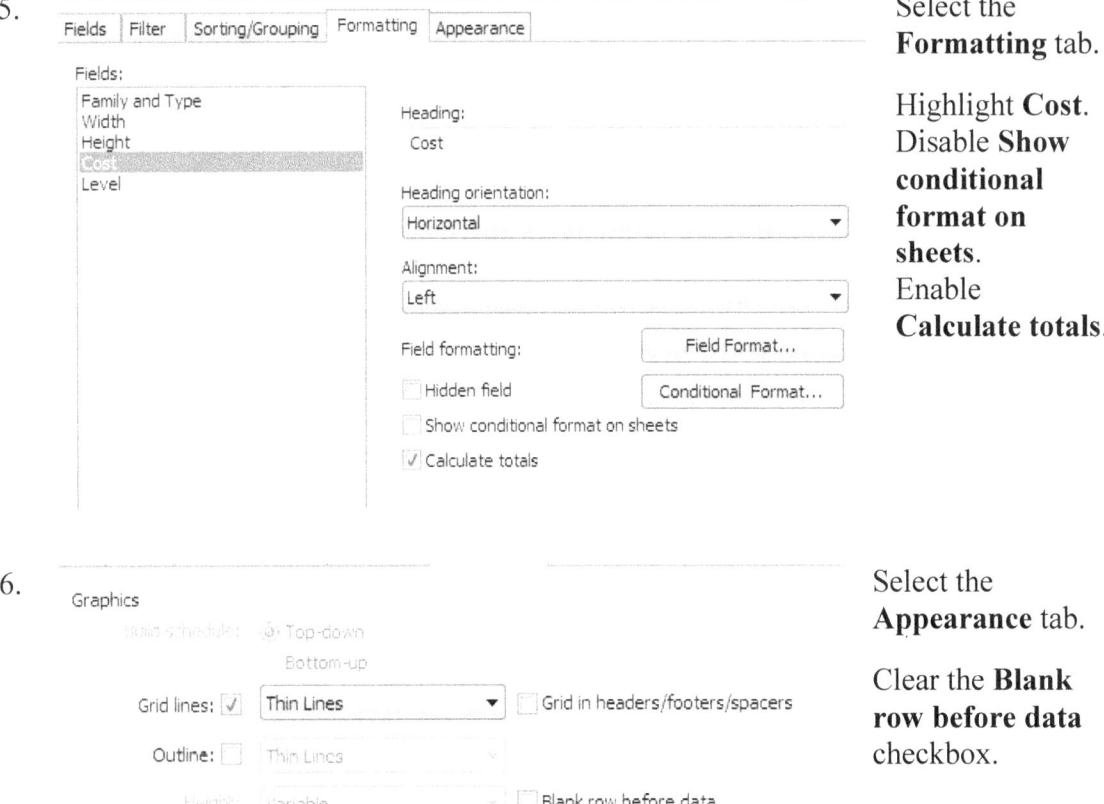

On the exam, there may be one question where you will be asked how many doors or windows of a specific type are located on a level. Repeat this exercise until you are comfortable with getting the correct answer.

9. Close without saving.

Construction Documentation

Command Exercise
Exercise 8-2 – Creating a Legend

Drawing Name: **i_Legends.rvt**
Estimated Time to Completion: 30 Minutes

Scope

Create a Legend.
Add to a sheet.
Export Legend.

Solution

1. Activate the **View** tab on the ribbon.

   Select **Legend** from the Create panel.

2. Set the Name to **Door and Window Legend**.

   Set the Scale to **¼" = 1'-0"**.

   Press **OK**.

3. The Legend is listed in the browser.

   An empty view window is opened.

4. In the browser, locate all the door families.

5. Expand each door family to see which types are available.

8-5

6. 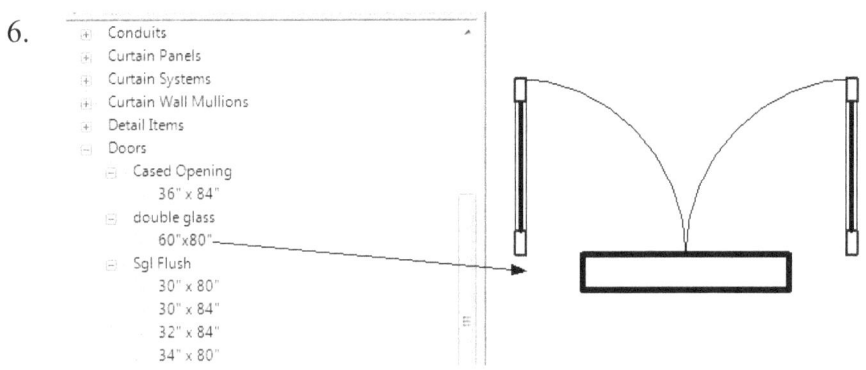 Highlight the **60″ x 80″ Double Glass** door.

   Drag and drop it into the Legend view.

7. 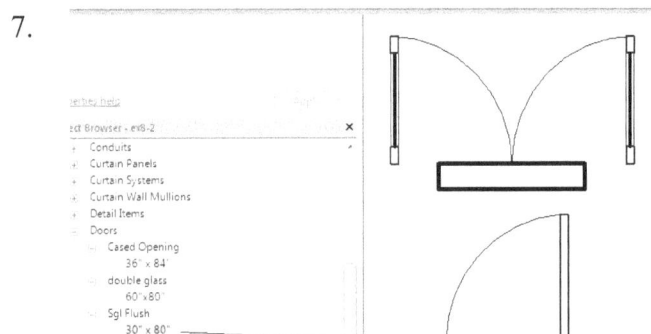 Highlight the **30″ x 80″ Sgl Flush** door.

   Drag and drop it into the Legend view.

8. 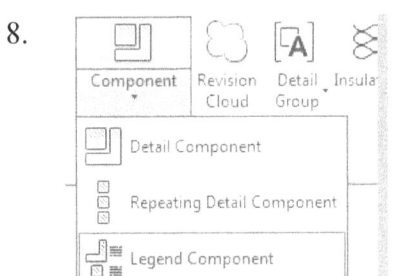 Activate the Annotate ribbon.

   Select the **Legend Component** tool from the Detail panel.

9. On the Options bar:

   Select the **Windows : archtop fixed : 36″ w x 48″ h** from the list.

10. Place the symbol below the other two symbols.

Construction Documentation

11. Drag and drop each window symbol into the Legend view.

    Symbols can be dragged and dropped from the Project Browser or using the Legend Component tool.

12. Activate the **Annotate** ribbon.

    Select the **Text** tool.

13. DOUBLE GLASS DOOR   Add text next to each symbol.

    SINGLE FLUSH DOOR

14. Mouse over a symbol to see what family type it is, if needed.

    Legend Components : Legend Component - Windows : archtop fixed : 36"w x 48"h (Floor Plan)

15. DOUBLE GLASS DOOR   Add text as shown.

    SINGLE FLUSH DOOR

    ARCHTOP FIXED

    CASEMENT

    DOUBLE CASEMENT WITH TRIM

    DOUBLE HUNG WITH TRIM

    FIXED

16. Select the **Detail Line** tool from the Detail panel on the Annotate ribbon.

17. 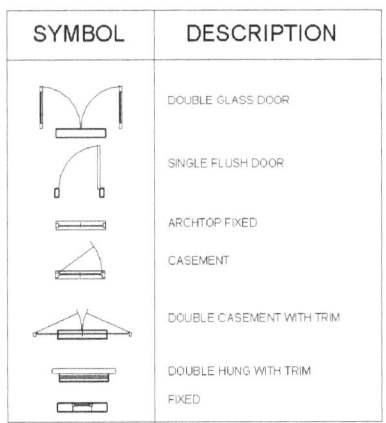 Add a rectangle.

    Add a vertical and horizontal line as shown.

    Add the header text.

18. Locate the **Sheets** category in the browser.

    Right click and select **New Sheet**.

19. 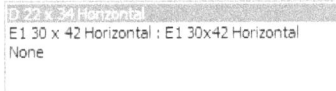 Select **Load**.

20. Locate the *Titleblocks* folder.

    Locate the **D 22 x 34 Horizontal** title block.

    Press **Open**.

21. Highlight the **D 22 x 34 Horizontal** title block.

    Press **OK**.

22. Drag and drop the Level 1 floor plan on to the sheet.

    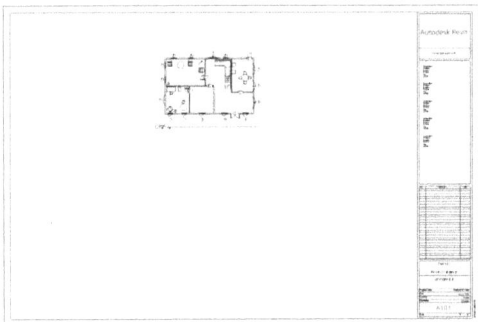

Construction Documentation

23. In the Properties pane:

    Change the View Scale to **¼" = 1'-0"**.

24. Drag and drop the Door and Window Legend onto the sheet.

    Place next to the floor plan view.

    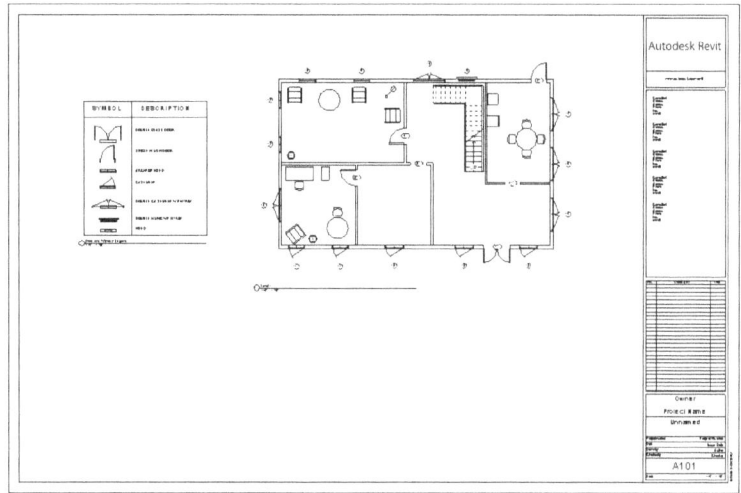

25. Select the legend so it highlights.

    Select **Edit Type** on the Properties pane.

26. Select **Duplicate**.

27. Type **Viewport with no Title**.

    Press **OK**.

28. Uncheck **Show Extension Line**.

    Set Show Title to **No**.

    Press **OK**.

8-9

29. 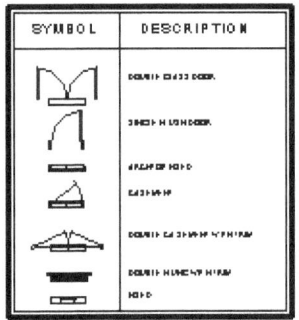 Zoom in to review the legend.

30. 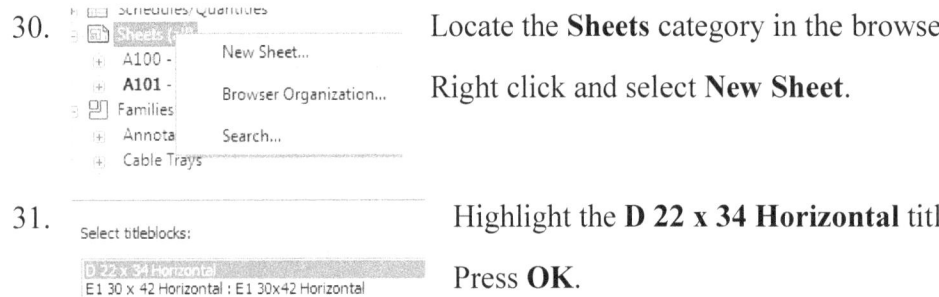 Locate the **Sheets** category in the browser.

    Right click and select **New Sheet**.

31. Highlight the **D 22 x 34 Horizontal** title block.

    Press **OK**.

32. Drag and drop the **Level 2** floor plan on to the sheet.

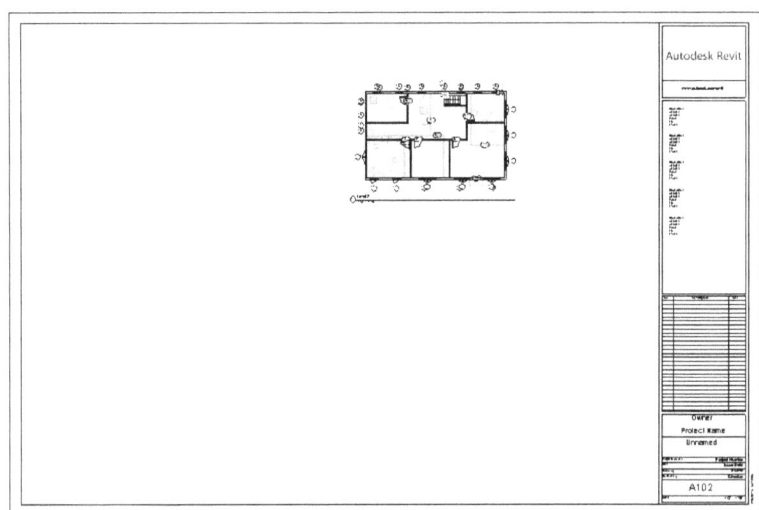

33. In the Properties pane:

    Change the View Scale to **¼" = 1'-0"**.

Construction Documentation

34. Drag and drop the Door and Window Legend onto the sheet.

    Place next to the view.

    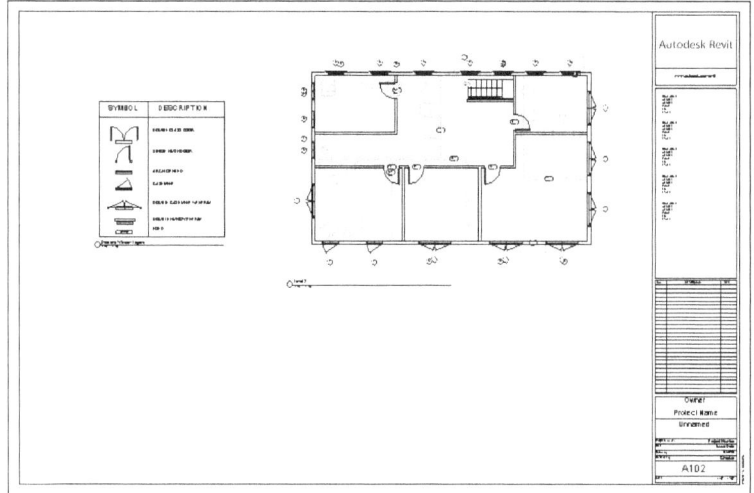

35. Select the legend view and then use the type selector in the Properties pane to set whether the view should have a title or not.

    Note that legends can be placed on more than one sheet.

36. Save as *ex8-2.rvt*.

Command Exercise

Exercise 8-3 – Adding Rooms to a Floor Plan

Drawing Name: **i_rooms.rvt**
Estimated Time to Completion: 10 Minutes

Scope

Add rooms to a floor plan.

Solution

1. Activate the 2nd **Floor** floor plan.

2. Select the **Room** tool from the Room & Area panel on the Architecture ribbon.

3.  Place a room in each enclosed boundary.

Construction Documentation

4. 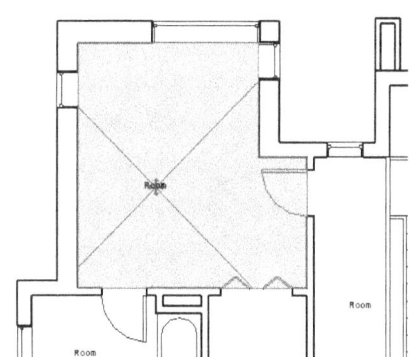 Select the room in the upper left corner.

5.

   In the Properties pane:

   Enter the data as shown.

   Press **Apply**.

6. Note that the Room Name updates to the name assigned.

7. Add the room names shown.

8. 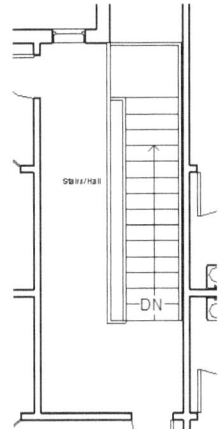 Continue adding names to the rooms.

9. 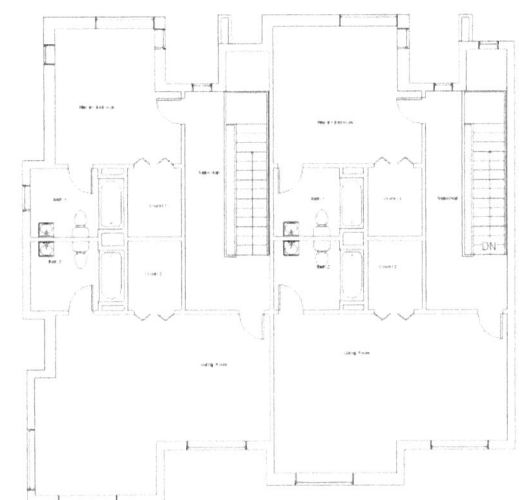 Verify that room names have been assigned to all rooms.

10. Save as *ex8-3.rvt*.

Command Exercise

Exercise 8-4 – Creating an Area Scheme

Drawing Name: **i_Color_Scheme.rvt**
Estimated Time to Completion: 15 Minutes

Scope:

Create an area plan using a color scheme.
Place a color scheme legend.

Solution

1. Activate the **01- Entry Level** floor plan.

2. In the Properties pane:

    Scroll down to the **Color Scheme** field.
    Press **<none>**.

3. Highlight **Name**.

4. In the Title field, enter **Area Legend**.

    Under Color: select the **Area** scheme.

5. Press **OK**.

6.

	Value	Visible	Color	Fill Pattern	Preview
1	47 SF	✓	RGB 156-185	Solid fill	
2	58 SF	✓	PANTONE 3	Solid fill	
3	64 SF	✓	PANTONE 6	Solid fill	
4	76 SF	✓	RGB 139-166	Solid fill	
5	90 SF	✓	PANTONE 6	Solid fill	
6	94 SF	✓	RGB 096-175	Solid fill	
7	95 SF	✓	RGB 209-203	Solid fill	
8	133 SF	✓	RGB 173-118	Solid fill	
9	168 SF	✓	RGB 194-161	Solid fill	

Note that different colors are applied depending on the square footage of the rooms.

Press **OK**.

7.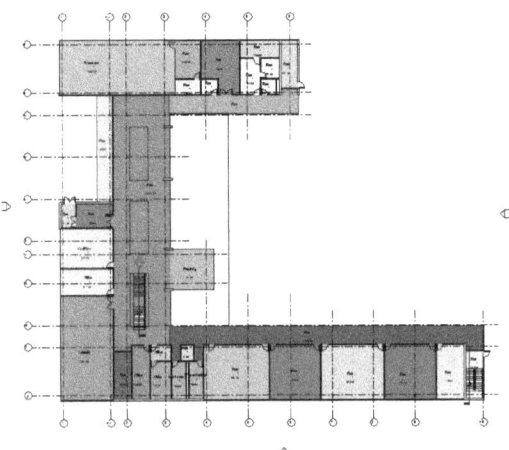

The rooms fill in according to the square footage.

8.

Activate the **Annotate** ribbon.

Select the **Legend** tool on the Color Fill panel.

9.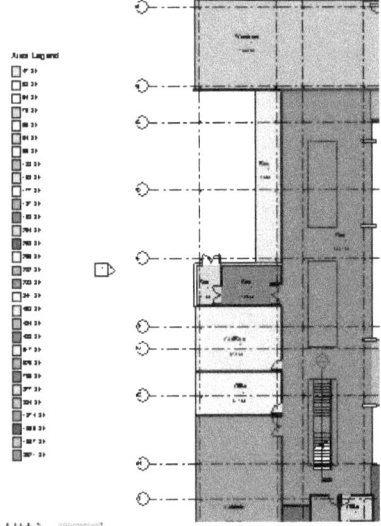

Place the legend in the view.

10.

Select the legend that was placed in the display window.

Select **Edit Scheme** from the Scheme panel on the ribbon.

11.

	Value	Visible	Color	Fill Pattern	Preview
1	47 SF	✓	RGB 156-185	Crosshatch-small	
2	58 SF	✓	PANTONE 3	Solid fill	
3	64 SF	✓	PANTONE 6	Solid fill	

Change the Fill Pattern for 47 SF **to Crosshatch- Small**.

12.

	Value	Visible	Color	Fill Pattern	Preview
1	47 SF	✓	RGB 156-185	Crosshatch-small	
2	58 SF	✓	PANTONE 3	Crosshatch	
3	64 SF	✓	PANTONE 6	Diagonal crosshatch	
4	76 SF	✓	RGB 139-166	Diagonal down	
5	90 SF	✓	PANTONE 6	Solid fill	
6	94 SF	✓	RGB 096-175	Solid fill	

Change the hatch patterns for the next few rows.

Press **OK**.

13. *Note that the legend updates.*

    Area Legend
    ☐ 47 SF
    ☐ 58 SF
    ☐ 64 SF
    ☐ 76 SF
    ☐ 90 SF

14. Select the Color Fill Legend. On the Properties pane: Select **Edit Type.**

15. Change the font for the Text to **Tahoma.**

Text	
Font	Tahoma
Size	3/16"
Bold	✓
Italic	
Underline	

    Enable **Bold**.

16. Change the font for the Title Text to **Tahoma.**

Title Text	
Font	Tahoma
Size	1/4"
Bold	✓
Italic	✓
Underline	

    Enable **Bold**.
    Enable **Italic.**

    Press **OK**.

17. Note how the legend updates.

    Area Legend
    ☐ 47 SF
    ☐ 58 SF
    ☐ 64 SF

18. Close without saving.

Command Exercise

Exercise 8-5 – Creating an Area Plan

Drawing Name: **i_Area_Plan.rvt**
Estimated Time to Completion: 30 Minutes

Scope:
Create an area plan.
Add areas to a floor plan.

Solution

1. 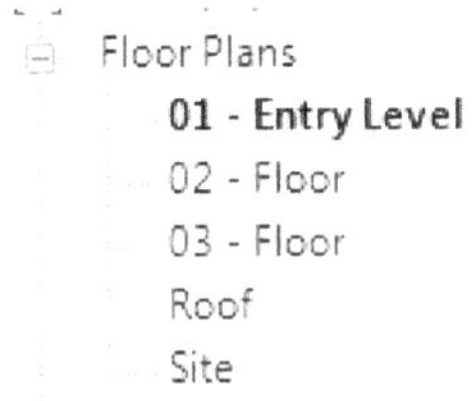 Activate the **01- Entry Level** floor plan.

2. 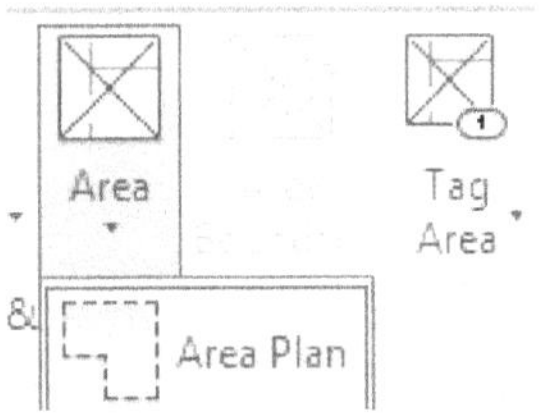 Select the **Area Plan** tool from the Room & Area panel on the **Architecture** ribbon.

3. 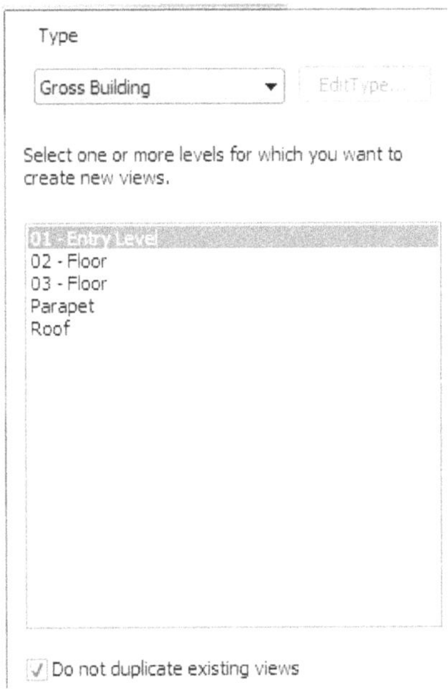 Select **Gross Building** under the Type drop-down.

    Select **01- Entry Level**.

    Enable **Do not duplicate existing views**.

    Press **OK**.

4. Automatically create area boundary lines associated with external walls and gross building area?   Press **Yes**.

5. Section 2
   Area Plans (Gross Building)
   01 - Entry Level
   Legends   A new area plan will be listed in the Project Browser.

6. 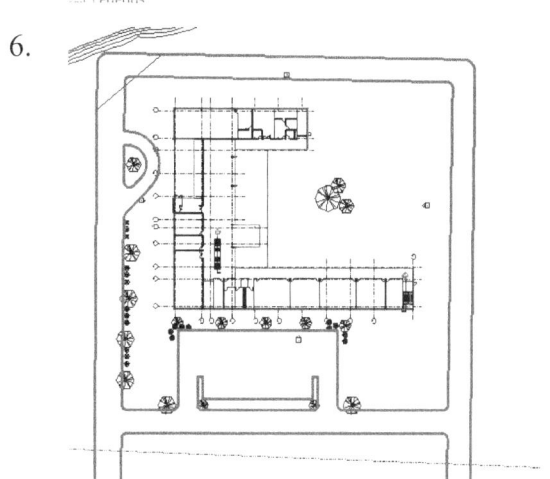   A new view will open in the graphics window.

7. Type **VV** to launch the Visibility/Graphics dialog.

8.    Clear the **Topography** box to turn off the visibility.

9. Clear the **Planting** box to turn off the visibility.

   Press **OK** to close the dialog.

10. Select the property line in the graphics window.   Right click and select **Hide in view → Category**.

    This will hide all property lines.

11.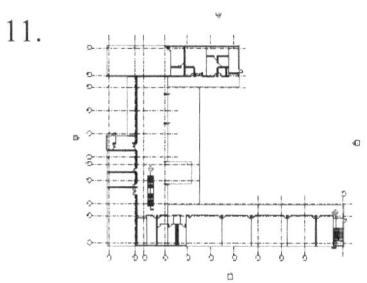

    The view now shows just the building with grid lines.

12. Activate the **Modify** ribbon.

    Select the **Trim** tool.

13. Select the two lines indicated.

14. The boundary lines trim to create an enclosed area.

15. Activate the **Architecture** ribbon.

    Select the **Area Boundary** tool.

16. Select the **Pick Lines** tool from the Draw panel.

17. Select the wall indicated to place an area boundary line.

    This divides the area into two different sections.

18. Activate the **Architecture** ribbon.

    Select the **Area** tool.

    Place an area in the area on the right.

Construction Documentation

19. You should see two tags. One is for the area on the left and one is for the area on the right.

You may need to move them to see them clearly.

20. Change the name of the area tag on the right to **Service**.

21. Change the label for the left area tag to **Administration**.

22. Left click anywhere in the window to clear any selections.

On the Properties pane:

Click on **<none>** next to Color Scheme.

23. Select **Gross Building Area** under Category.

24. Under Color: select **Name**.

25. Press **OK**.

26. The two named areas are listed.

Colors are automatically assigned.

27. Press **OK**.

28. Save as *ex8-5.rvt*.

Command Exercise

Exercise 8-6 – Creating a Room Schedule

Drawing Name: **i_Finish_Schedule.rvt**
Estimated Time to Completion: 10 Minutes

Scope

Create a room finish schedule.

Solution

1. Activate the **View** tab on the ribbon.

   Select the **Schedule/Quantities** tool from the Create panel.

2. 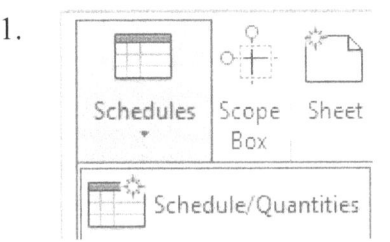 Select **Rooms** in the Category list box.

   Change the Name to **Room Finish Schedule**.

   Press **OK**.

3. Add the fields shown to the schedule.

   Scheduled fields (in order):
   - Number
   - Name
   - Base Finish
   - Floor Finish
   - Wall Finish
   - Ceiling Finish
   - Area
   - Comments
   - Level

8-22

4. Select the **Sorting/ Grouping** tab.

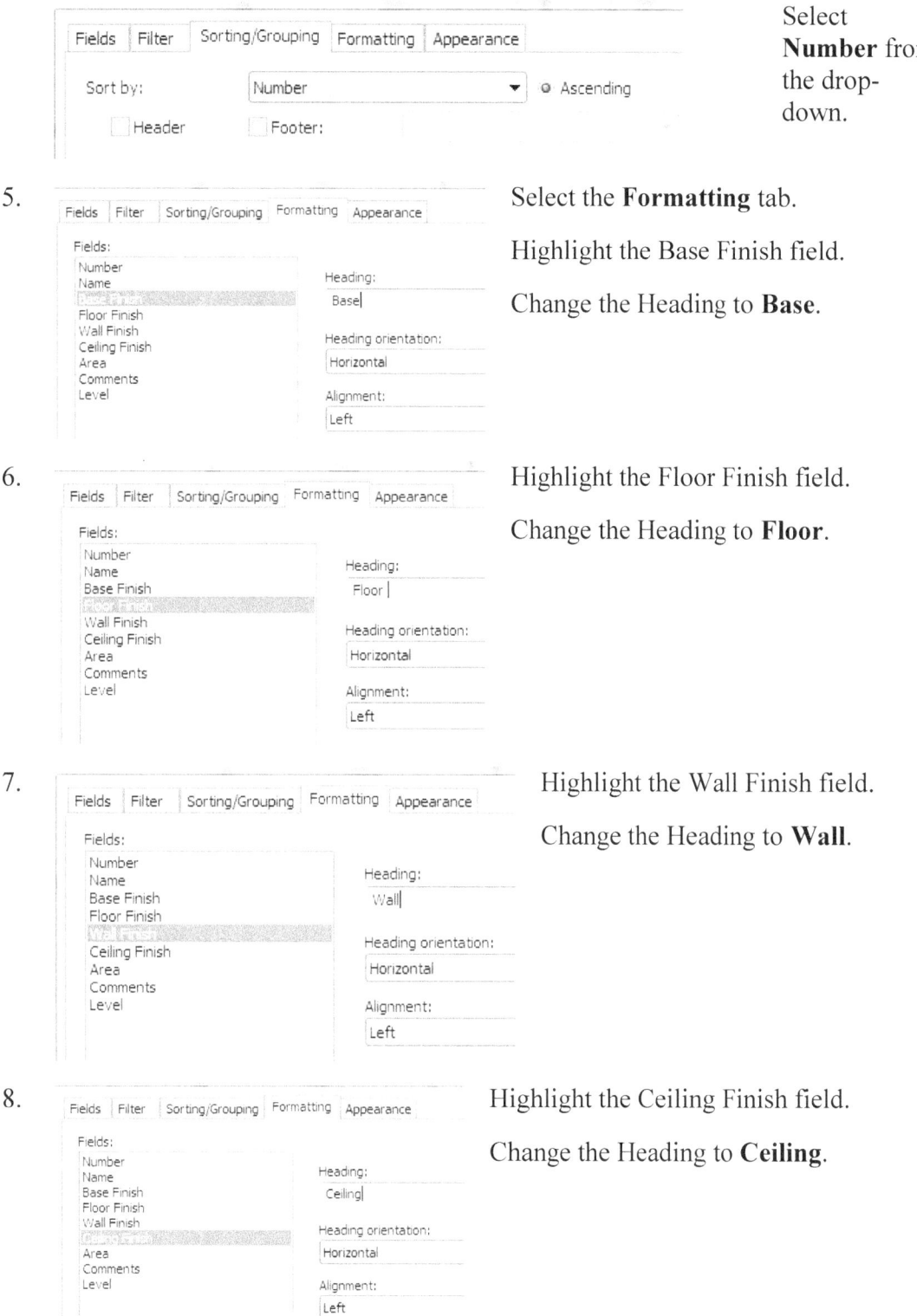

Select **Number** from the drop-down.

5. Select the **Formatting** tab.

Highlight the Base Finish field.

Change the Heading to **Base**.

6. Highlight the Floor Finish field.

Change the Heading to **Floor**.

7. Highlight the Wall Finish field.

Change the Heading to **Wall**.

8. Highlight the Ceiling Finish field.

Change the Heading to **Ceiling**.

9. Press **OK** to finish the schedule.

10. The schedule view opens.

	A	B	C	D	E	F	G	H	I
	Number	Name	Base	Floor	Wall	Ceiling	Area	Comments	Level
	101	Vest.					406 SF		01 - Entry Level
	102	Lobby					3537 SF		01 - Entry Level
	103	Conference					359 SF		01 - Entry Level
	104	Instruction					712 SF		01 - Entry Level
	105	Instruction					891 SF		01 - Entry Level
	106	Instruction					712 SF		01 - Entry Level
	107	Corridor					1504 SF		01 - Entry Level
	108	Instruction					900 SF		01 - Entry Level
	109	Women					135 SF		01 - Entry Level

11. **Group**   Drag your cursor to select the Base, Floor, Wall, and Ceiling headers.

    Select the **Group** tool.

12. [schedule with FINISH header above Base, Floor, Wall, Ceiling columns]   Add the word **FINISH** above the grouped fields.

13. [schedule showing Vest. row with PRIMER, GRANITE, SW 6136, SW 6136 values]   To add finish information, just type in the cell.

14. Close the file without saving.

Construction Documentation

Command Exercise
Exercise 8-7 – Creating a Drawing List

Drawing Name: **sheets.rvt**
Estimated Time to Completion: 15 Minutes

Scope

Create a sheet list schedule.
Add to a sheet.

Solution

1. Highlight **Sheets** in the Project Browser.

   Right click and select **New Sheet**.

2. Select **Load**.

3. Browse to the *Titleblocks* folder.

   Select the *B 11 x 17 Horizontal Titleblock* and press **Open**.

4. The B size Titleblock is listed in the dialog list.

   Highlight and press **OK**.

8-25

5. 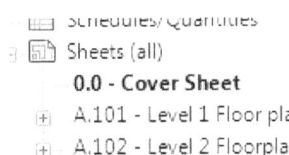 In the Properties pane:

Type **0.0** for the Number.

Type **Cover Sheet** for the Sheet Name.

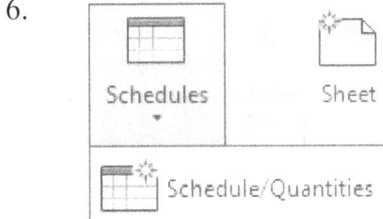

Notice that sheet name and number update in the browser.

6. 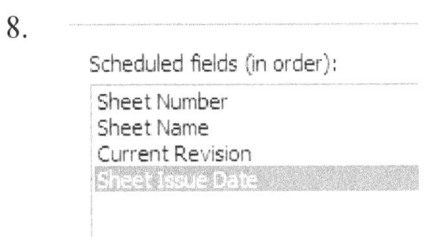 Activate the View ribbon.

Select the **Sheet List** tool under Schedules on the Create panel.

8. 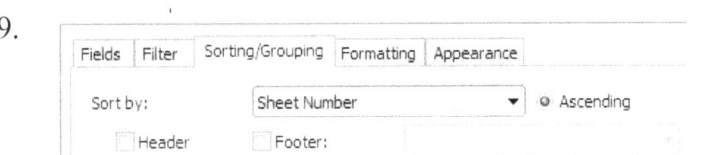 Add the following fields to the schedule:

Sheet Number
Sheet Name
Current Revision
Sheet Issue Date

9. Select the Sorting/Grouping tab.
Sort by **Sheet Number**.

10. Press **OK** to create the schedule.

Construction Documentation

11. The drawing list schedule view is displayed.

12. Activate the **0.0- Cover Sheet**.

13. Drag and drop the sheet list onto the cover sheet.

14. Activate the Cover Sheet.

15. In the Properties pane:

    Uncheck **Appears in Sheet List**.

16. The schedule will automatically update.

    Close without saving.

8-27

Command Exercise

Exercise 8-8 – Create a Note Symbol

Drawing Name: **none**
Estimated Time to Completion: 30 Minutes

Scope

Create Note Symbol.

Solution

1. Go to **New→Annotation Symbol** on the Applications Menu.

2. Select *Generic Annotation*.
   Press **Open**.

3. Locate the Word document titled *Notes* in the exercise folder and open it.

4. Highlight the Floor Plan Notes.
   Use Ctl-C to copy.

   FLOOR PLAN NOTES:

   1. INTERIOR DIMENSIONS SHOW CENTERLINE.

   2. EXTERIOR PLAN DIMENSIONS RECESSED BRICK FEATURES.

5. Switch to Revit.

6. Select the **Text** tool from the Text panel.

   Place your cursor inside the text box.
   Enter Ctl-V to paste.

7. Use the Format settings to add an underline and bold the heading for the notes.

8. Left click anywhere in the drawing window to exit the Text command. Right click and select **Cancel**.

9. Select the note that was included in the template.
Right click and select **Delete**.

10. Position the note so it aligns with the intersection of the reference planes.

    This is the insertion point.

11. Save the family as *Floor Plan Notes - Embry Home*.

12. Close the file.

13. 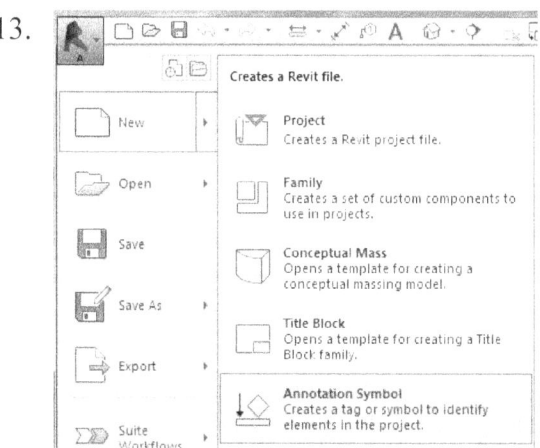 Go to **New→Annotation Symbol** on the Applications Menu.

14. Select *Generic Annotation*.

    Press **Open**.

15. Locate the Word document titled *Notes* in the exercise folder and open it.

16. Highlight the General Project Notes.

    Use Ctl-C to copy.

17. Switch to Revit.

    Select the **Text** tool from the Text panel.

18. Place your cursor inside the text box.
    Enter Ctl-V to paste.

19. Use the Format settings to add an underline and bold the heading for the notes.

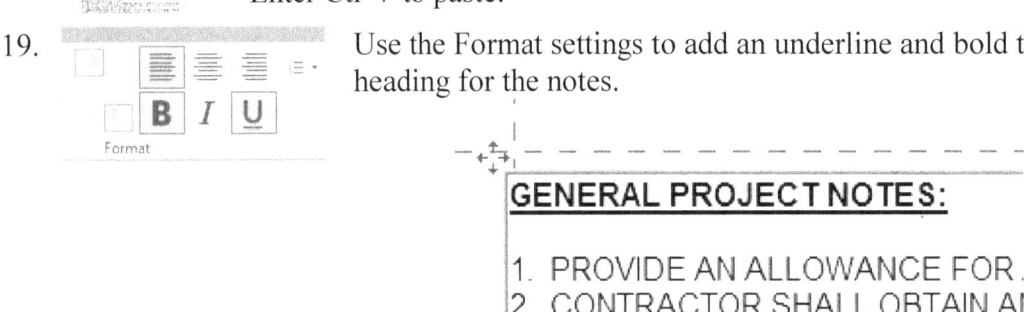

20. Left click anywhere in the drawing window to exit the Text command.
    Right click and select **Cancel**.

21. 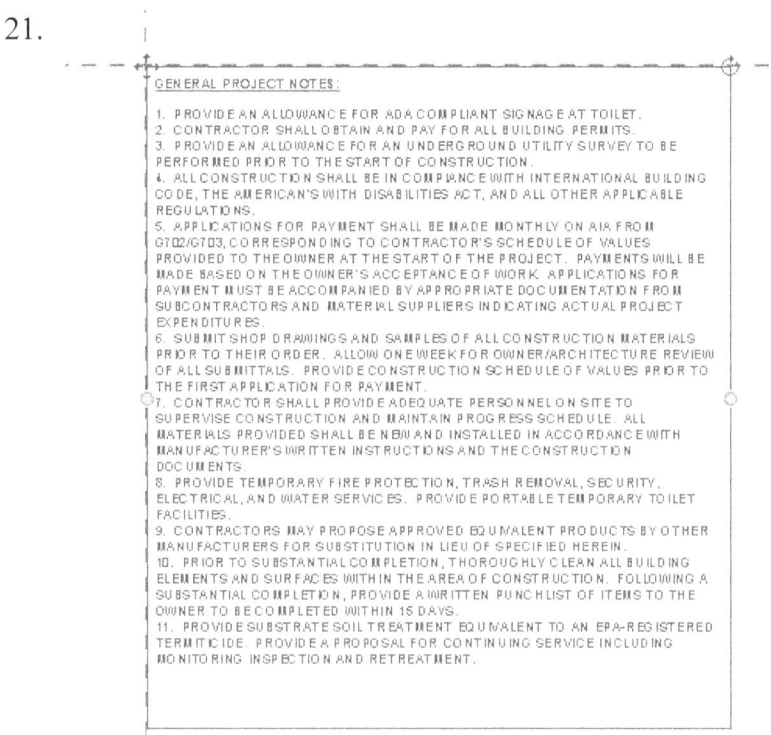 Use the grips on the sides to resize the text box.

22. Use the **Numbers** formatting tool on the Format panel to clean up the notes.

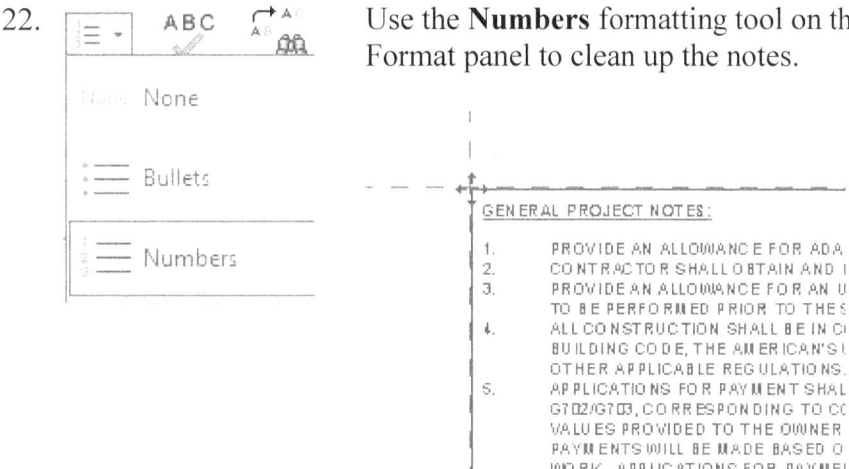

23. Select the template note.
Right click and select **Delete**.

24. [image of general project notes]   The notes should look like this.

The width of the note should be no more than 5" in order to fit properly in the title block.

25. File name: General Project Notes
    Files of type: Family Files (*.rfa)

    Save the family as *General Project Notes*.

26. Close the file.

Construction Documentation

Command Exercise
Exercise 8-9 – Add Notes

Drawing Name: **notes.rvt**
Estimated Time to Completion: 10 Minutes

Scope

Use Annotation Symbols.

Solution

1. Activate the **Insert** ribbon.

   Select **Load Family** from the Load from Library panel.

   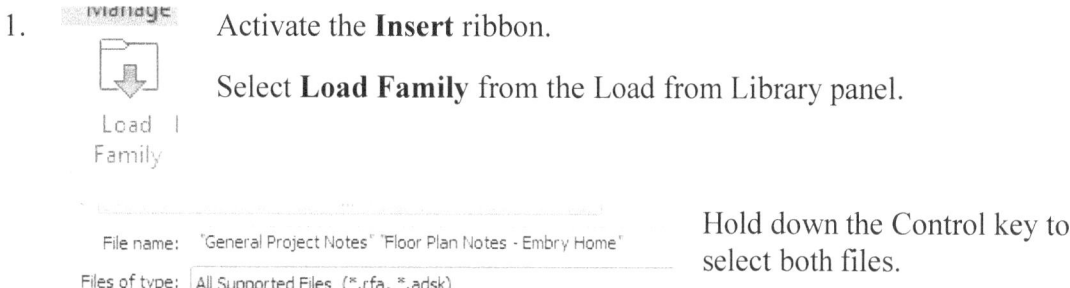

   Hold down the Control key to select both files.

2. Locate the two annotation symbols created called *General Project Notes* and *Floor Plan Notes - Embry Home*.

   Press **Open**.

5. Activate the **Ground Floor Plan** sheet.

   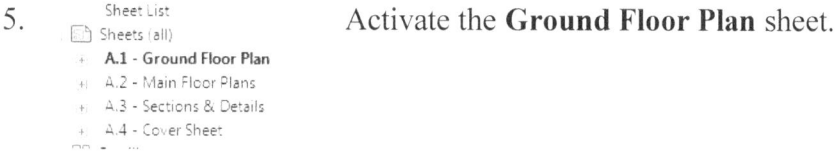

6. Activate the **Annotate** ribbon.

   Select the **Symbol** tool on the Symbol panel.

8-33

7. Select the **General Project Notes** from the Properties pane.

   Place in the title block.

   Right click and select **Cancel**.

   To modify the notes, double left click on them, adjust, save, and reload.

8. Activate the **Annotate** ribbon.

   Select the **Symbol** tool on the Symbol panel.

9. Select the **Floor Plan Notes** using the Type Selector on the Properties pane.

10. Place on the sheet above the title mark.

11. Close without saving.

Construction Documentation

Command Exercise

Exercise 8-10 – Create a Material TakeOff Schedule

Drawing Name: **i_materials.rvt**
Estimated Time to Completion: 10 Minutes

Scope

Create a material takeoff schedule

Solution

1. 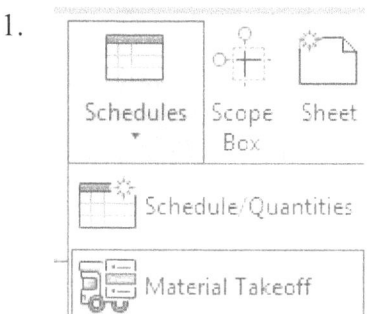 Activate the View ribbon.

   Select **Schedules→Material Takeoff** from the Create panel.

2.  Highlight **Walls**.

   Press **OK**.

3. On the **Fields** tab:

   Add the following fields:
   - Family and Type
   - Material: Name
   - Material: Area

8-35

4. Select the **Sorting/Grouping** tab.

   Sort by: **Material Name**
   Enable **Footer: Title, count and totals**.
   Then by: **Family and Type**

   Enable **Itemize every instance**.

5. Activate the **Formatting** tab.

   Highlight **Family and Type**.
   Change the Heading to **Wall Style**.

6. Activate the **Formatting** tab.

   Highlight **Material: Name**.
   Change the Heading to **Material**.

7. Activate the Formatting tab.

   Highlight **Material: Area**.
   Change the Heading to **Area**.

   Enable **Calculate totals**.

8.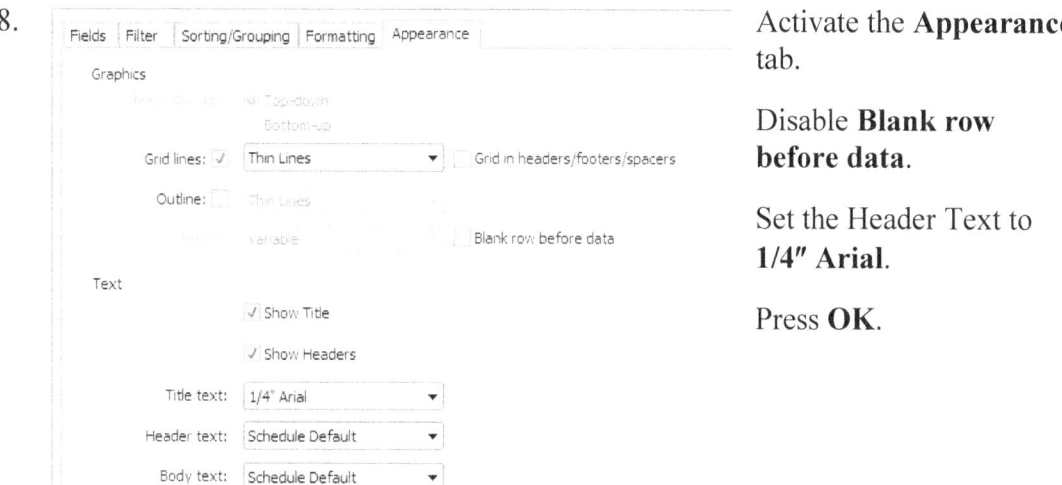

    Activate the **Appearance** tab.

    Disable **Blank row before data**.

    Set the Header Text to **1/4″ Arial**.

    Press **OK**.

9. A window opens with the new schedule.

10. Close without saving.

Command Exercise

Exercise 8-11 – Create a Mass Floor Schedule

Drawing Name: **mass_schedule.rvt**
Estimated Time to Completion: 10 Minutes

Scope

Create a mass floor schedule

Solution

1. Activate the **View** ribbon.

   Select **Schedules→Schedule/ Quantities** from the Create panel.

2. Highlight **Mass**.
   Press **OK**.

3. Add the following fields:
   - **Family and Type**
   - **Gross Floor Area**
   - **Gross Volume**
   - **Description**

4. Select the Formatting tab.

   Select the **Gross Floor Area**.

   Enable **Calculate totals**.

5. Select the Appearance tab.
   Disable **Blank row before data**.

   Press **OK**.

6.

<Mass Schedule>			
A	B	C	D
Family and Type	Gross Floor Area	Gross Volume	Description
Building 1: Building 1	40724 SF	475111.08 CF	
Building 2: Building 2	4973 SF	58017.48 CF	
Building 3: Building 3	59395 SF	695765.31 CF	
Building 4: Building 4	66309 SF	779134.58 CF	
Towers: Towers	28104 SF	331007.33 CF	
Building 5: Building 5	15398 SF	174509.62 CF	

7. Close the file without saving.

Practice Exam

1. Legend views:
    - A. Are placed on only one sheet at a time
    - B. Contain model and annotation elements
    - C. Have model components tagged
    - D. Have dimensions added to system families

2. To add or remove fields from a revision schedule:
    - A. Select the revision schedule in the browser, right click and select Properties
    - B. Select the Schedules tool on the View ribbon.
    - C. First you must open the title block family for editing
    - D. Select the Revision Table in the graphics window, right click and select Element Properties

3. To add revisions to the Sheet Issues/Revisions dialog, you must access this ribbon:
    - A. View
    - B. Manage
    - C. Annotate
    - D. Insert

4. On which tab of the Schedule Properties dialog can hidden fields be enabled?
    - A. Fields
    - B. Filter
    - C. Formatting
    - D. Appearance
    - E. Sorting/Grouping

5. Color Schemes can be placed in the following views (Select 3):
    - A. Floor plan
    - B. Ceiling Plan
    - C. Section
    - D. Elevation
    - E. 3D

6. To define the colors and fill patterns used in a color scheme legend, select the legend, and Click Edit Scheme on the:
    - A. Properties palette
    - B. View Control Bar
    - C. Ribbon
    - D. Options Bar

Room Schedule			
Number	Name	Area	Level
4	Accounting	49.29	Level 1
3	CEO	117.84	Level 1
9	Common Area	391.84	Level 1
6	Conference Room	146.21	Level 1
7	Copy/Mail Room	71.73	Level 1
1	Engineering	136.75	Level 1
8	Lobby	236.81	Level 1
5	Operations	49.29	Level 1
2	Sales/Marketing	119.02	Level 1
Grand total: 9		1318.77	

7. The Room Schedule shown sorts by Level and by _____.

   A. Name
   B. Area
   C. Number
   D. Total

8. A legend is a view that can be placed on:

   A. A plan view
   B. A drafting view
   C. Multiple Plan Regions
   D. Multiple Sheets

9. The calculated total for a mass floor area can be displayed in:

   A. Project Browser
   B. Properties pane
   C. Mass Floor Schedule
   D. Floors Schedule

10. To compute the total cost of a material in a schedule, you need to create a _____.

    A. Project parameter for cost
    B. Shared Parameter
    C. Calculated Value
    D. Project Filter

11. A _____ is a view that displays information about a building project in tabular form.

    A. Schedule
    B. Graph
    C. Legend
    D. Chart

12. Schedules can be filtered by the following disciplines EXCEPT:

    A. Architecture
    B. Structure
    C. Mechanical
    D. Electrical
    E. Piping
    F. Civil

13. To change the font used in a Color Scheme Legend:

    A. Select the legend, then select Edit Scheme from the ribbon
    B. Select the text in the legend, double click to edit
    C. Select the legend, then select Edit Type from the Properties pane.
    D. Select the legend, right click and select Edit Family.

Answers
1) B; 2) C; 3) A; 4) C; 5) A, C, D; 6) C; 7) A; 8) D; 9) C; 10) C; 11) A; 12) F; 13) C

Lesson Nine

Presenting the Building Model

This lesson addresses the following User and Professional exam questions:

- Sun and Shadow Settings
- Rendering
- Decals
- Hidden Views
- Graphic Display Overrides

Revit Architecture can create photorealistic images of both exterior and interior views of your model. Users can create lighting, plants, decals, and place people in their model.

If you need to monitor how much memory the rendering process is using in the Windows Task Manager, the rendering process is named fbxooprender.exe. When you render an image, the rendering process may use up to 4 CPUs.

Because rendering can use up system resources, turn off active screen savers, and shut down any non-essential processes. (For example, close your email program and don't browse the internet during rendering.) By closing applications, more CPU capacity will be made available to the rendering process and can reduce render time. Many users will have two workstations: one for rendering and one for regular office work if they create a lot of renderings.

If you experience long delays when rendering, use the Windows Task Manager to monitor processes. If fbxooprender.exe is not using close to 99% of processor power, other active processes may be interfering with the rendering process. Shut down non-essential tasks to make more processor power available for the rendering process.

Autodesk now allows users the option of rendering on their cloud server. This allows you to keep working during the rendering process.

Command Exercise

Exercise 9-1 – Creating a Toposurface

Drawing Name: **C_Condo_complex.rvt**
Estimated Time to Completion: 40 Minutes

Scope

Create a toposurface.
Add site components.
Add entourage.

Solution

1. Activate the **Site** view.
   Turn off the visibility of grids, elevations, and sections.

2. Activate the **Massing & Site** ribbon.

   Select the **Toposurface** tool.

3. Use the **Place Point** tool to create an outline of a lawn surface.

4.  Pick the points indicated to create a lawn expanse.

   You can grab the points and drag to move into the correct position.

5. Click on the **Modify** tool on the ribbon to exit out of the add points mode.

Presenting the Building Model

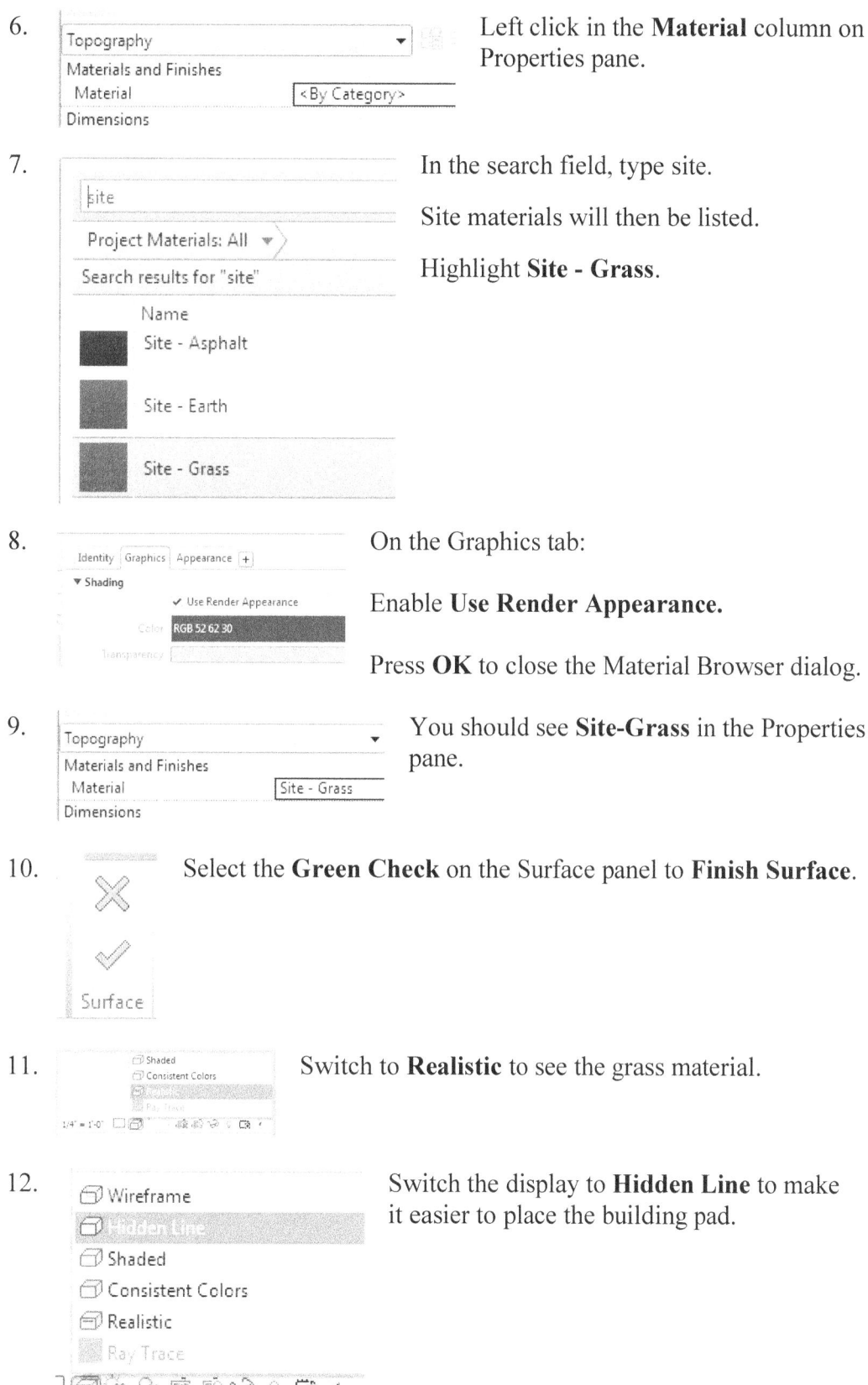

6. Left click in the **Material** column on the Properties pane.

7. In the search field, type site.

   Site materials will then be listed.

   Highlight **Site - Grass**.

8. On the Graphics tab:

   Enable **Use Render Appearance**.

   Press **OK** to close the Material Browser dialog.

9. You should see **Site-Grass** in the Properties pane.

10. Select the **Green Check** on the Surface panel to **Finish Surface**.

11. Switch to **Realistic** to see the grass material.

12. Switch the display to **Hidden Line** to make it easier to place the building pad.

9-3

13. Activate the **Massing & Site** ribbon.

    Select the **Toposurface** tool.

14. Use the **Place Point** tool to create an outline of a lawn surface.

15. Select the four corners of the building to form a rough rectangle.

16. Click on the **Modify** tool on the ribbon to exit out of the add points mode.

17. Left click in the **Material** column on the Properties pane.

18. In the search field, type **concrete**.

    Highlight **Concrete - Cast In-Place Concrete**.

19. Enable **Use Render Appearance** in the Material Editor dialog.

    Press **OK** to close the Material Browser dialog.

20. You should see **Concrete: Cast-In-Place Concrete** in the Properties pane.

21. Press **OK**.

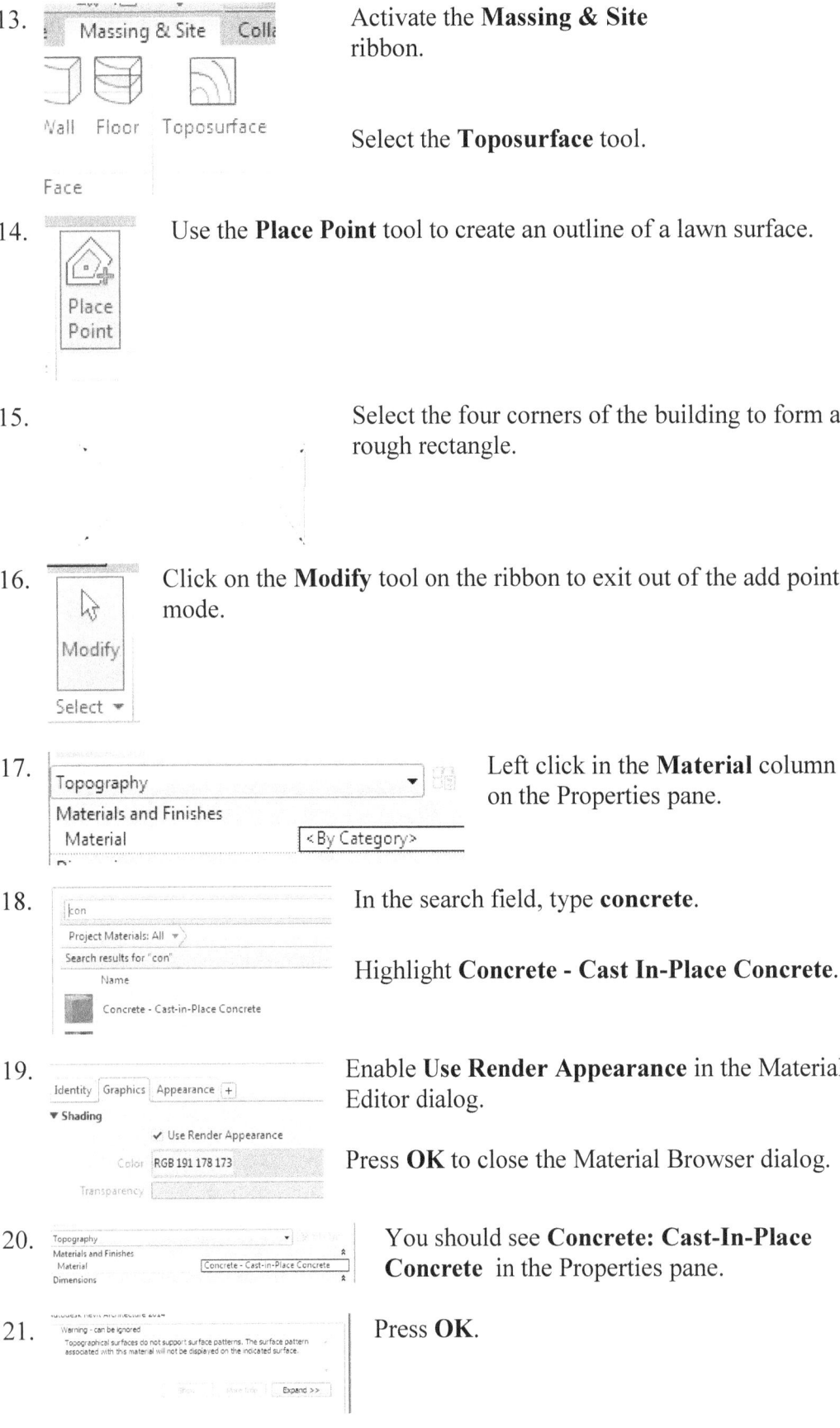

Presenting the Building Model

22. Select the **Green Check** on the Surface panel to **Finish Surface**.

23. Select the **Building Pad** tool from the Model Site panel.

24. Select the **Rectangle** tool from the Draw panel

25. Use **Rectangle** to create a sidewalk up to the left entrance of the building.

The rectangle must be entirely within the toposurface or you will get an error message.

26. On the Properties pane: Select **Edit Type**.

27. Select **Duplicate**.

28. Enter **Walkway** in the Name field. Press **OK**.

29. Select **Edit** under Structure.

9-5

30. 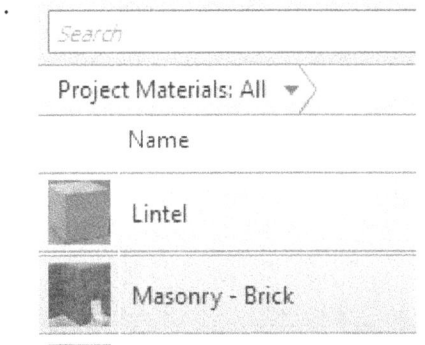 Assign the Masonry- Brick material to the structure layer.

31. 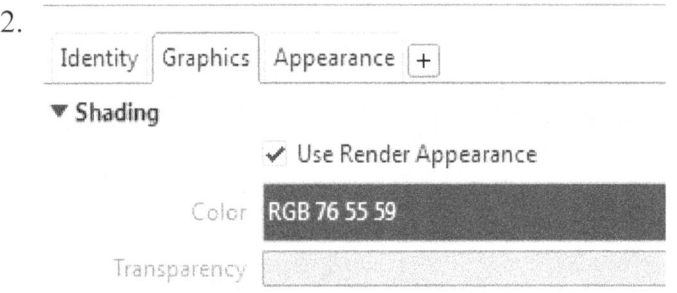 Select Masonry-Brick as the material.

32. 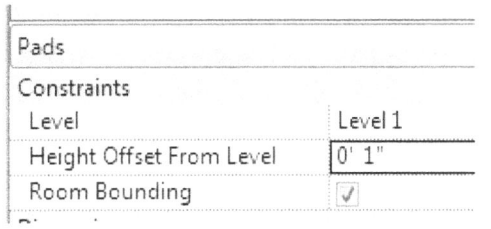 Enable **Use Render Appearance for Shading**.

    Press **OK** to exit all the material dialogs.

33. In the Properties pane:

    Set the Height Offset from Level to **1″**.

    This protrudes the walkway 1″ above the toposurface so it appears better.

34.  Select the **Green Check** on the Mode panel to **Finish Building Pad**.

35. Switch to a **3D** view.

Presenting the Building Model

36. 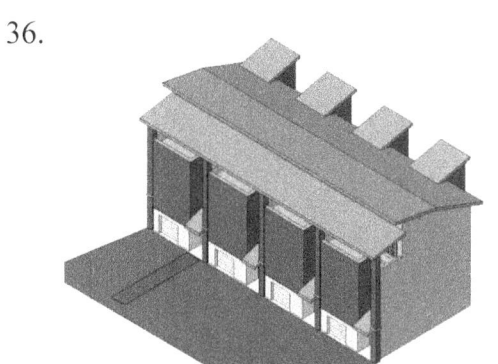 Use View Properties (VP) to set the style to **Shaded**.

    Rotate the view so you can see the topo surface.

37. Switch to the **Site** view.

38. Switch the display to **Hidden Line**.

39. Activate the **Massing & Site** ribbon.

40. Select the **Site Component** tool on the Model Site panel.

41. Select the **Load Family** tool from the Mode panel.

42. Select the *Planting* folder.

43. Select the *RPC Tree – Fall.rfa*.

    Press **Open**.

9-7

44.  Select **Japanese Maple – 10'** from the Type Selector drop-down list.

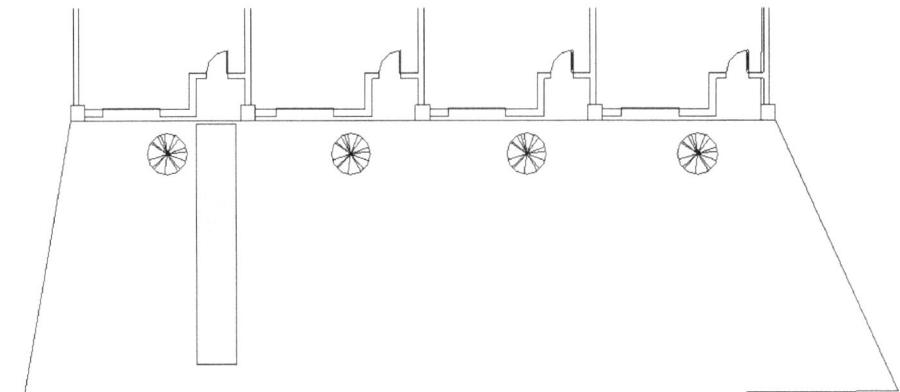

45. Place a tree at each entrance.

46. Select the **Building Pad** tool from the Model Site panel.

47. Select the **Rectangle** tool from the Draw panel

48.  Use **Rectangle** to create a drive up to the garage door of the building.

The rectangle must be entirely within the toposurface or you will get an error message.

Presenting the Building Model

49. On the Properties pane:
Select **Edit Type**.

50. Select **Duplicate**.

51. Enter **Driveway** in the Name field.
Press **OK**.

52. Select **Edit** under Structure.

53. Assign the **Site-Asphalt** material to the structure layer.

54. Enable **Use Render Appearance** on the Graphics tab.

Press **OK**.

55. Verify that the correct material has been assigned.

Press **OK**.

56. In the Properties pane:

Set the Height Offset from Level to **1"**.

This protrudes the walkway 1" above the toposurface so it appears better.

9-9

57. 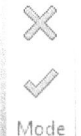 Select the **Green Check** on the Mode panel to **Finish Building Pad**.

58. Select the **Site Component** tool on the Model Site panel.

59. Select the **Load Family** tool from the Mode panel.

60. Browse to the *Entourage* folder.

61. Locate the **RPC Female [M_RPC Female.rfa]** file. Press **Open**.

62.  Set the Female to **Cathy** in the Properties pane.

63. 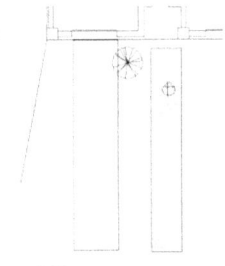 Place the person on the walkway.

64. If you zoom in on the person, you only see a symbol – you don't see the real person until you perform a Render. You may need to switch to Hidden Line view to see the symbol for the person.

    The point indicates the direction the person is facing.

65. Rotate your person so she is facing the building.

66.    Use the site component tool to place a VW beetle on the driveway.

67. Save the file as *ex9-1.rvt*.

Tips Tricks

Make sure **Level 1** or **Site** is active, or your trees could be placed on Level 2 (and be elevated in the air). If you mistakenly placed your trees on the wrong level, you can pick the trees, right click, select Properties, and change the level.

Command Exercise

Exercise 9-2 – Defining Camera Views

Drawing Name: **ex9-1.rvt**
Estimated Time to Completion: 30 Minutes

Scope

Create a camera view.
Rename the view.
Duplicate the view
Set view properties.
Compare different view display setttings

Solution

1. Activate the **Site** floor plan.

2. Activate the **View** ribbon.

   Select the **3D View→Camera** tool.

3. Your cursor icon changes to a camera.

   In the lower left prompt area: the prompt says "Click to place the eye position at the cursor location."

Presenting the Building Model

4. Aim the camera towards the front entrance to the building.

5. A window opens with the camera view of your model.

   The person and the tree appear as stick figures because the view is not rendered yet.

6. Switch to a **Realistic** display.

7. Change the plant to a Gray Birch.

9-13

8.  Our view changes to a colored view.

9.   ⊞ Ceiling Plans
      ⊟ 3D Views
          3D View 1
          {3D}

   If you look in the browser, you see that a view has been added to the 3D Views list.

10. Name: 3D Realistic

    Highlight the **3D View 1**.

    Right click and select **Rename**.

    Rename to **3D Realistic**.

The view we have is a perspective view – not an isometric view. Isometrics are true scale drawings. The Camera View is a perspective view with vanishing points. Isometric views have no vanishing points. Vanishing points are the points at which two parallel lines appear to meet in perspective.

11. | Crop View | |
    | Camera | |
    | Rendering Settings | Edit... |
    | Locked Orientation | |
    | Perspective | |
    | Eye Elevation | 4' 6" |
    | Target Elevation | 5' 10" |
    | Camera Position | Explicit |

    In the Properties pane:

    Change the Eye Elevation to **4' 6"** **[1370 mm]**.

    Change the Target Elevation to **5' 10"** **[3200 mm]**.

12. Press **Apply**.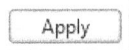

    Your view shifts slightly.

13. 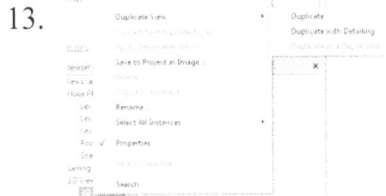 Highlight the 3D Realistic view in the Project Browser. Right click and select **Duplicate View→Duplicate**.

9-14

Presenting the Building Model

14. 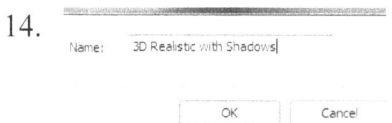 Highlight the **Copy of 3D Realistic**.

    Right click and select **Rename**.

    Rename to **3D Realistic with Shadows**.

15.  Enable **Shadows**.

16.  Highlight the 3D Realistic with Shadows view in the Project Browser.

    Right click and select **Duplicate View→Duplicate**.

17. 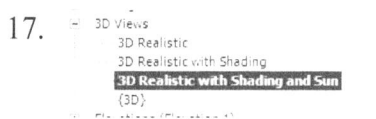 Highlight the **Copy of 3D Realistic with Shadows**.

    Right click and select **Rename**.

    Rename to **3D Realistic with Shadows and Sun**.

18. Turn **Sun Path On**.

    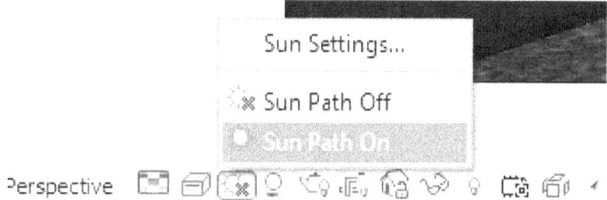

19. Select **Use the specified project location, date and time instead**.

20. 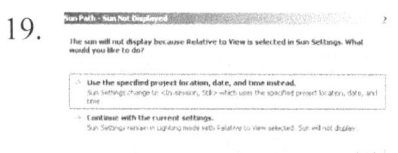 Highlight the **3D Realistic with Shadows and Sun** view in the Project Browser.
    Right click and select **Duplicate View→Duplicate**.

21.  Highlight the **Copy of 3D Realistic with Shadows and Sun**.

    Right click and select **Rename**.

    Rename to **3D Realistic with Shadows and Sun - Graphics**.

22.  Click **Edit** next to Graphic Display Options.

9-15

23. Change the Silhouettes to **Medium Lines**.

    Under Shadows:
    Enable **Cast Shadows**.
    Enable **Show Ambient Shadows**.

    Set the Background to **Sky**.

    Click **Apply** to see how the view changes.

24. Under Lighting:

    Set the Sun to **30**.
    Set the Ambient Light to **36**.
    Set the Shadows to **50.**
    Click **Apply** to see how the view changes.

    Press **OK** to close the dialog.

25. Highlight the **3D Realistic with Shadows and Sun - Graphics** view in the Project Browser.
    Right click and select **Duplicate View→Duplicate**.

26. Highlight the **Copy of 3D Realistic with Shadows and Sun -Graphics**.

    Right click and select **Rename**.

    Rename to **3D Realistic with Shadows and Sun - Photo**.

27. Click **Edit** next to Graphic Display Options.

Presenting the Building Model

28. Under Photographic Exposure:
Check **Enabled**.
Enable **Manual**.
Set the value to **16**.
Click the **Color Correction** button.

29. Set the Highlights to **0.35**.
Set the Shadow Intensity to **0.50**
Set the Color Saturation to **2.50**.
Set the Whitepoint to **6587**.

Press **OK** twice to exit the dialogs.

30. Highlight **Sheets** in the Project Browser.
Right click and select **New Sheet.**

31. Highlight the **E1** titleblock.

Press **OK**.

32. Rename the sheet - **Camera Views-Exterior**.

33. Drag and drop the 3D camera views you created on to the sheet.

34. Compare the different views and how the settings affect the displays.

35. Save the file as *ex9-2.rvt*.

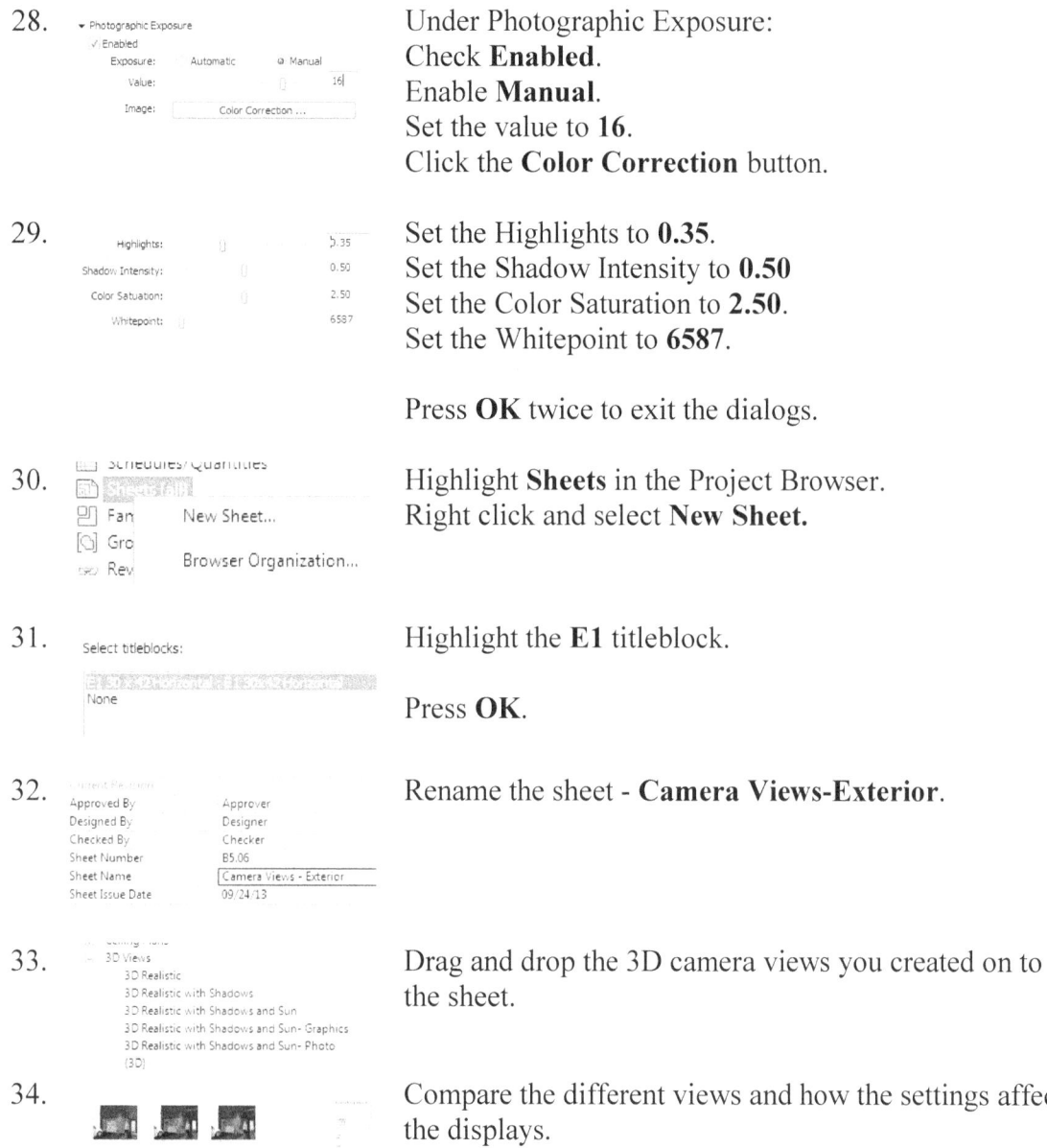

Tips & Tricks

Additional material libraries can be added to Revit. You may store your material libraries on a server to allow multiple users access to the same libraries. You must add a path under Settings → Options to point to the location of your material libraries. There are many online sources for photorealistic materials; www.accustudio.com is a good place to start.

Presenting the Building Model

Command Exercise

Exercise 9-3 – Sun Settings

Drawing Name: **ex9-2.rvt**
Estimated Time to Completion: 15 Minutes

Scope

Sun Settings

Solution

1. Activate the **3D Realistic with Shadows and Sun** view.

3.  On the View Display bar:
   Select **Sun Settings**.

4. Enable **Still** under Solar Study.

5. Select the browse button next to Location.

6. Enter an address into the Project Address field.

   Press **Search** to locate the address.
   Type in the address shown.

   Enable **Use Daylight Savings time**.

   Press **OK**.

7.  Enable **Ground Plane at Level: Level 1**.

   Press **OK**.

8. Activate the 3D view.

9. Set **Sun Path On**.

10. Select **Use the specified project location, date, and time instead**.

11. 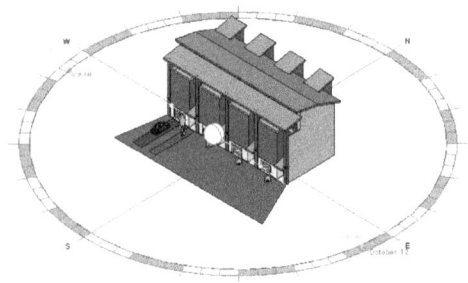 Zoom out to see the sun path.

    Place the cursor over the sun.

    Hold down the left mouse button and move the sun.

12. 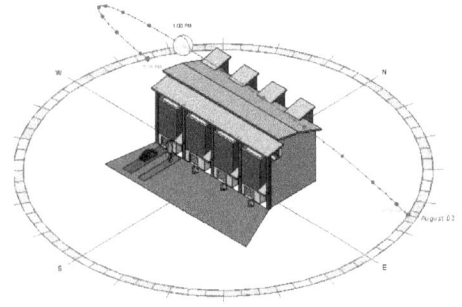 See if you can move the sun to a position of 1:00 PM and August.

Presenting the Building Model

13.  Activate the **3D Realistic with Shadows and Sun** view.

14. 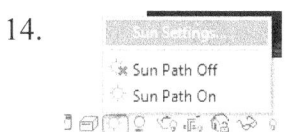 On the View Display bar: Select **Sun Settings**.

15. Note that the sun settings for the view are specific to each view.

    Change the date to **August** and the time to **1:00 PM**.

    Press **Apply** and **OK**.

16. Save as *ex9-3.rvt*.

Command Exercise

Exercise 9-4 – Rendering Settings

Drawing Name: **ex9-3.rvt**
Estimated Time to Completion: 15 Minutes

Scope

Create a rendering.
Control rendering options.
Save a rendering to the project.

Solution

1. Activate the **3D Realistic with Shadows and Sun** view.

2. On the Properties pane:

   Under Camera:
   Select **Edit** for Rendering Settings.

3. Set the Quality Setting to **Medium**.

   Press **OK**.

4. 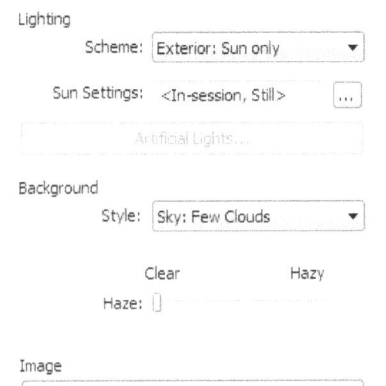 Select the **Rendering** tool located on the Display Control bar.

9-22

Presenting the Building Model

5. Select the **Render** button.

6. Your window will be rendered.

7. Select **Save to Project**.

8. Press add **Rendering 1 to** the default name.

9. Under Renderings, we now have a view called **3D Realistic with Shadows and Sun _Rendering_1**.

10. Select the **Show the model** button. Our window changes to – not Rendered – mode.

    Close the Rendering dialog.

11. Activate the sheet with the Camera Views.

12. Add the new rendering view to the sheet.

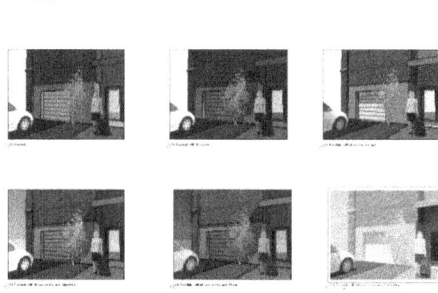

13. Save the file as *ex9-4.rvt*.

Command Exercise
Exercise 9-5 – Render in Cloud

Drawing Name: **ex9-4.rvt**
Estimated Time to Completion: 10 Minutes

Scope

Render in Cloud

In order to use the Cloud, you must have an internet connection and an Autodesk 360 account. Students and teachers qualify for a free Autodesk 360 account. If you have an Autodesk subscription, you also qualify for a free account. Check with your local reseller if you are not sure about a subscription.

Solution

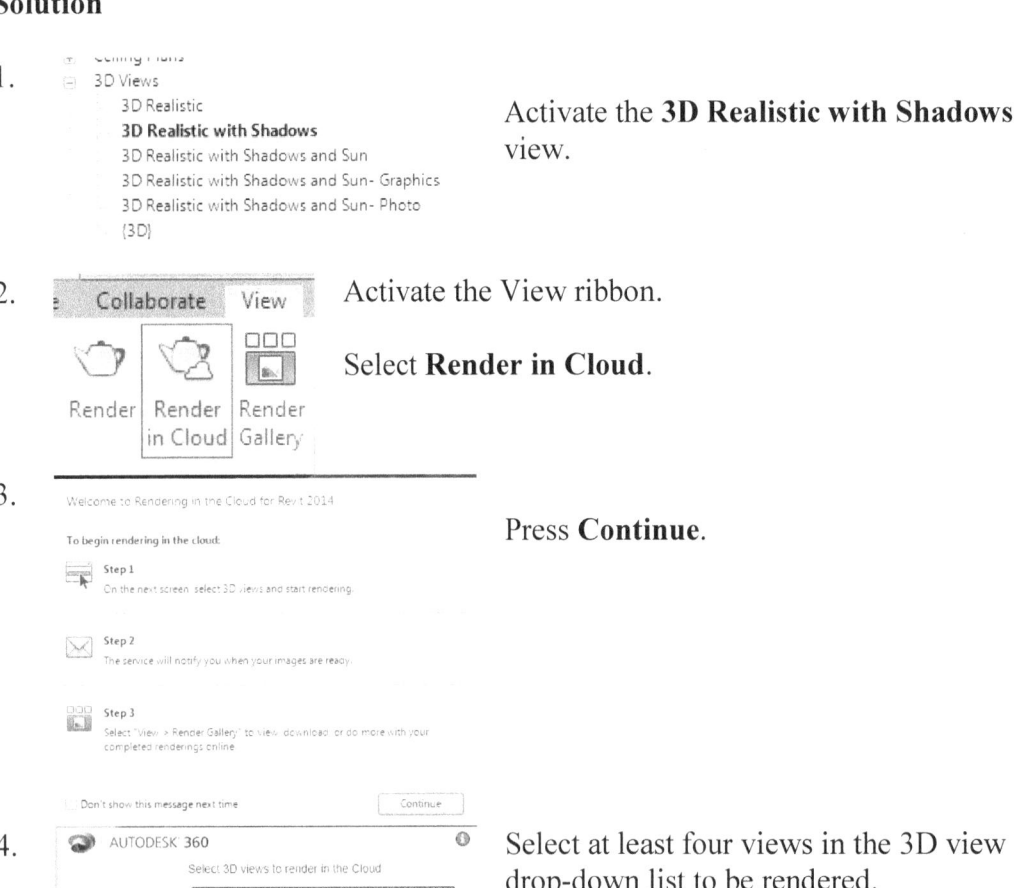

1. Activate the **3D Realistic with Shadows** view.

2. Activate the View ribbon.

   Select **Render in Cloud**.

3. Press **Continue**.

4. Select at least four views in the 3D view drop-down list to be rendered.

5.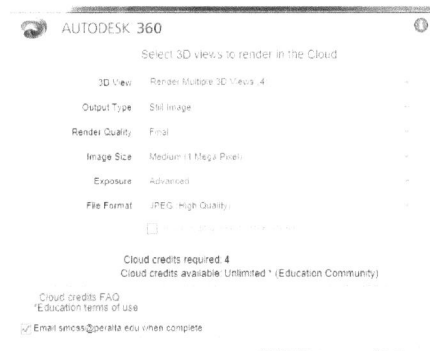

Set the Output Type to **Still Image**.
Set the Render Quality to **Final.**
Set the Image Size to **Medium**.
Set the Exposure to **Advanced**.
Set the File Format to **JPEG**.

Press **Start Rendering**.

6.

Select **Render Gallery** to view the completed renderings.

7. You should see a list of images to be reviewed.

8.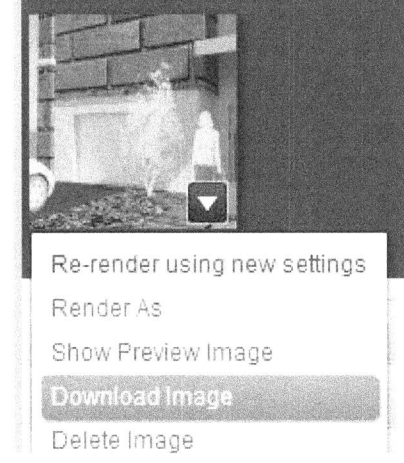

Select the image you like best.

Left click on the image and select **Download Image** to save the image to your computer.

The image will most likely be saved to the Downloads folder.

9. 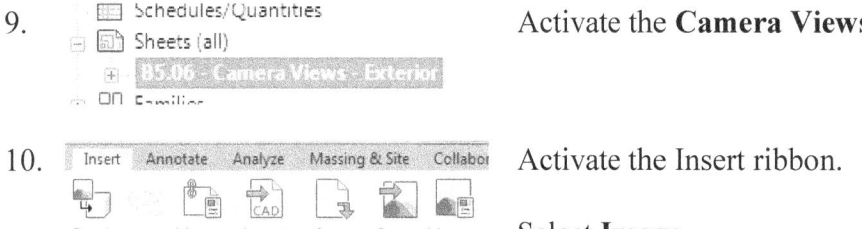 Activate the **Camera Views** sheet.

10. 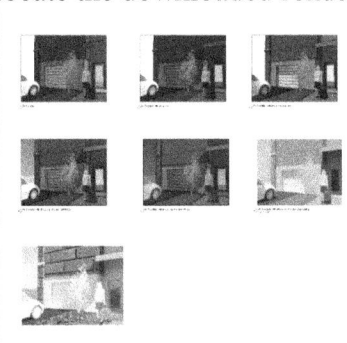 Activate the Insert ribbon.

    Select **Image.**

11. Locate the downloaded rendering file and select.

12. Place on the sheet.

    Compare the image created on the Cloud against the other images.

13. Save as *ex9-5.rvt*.

Command Exercise

Exercise 9-6 – Place a Decal

Drawing Name: **ex9-5.rvt**
Estimated Time to Completion: 30 Minutes

Scope

Create a decal type.
Place a decal.
Add model text.
Join Geometry
Render.

Solution

1. 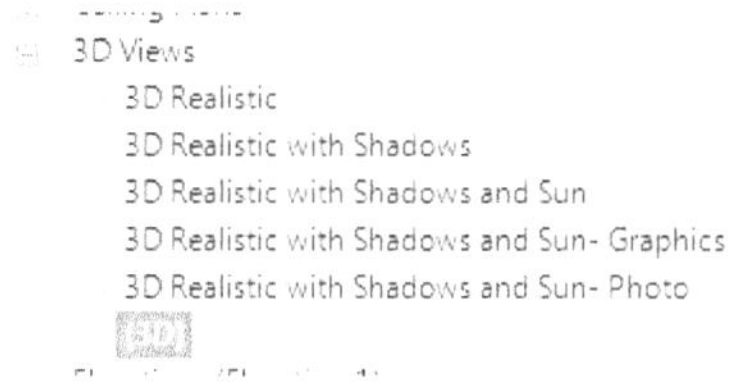 Activate the **3D** view by selecting the House icon.

2.  Set the Visual Style to **Realistic**.

3.  Turn the **Sun Path Off**.

4.  Turn **Shadows Off**.

5. 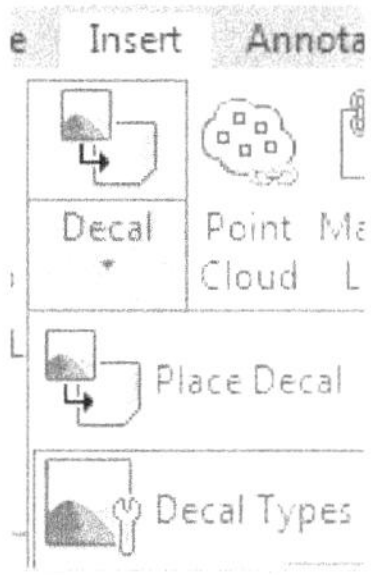 Activate the **Insert** ribbon.

   Select **Decal Types** on the Link panel.

Presenting the Building Model

9-27

6. Select **Create New Decal Type**.
   This tool is located on the lower left of the dialog.

7. Enter **Parking Sign** in the Name field.
   Press **OK**.

8. Select the Browse button to select the image file to be used.

9. Locate the *parking sign.png* file.
   You can use a different image file if you prefer.
   Press **Open**.

10. You will see a preview of the image file.
    Press **OK**.

11. Select the **Place Decal** tool from the Link panel

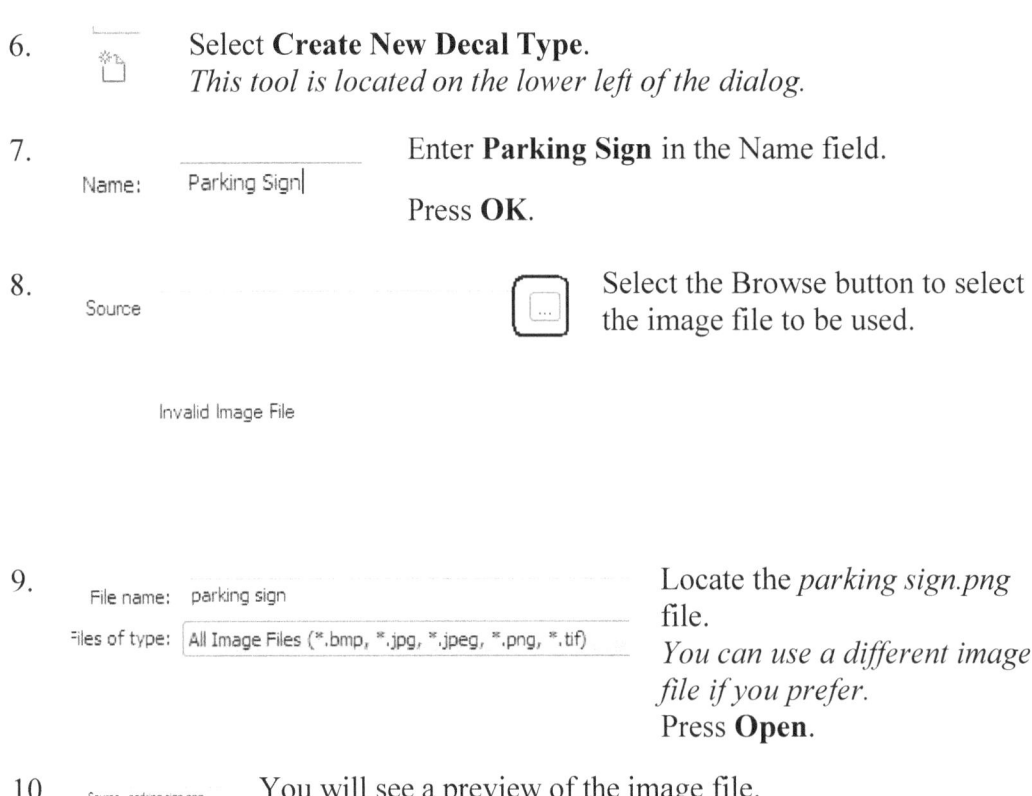

12.  Place the decal on the side of the building.

Presenting the Building Model

13. Select the Modify tool and then pick the decal that was placed.

14.  Use the grips located on the corners to enlarge the decal.

    The grips can also be used to move the decal into the desired position.

15. Activate the **Architecture** ribbon.

    Select the **Set Work plane** tool from the Work Plane panel.

16. Enable **Pick a plane**.

    Press **OK**.

17.  Select the wall above where the decal is placed.

    If you wish to verify that you have selected the correct work plane, use the Show Work Plane tool.

18. Select the **Model Text** tool from the Model panel on the Architecture ribbon.

9-29

19. Enter **Crestview Commons** into the Edit Text dialog.
    Type it as two lines as shown.

    Press **OK**.

20. 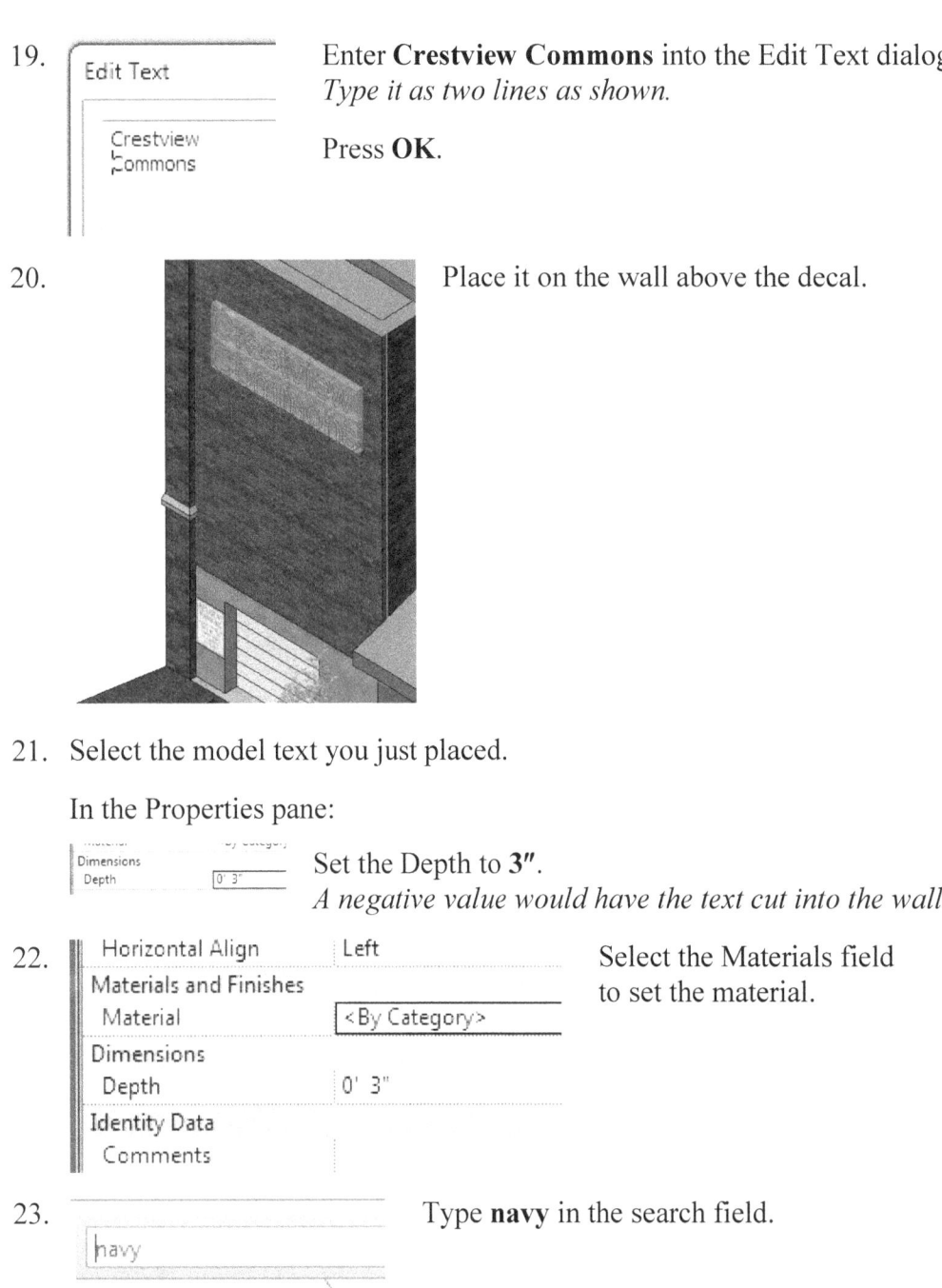 Place it on the wall above the decal.

21. Select the model text you just placed.

    In the Properties pane:

    Set the Depth to **3"**.
    A negative value would have the text cut into the wall.

22. Select the Materials field to set the material.

23. Type **navy** in the search field.

Presenting the Building Model

24. A Laminate, Navy material will be listed in the lower pane.

Click on the **Add material to document** icon.

25. Highlight the material and press **OK**.

26. With the Model Text selected:
Set the Horizontal Align to **Center** in the Properties pane.
Left click in the window to release the selection.

27.

Save the file as *ex9-6.rvt*.

9-31

Command Exercise

Exercise 9-7 – Adding a Background

Drawing Name: **background.rvt**
Estimated Time to Completion: 30 Minutes

Scope

Modify Graphic Display Options
Add a background to a view
Render a view
Compare realistic and rendered views

Solution

1. Activate the **Day View with Background** view.

2. Select **Edit** next to Graphic Display Options.

3.
   Set the Style to **Realistic**.
   Enable **Show Edges**.
   Set Silhouettes to **Medium Lines**.
   Enable **Cast Shadows**.
   Enable **Show Ambient Shadows**.
   Set Scheme to **Exterior: Sun Only**.
   Set Sun to **30**.
   Set Ambient Light to **40**.
   Set Shadows to **50**.

Presenting the Building Model

4. Under Background:

    Select **Image**.
    Select **Customize Image**.

5. Select the **Image** button.

6. Select *sky1.png*.

    Press **Open**.

    I have included several images you can use for backgrounds, so you can try out different backgrounds to get different effects.

7. The background will preview.

    Press **OK**.

8. Press **Apply**.

    Press **OK**.

    The image will update with the background.

    Repeat steps 5 through 9 and use *sky2.jpeg*.

9. Verify that both sun and shadows are set to **ON**.

10. 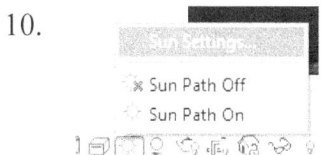 Select **Sun Settings** on the View Display bar.

11. Set the Sun Settings to **8/1/2010**.
1:00 PM.

Press **OK.**

12. Select the **Render** tool on the View ribbon.

    You can also select the Render tool on the display bar at the bottom of the screen.

13. Set the Quality to **Medium**.
    Set the Scheme to **Exterior: Sun Only**.
    Set the Background to **Image**.
    Select **Customize Image** and select *sky2.jpg* for the background.

    Press **Render**.

14. Select **Save to Project**.

15. Press **OK** to accept the name.

    Close the rendering dialog.

16. Activate **Night View with background**.

17. Verify that both sun and shadows are set to **ON**.

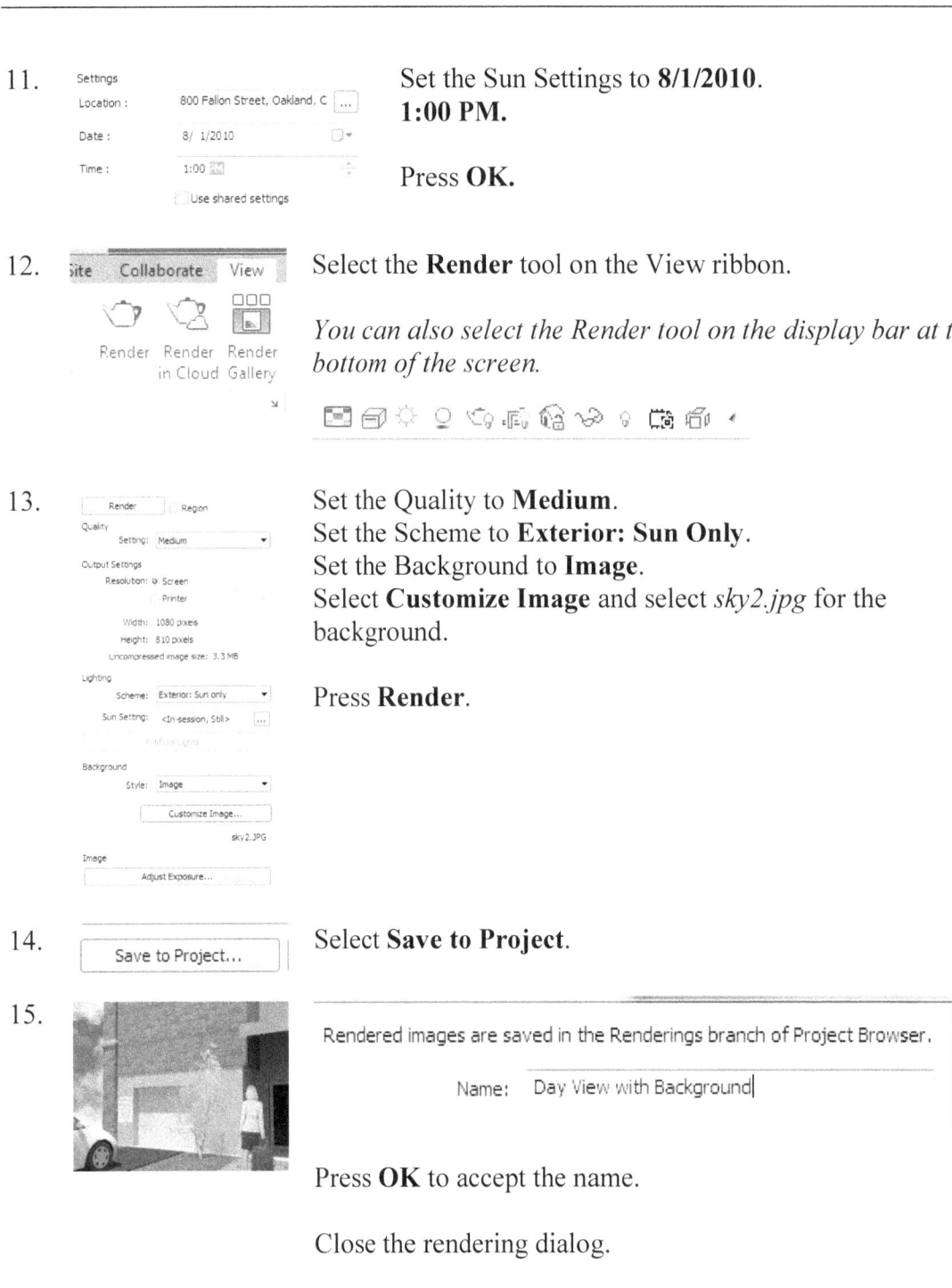

Presenting the Building Model

18. Select **Sun Settings** on the View Display bar.

19. Set the Sun Settings to **8/1/2010**.
6:00 PM.

Press **OK.**

20. Select **Edit** next to Graphic Display Options.

21. Set the Style to **Realistic**.
Enable **Show Edges**.
Set Silhouettes to **Medium Lines**.
Enable **Cast Shadows**.
Enable **Show Ambient Shadows**.
Set Scheme to **Exterior: Sun and Artificial**.
Set Sun to **8**.
Set Ambient Light to **30**.
Set Shadows to **50**.
Press **Apply**.

22. Under Background:

Select **Image**.
Select **Customize Image**.

23. Select the **Image** button.

24. Select *night4.jpg*.

Press **OK**.

9-35

25.  The background will preview.

Press **OK**.

26.  Press **Apply**.

The image will update with the background.

Press **OK**.

27.  Select the **Render** tool in the View Display bar.

28. Set the Quality to **Medium**.
Set the Scheme to **Exterior: Sun and Artificial Lights**.
Set the Background to **Image**.
Select **Customize Image** and select *night4.jpg* for the background.

Press **Render**.

29. Select **Save to Project**.

30. 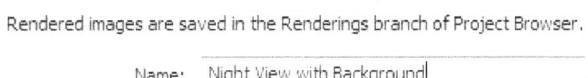 Press **OK** to accept the default name.

Close the rendering dialog.

31.  Activate the **Exterior-Background Views** sheet.

32. 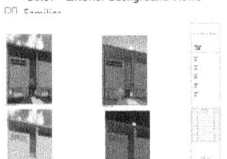 Add the realistic views and the rendering views to the sheet so you can compare the different views.

Save as *ex9-7.rvt*.

Command Exercise

Exercise 9-8 – Using Transparency Settings

Drawing Name: **office.rvt**
Estimated Time to Completion: 30 Minutes

Scope

Apply transparency to elements

Solution

1. Activate the **3D Section** view.

2. Select the walls for the first cubicle.

3.  Right click and select **Override Graphics in View→By Element.**

4. Set the Surface Transparency to **80**.

   Press **OK**.

5. You can now see clearly inside the cubicle.

   Close without saving.

Command Exercise

Exercise 9-9 – Custom Render Settings

Drawing Name: **office.rvt**
Estimated Time to Completion: 15 Minutes

Scope
Create custom rendering settings

If the link for the decal needs to be reset: Select the Decal, Select Edit Type, Edit Decal Attributes and select the decal – street scene.

Solution

1. Activate the **Inside Cubicle 3D** view.

2. Select the **Render** tool from the View Control bar.

3. Under Quality:
   Setting: Select **Edit**.

4. Set the Setting to **Custom**.

5. Set Image Precision to **5**.

   Set Maximum Number of Reflections to **22**.

   Set Maximum Number of Refractions to **14**.

9-38

Presenting the Building Model

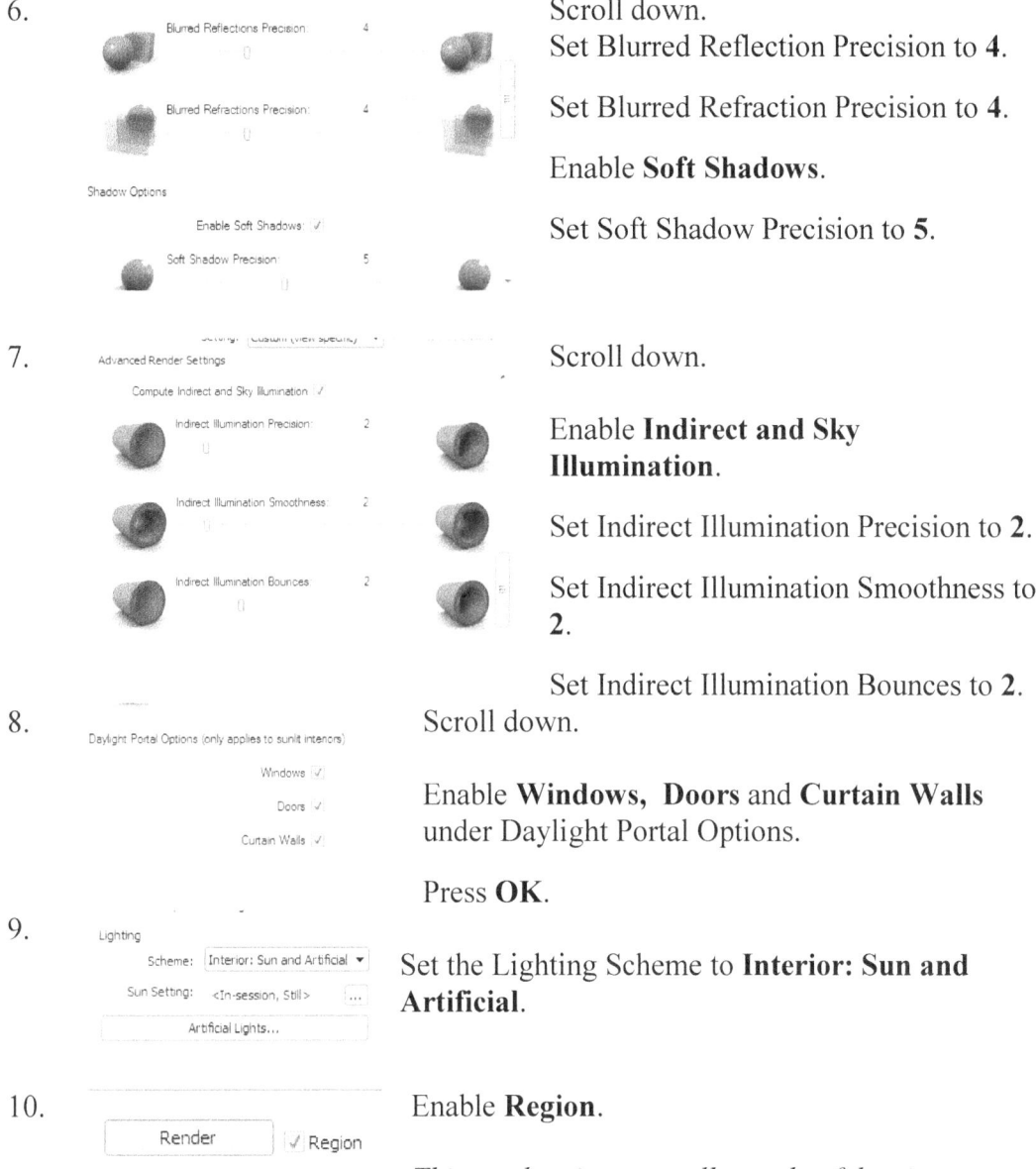

6. Scroll down.

    Set Blurred Reflection Precision to **4**.

    Set Blurred Refraction Precision to **4**.

    Enable **Soft Shadows**.

    Set Soft Shadow Precision to **5**.

7. Scroll down.

    Enable **Indirect and Sky Illumination**.

    Set Indirect Illumination Precision to **2**.

    Set Indirect Illumination Smoothness to **2**.

    Set Indirect Illumination Bounces to **2**.

8. Scroll down.

    Enable **Windows, Doors** and **Curtain Walls** under Daylight Portal Options.

    Press **OK**.

9. Set the Lighting Scheme to **Interior: Sun and Artificial**.

10. Enable **Region**.

    This renders just a small sample of the view.

9-39

11. Select the region and use the grips to position and size the desired region for rendering.

12. Press **Render**.

13. Select **Adjust Exposure**.

14. Increase the Exposure Value to 14.5 or greater.

    Note how the rendering adjusts.

15. Close the file without saving.

Command Exercise

Exercise 9-10 – Using Element Graphic Overrides in a Hidden View

Drawing Name: **hidden_views.rvt**
Estimated Time to Completion: 10 Minutes

Scope

Apply Graphic Overrides to Enhance a Hidden View

Solution

1. Activate the **3D Hidden View**.

2. Select **Edit** next to Graphic Display Options in the Properties pane.

3. Set the Style to **Hidden Line.**

   Set the Silhouettes to **Medium Lines.**

   Enable **Cast Shadows**.
   Enable **Show Ambient Shadows**.
   Select **Image** under the Background drop-down list.

4. Select **Customize Image**.

5. Select the **Image** button to browse for the image file.

6. Select *sky4.jpg*.

   Press **OK** twice.

		The image is displayed in the view.
7.		Select the far left column. Right click and select **Override Graphics in View→By Element.**
8.		Under Surface Patterns: 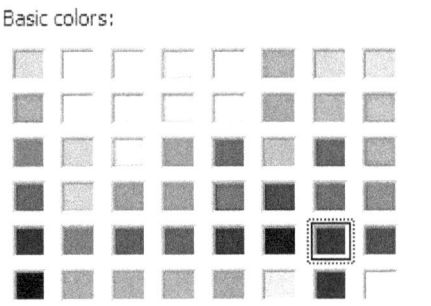 Set the Color to the Color Swatch indicated. Set the Pattern to **Solid Fill**. Press **OK**. Left click in the window to release the selection.
9.		Select the model text on the building. Right click and select **Override Graphics in View→By Element.**
10.		Under Surface Patterns: Set the Color to **Blue**. Set the Pattern to **Solid Fill**. Press **OK**.

11.  Try changing other elements to see what effects you can create.

Save as *ex9-10.rvt*.

Practice Exam

1. Renderings can be created in which TWO view types:
   - A. Elevation
   - B. Plan
   - C. Section
   - D. 3D
   - E. Camera

2. THREE items important for accurate shadow studies:
   - A. Shared parameters
   - B. Toposurface
   - C. Color fill schemes
   - D. Project location
   - E. Sun position (date and time)

3. In the *default.rte* template, what level is the Ground Plane set in the Sun and Shadow Settings?
   - A. Site
   - B. Level 1
   - C. Level 2

4. The Place Decal tool is located on which ribbon:
   - A. Architecture
   - B. Insert
   - C. View
   - D. Annotations

5. Before you add a building pad, you need a:
   - A. floor
   - B. toposurface
   - C. wall
   - D. isolated foundation

6. You can create a toposurface by placing points OR:
   - A. creating from import
   - B. drawing contour lines
   - C. drawing a path
   - D. picking points

7. When creating a rendering, you can control the Maximum Number of Reflections by adjusting the _____ settings.

   A. Image
   B. Lighting
   C. Quality
   D. Output

8. Select the display setting that allows you to preview a decal.

   A. Wireframe
   B. Hidden line
   C. Consistent Colors
   D. Realistic

9. In a view, you want to use the Visibility/Graphics dialog to override the graphics of an element category, but the category displays in gray, and you cannot change it. What is the problem?

   A. The visibility of the category may be controlled by a view template.
   B. The element category is locked
   C. The view display is set to Hidden.
   D. The view is a Section.

10. Decals can be placed on _____ and _____ surfaces.

    A. Flat
    B. Slanted
    C. Round
    D. Visible

11. Backgrounds can be added in ALL of the following view types EXCEPT:

    A. Elevation
    B. Section
    C. Detail
    D. Isometric
    E. Camera (Perspective) 3D

12. Types of backgrounds are ALL of the following EXCEPT:

    A. Sky
    B. Gradient
    C. Image
    D. Region

Answers
1) D & E; 2) B, D & E; 3) B; 4) B; 5) B; 6) A; 7) C; 8) D; 9) A; 10) A & C; 11) C; 12) D

Lesson Ten

Collaboration

This lesson addresses the following User and Professional exam questions:

- Demonstrate how to copy and monitor elements in a linked file
- Apply interference checking in Revit
- Using Shared Coordinates
- Worksets
- Linked files

In most building projects, you need to collaborate with outside contractors and with other team members. A mechanical engineer uses an architect's building model to layout the HVAC (heating and air conditioning) system. Proper coordination and monitoring ensures that the mechanical layout is synchronized with the changes that the architect makes as the building develops. Effective change monitoring reduces errors and keeps a project on schedule.

Worksets are used in a team environment when you have many people working on the same project file. The project file is located on a server (a central file). Each team member downloads a copy of the project to their local machine. The person is assigned a workset consisting of building elements which they can change. If you need to change an element that belongs to another team member, you issue an Editing Request which can be granted or denied. Workers check in and check out the project, updating both the local and server versions of the file upon each check in/out.

Project sharing is the process of linking projects across disciplines. You can share a Revit Structure model with an MEP engineer.

You can link different file formats in a Revit project, including other Revit files (Revit Architecture, Revit Structure, Revit MEP), CAD formats (DWG, DXF, DGN, SAT, SKP), and DWF markup files. Linked files act similarly as external references (XREFs) in AutoCAD. You can also use file linking if you have a project which involves multiple buildings.

It is recommended to use linked Revit models for

- Separate buildings on a site or campus

- Parts of buildings which are being designed by different design teams or designed for different drawing sets

- Coordination across different disciplines (for example, an architectural model and a structural model)

Linked models may also be appropriate for the following situations:

- Townhouse design when there is little geometric interactivity between the townhouses

- Repeating floors of buildings at early stages in the design, where improved Revit model performance (for example, quick change propagation) is more important than full geometric interactivity or complete detailing

You can select a linked project and bind it within the host project. Binding converts the linked file to a group in the host project. You can also convert a model group into a link which saves the group as an external file.

I have included a portion of the worksets exercise I do in my classroom. This exercise requires that there is a shared location on a server where students have read/write access. In many classroom settings, the IT department only provides students with read access and they can only write to a local flash drive. This is to prevent file corruption and minimize exposures to computer viruses. Instructors and users should keep this in mind during the workset exercise.

Command Exercise

Exercise 10-1 – Monitoring a Linked File

Drawing Name: **i_multiple_disciplines.rvt**
Estimated Time to Completion: 40 Minutes

Scope

Link a Revit Structure file
Monitor the levels in the linked file
Reload the modified Structure file
Perform a coordination review
Create a Coordination Review report

Solution

1. 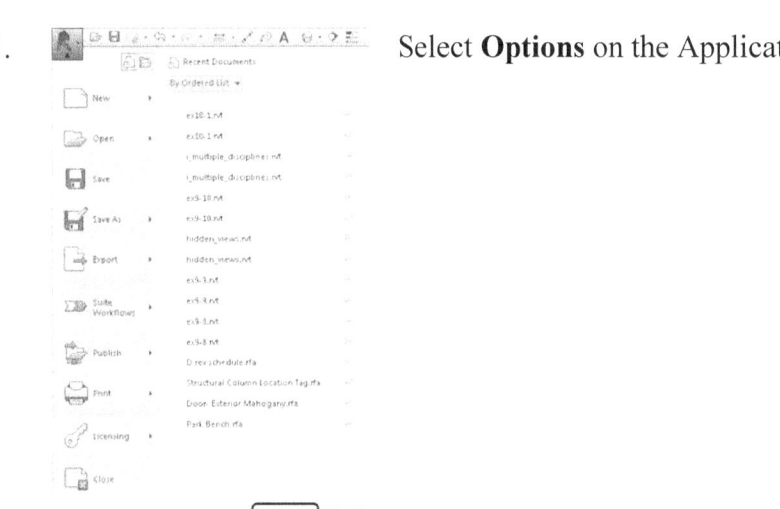 Select **Options** on the Application menu.

2. In the Username field, type in your first initial and last name.

   Press **OK**.

   Revit will auto-fill the username in reports and the title block.

   If you have a 360 Account, the sign in for the user account is used for your user name.

3. Open the *i_multiple_disciplines.rvt* file.

4. 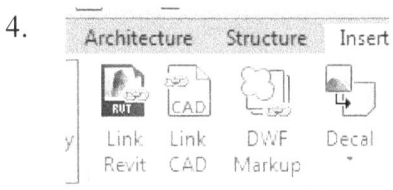 Activate the **Insert** ribbon.

   Select the **Link Revit** tool on the Link panel.

5. 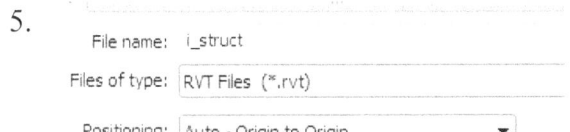 Locate the *i_struct* file.

   Set the Positioning to **Auto-Origin to Origin**.

   Press **Open**.

6. 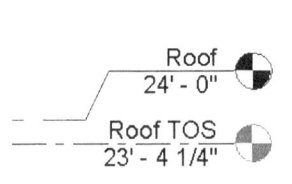 Zoom into the left side of the screen where the level markers are.

   There are two sets of level lines. One is part of the original file and the other is from the linked file.

7.  Activate the Collaborate ribbon.

   Select **Copy/Monitor→Select Link** from the Coordinate panel.

   Pick in the window to select the linked file.

8. 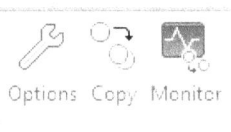 Select the **Monitor** tool from the Tools panel.

9.  Select the Level 1 level line twice.
   The first selection will be the host file.
   The second selection will be the linked file.
   This is because the level lines overlap.

10. You should see a symbol on Level 1 indicating that Level 1 is currently being monitored for changes.

Collaboration

11.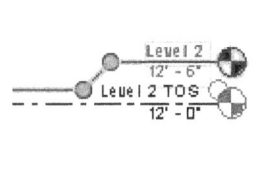

    Select the Level 2 level line.
    The first selection will be the host file.
    The second selection will be the linked file.
    You should see a tool tip indicating a linked file is being selected.

12. Zoom out.

    You should see a symbol on Level 2 indicating that Level 2 is currently being monitored for changes.

13.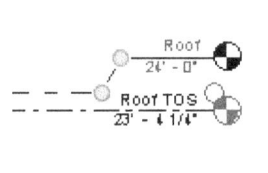

    Select the Roof level line.
    The first selection will be the host file.
    The second selection will be the linked file.
    You should see a tool tip indicating a linked file is being selected.

14. 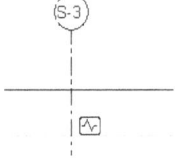 Zoom out.

    You should see a symbol on the Roof level indicating that it is currently being monitored for changes.

15. 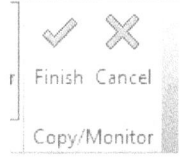 If you try to select elements which have already been set to be monitored, you will see a warning dialog.
    Simply close the dialog and move on.

16.  Select **Finish** on the Copy/Monitor panel.

17. Activate the **Insert** ribbon.

    Select **Manage Links** from the Link panel.

10-5

18. 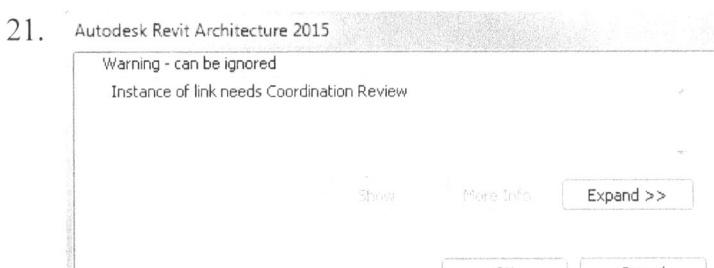 Select the **Revit** tab.
Highlight the *i_struct.rvt* file.

19. Select **Reload From**.

20. Locate the *i_struct_revised* file.
Press **Open**.

21. 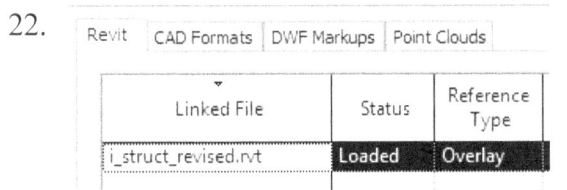 *A dialog will appear indicating that the revised file requires Coordination Review.*

Press **OK**.

22. Note that the revised file has replaced the previous link.

This is similar to when a sub-contractor or other consultant emails you an updated file for use in a project.

Press **OK**.

23. 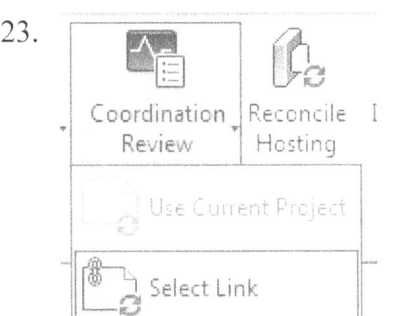 Activate the **Collaborate** ribbon.

Select **Coordination Review→Select Link** from the Coordinate panel.

Select the linked file in the drawing window.

24. 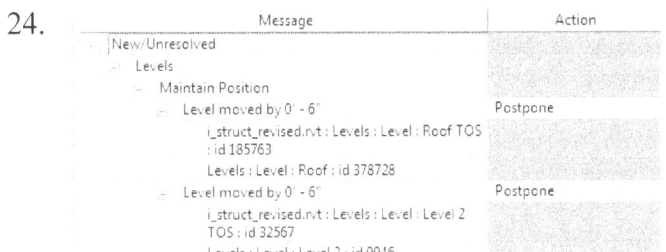 A dialog appears.

Expand the notations so you can see what was changed.

10-6

Collaboration

25. Highlight first change. Select **Move Level Roof** from the drop-down list.

26. Select the **Add Comment** button.

27. Type **Approved.** Press **OK**.

Note that Revit automatically adds your user name and a time stamp.

28. Highlight the second change. Select **Move Level 'Level 2'**.

29. Select the **Add Comment** button.

30. Type **Approved.** Press **OK**.

31. Select **Create Report** on the bottom left of the dialog.

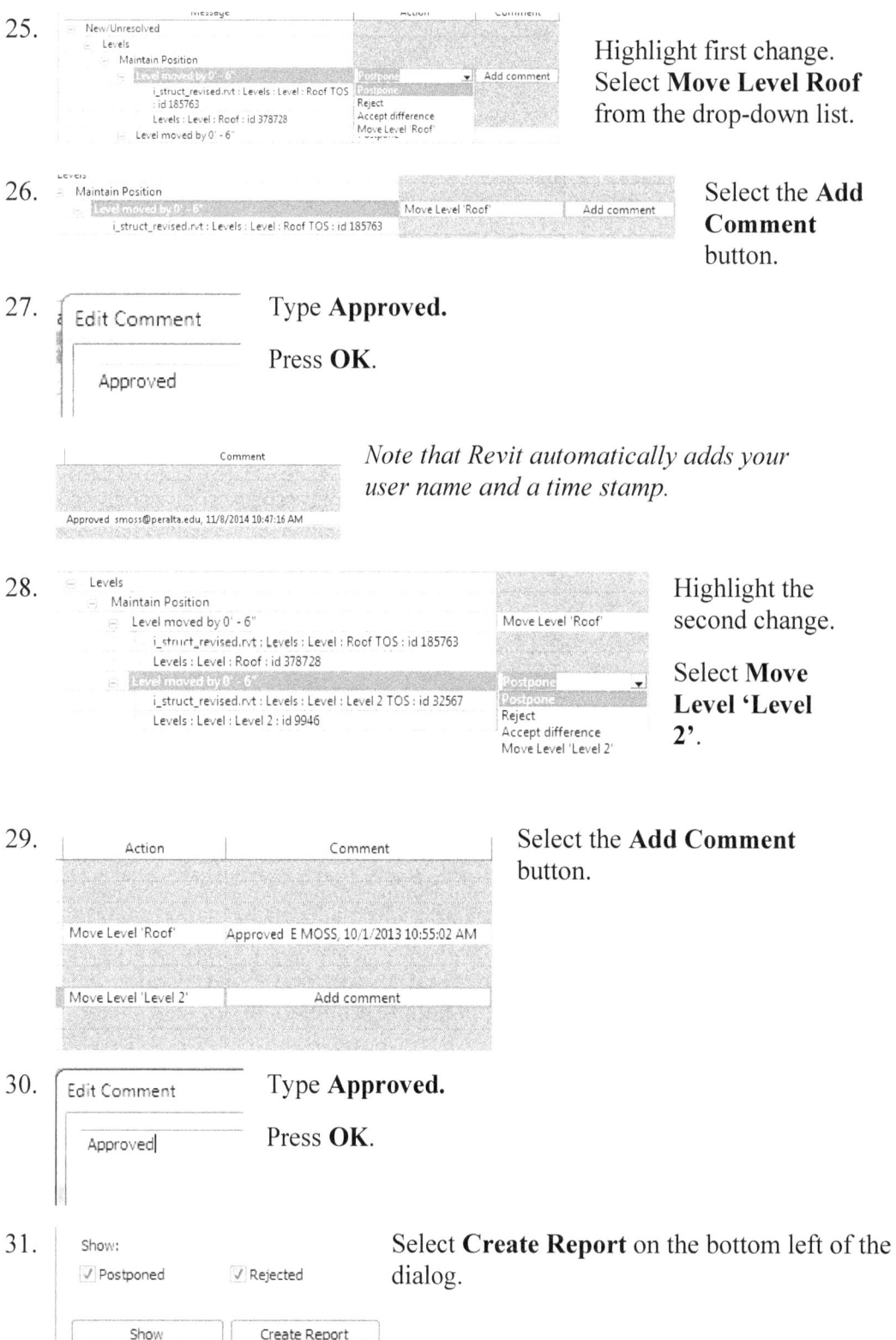

10-7

32.  Browse to your exercise folder.
    Name the file *review*.

    Press **Save**.

33. Press **OK** to close the Coordination Review dialog box.

    If you close the Coordination Review dialog before you create the report and then re-open it to create the report, the report will be blank because all the issues will have been resolved.

34. Locate the report you created and double click on it to open.

35.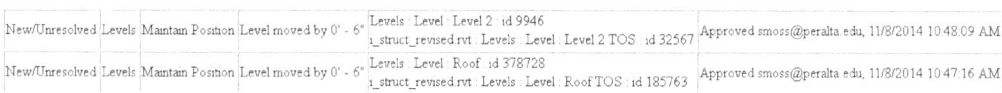

    This report can be emailed or used as part of the submittal process.

36. Close the file without saving.

Command Exercise
Exercise 10-2 – Interference Checking

Drawing Name: **c_interference_checking.rvt**
Estimated Time to Completion: 30 Minutes

Scope

Describe Interference Checks
Check and fix interference conditions in a building model
Generate an interference report

Solution

1. Activate the **Collaborate** ribbon.

    Select the **Interference Check→ Run Interference Check** tool on the Coordinate panel.

2. 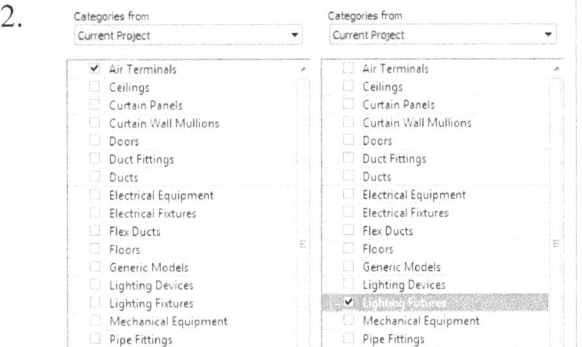 Enable **Air Terminals** in the left panel.

    Enable **Lighting Fixtures** in the right panel.

    If you select all in both panels, the check can take a substantial amount of time depending on the project and the results you get may not be very meaningful.

3. Press **OK**.

4.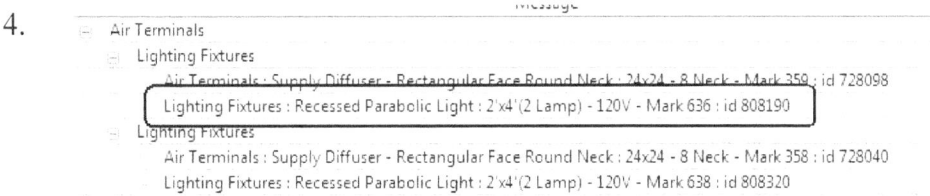

Highlight the Lighting Fixture in the first error.

5. 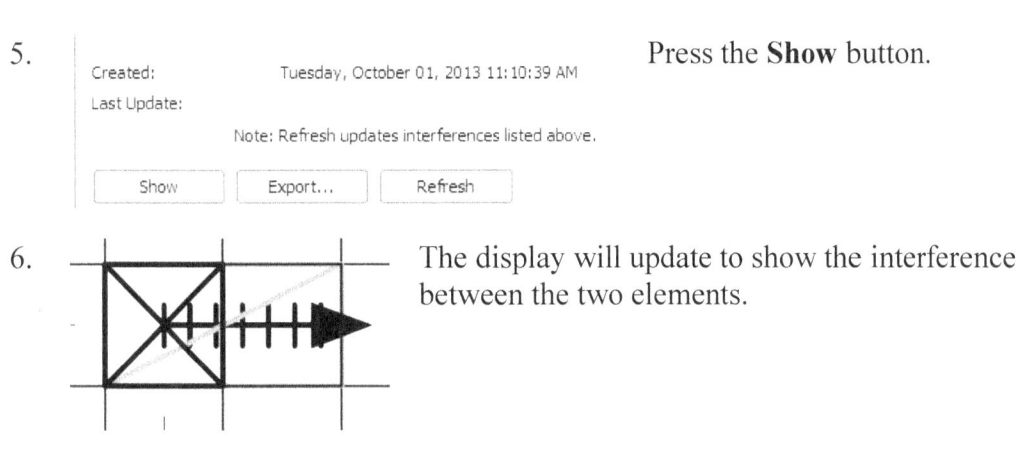 Press the **Show** button.

6.  The display will update to show the interference between the two elements.

7. 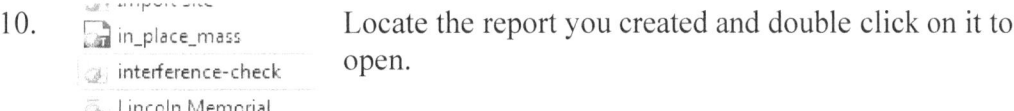 Select **Export**.

8. 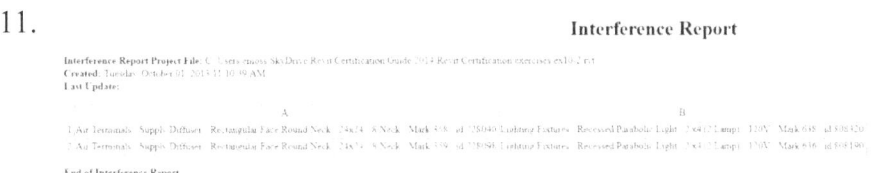 Browse to your exercise folder.

Press **Save**.

9. Press **Close** to close the dialog box.

10. Locate the report you created and double click on it to open.

11.

This report can be emailed or used as part of the submittal process.

12. 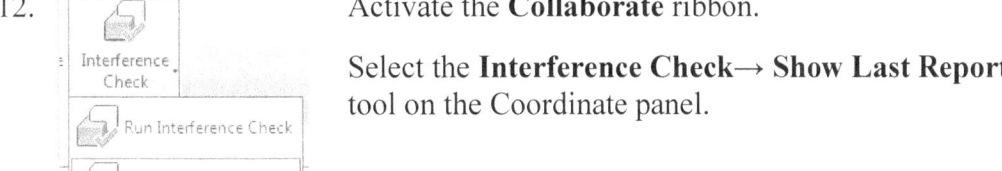 Activate the **Collaborate** ribbon.

Select the **Interference Check→ Show Last Report** tool on the Coordinate panel.

10-10

13.

Highlight the first lighting fixture.

Verify that you see which fixture is indicated in the graphics window.

Close the dialog.

14.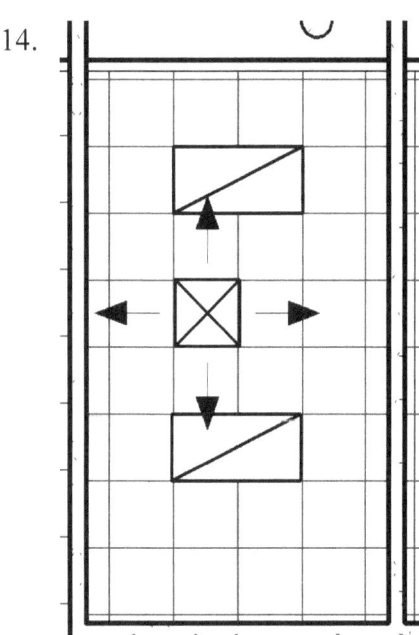

Move the lighting fixture to a position above the air terminal.

Arrange the lighting fixtures and the air diffuser so there should be no interference.

15.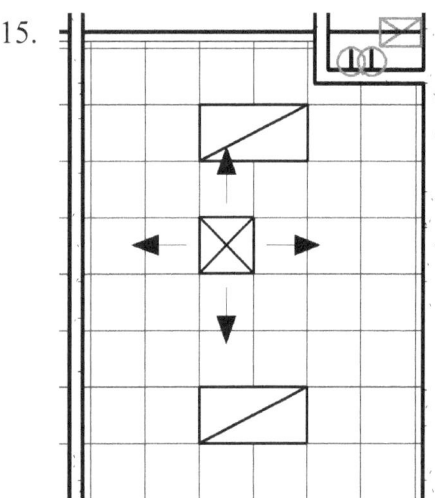

Move the second lighting fixture on the right so it is no longer on top of the air terminal.

Rearrange the fixtures in the room to eliminate any interference.

16. Activate the **Collaborate** ribbon.

    Select the **Interference Check→ Show Last Report** tool on the Coordinate panel.

17. Select **Refresh**.

18. The message list is now empty. Close the dialog box.

19. 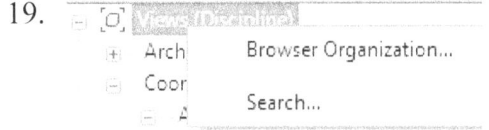 Highlight **Views** in the Project Browser. Right click and select **Search**.

20. 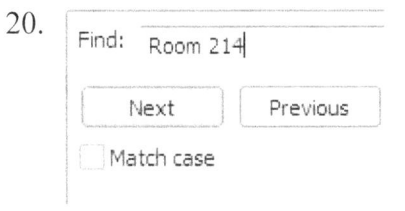 Type **Room 214**.
    Press **Next**.

21.  Press **Close** when the view is located.

    Activate the **Room 214 3D Fire Protection view** located under Mechanical.

22. 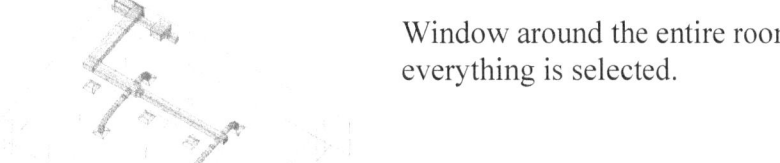 Zoom in so you can see the room.

    Window around the entire room so everything is selected.

10-12

23. Activate the **Collaborate** ribbon.

Select the **Interference Check→ Run Interference Check** tool on the Coordinate panel.

24. Note that the interference check is being run on the current selection only and not on the entire project.

Press **OK**.

25. Review the report.

Press **Close**.

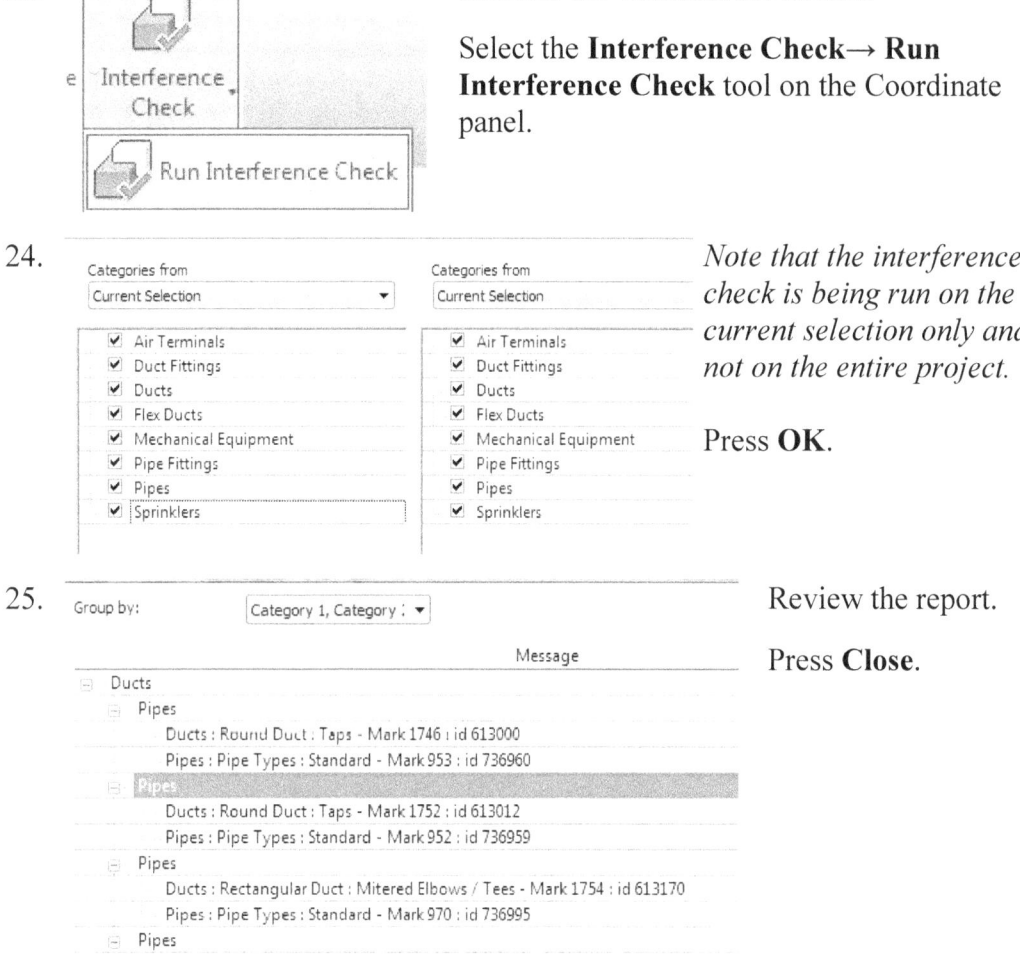

26. Close all files without saving.

In the Professional exam, you might be asked to run an interference check and identify the id number of the element affected by interference. Can you locate the id number for the element with Mark Number 970 in the report?

Command Exercise

Exercise 10-3 – Using Shared Coordinates

Drawing Name: **Import Site.Dwg**
Estimated Time to Completion: 40 Minutes

Scope

Link an AutoCAD file
Set Shared Coordinates
Set Project North

Every project has a project base point ⊗ and a survey point △, although they might not be visible in all views, because of visibility settings and view clippings. They cannot be deleted.

The project base point defines the origin (0,0,0) of the project coordinate system. It also can be used to position the building on the site and for locating the design elements of a building during construction. Spot coordinates and spot elevations that reference the project coordinate system are displayed relative to this point.

The survey point represents a known point in the physical world, such as a geodetic survey marker. The survey point is used to correctly orient the building geometry in another coordinate system, such as the coordinate system used in a civil engineering application.

Solution

1. Start a new project file using the Architectural template.

2. Activate the **Site** floor plan.

3. Activate the **Insert** ribbon.

    Select the **Link CAD** tool on the Link panel.

Collaboration

4.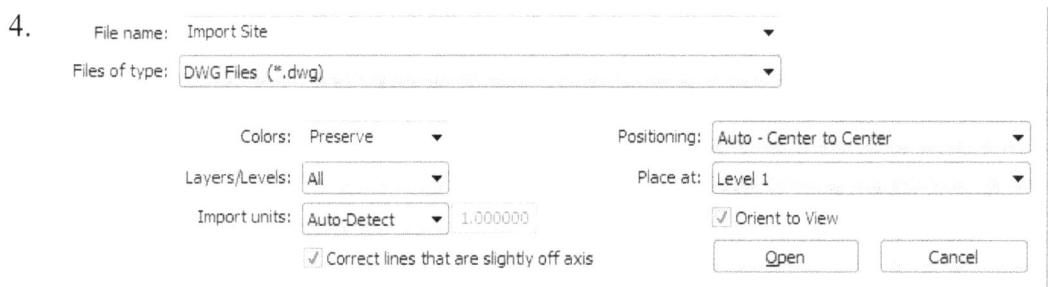

Select the *Import Site* drawing.

Set Colors to **Preserve**.
Set Import Units to **Auto-Detect**.
Set Positioning to **Auto - Center to Center**.
Press **Open**.

5.

Select the imported site plan.

In the Properties pane:

Select the Shared Site button.

6. Enable **Acquire the shared coordinate system…**

Select **Change**.

7. Activate the Location tab.

Enter the Project Address.

Press **Search**.
Then, press **OK**.

10-15

8. Press **Reconcile**.

9. Note that the survey point and project base point shift position.

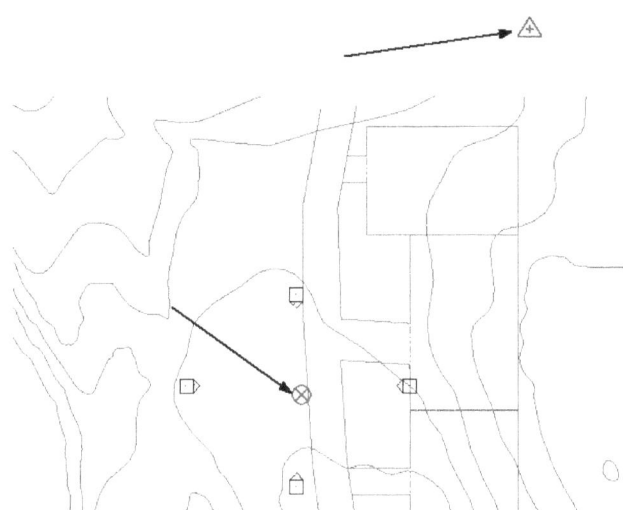

10. Select the Project Base point.

    This is the symbol that is a circle with an X inside.

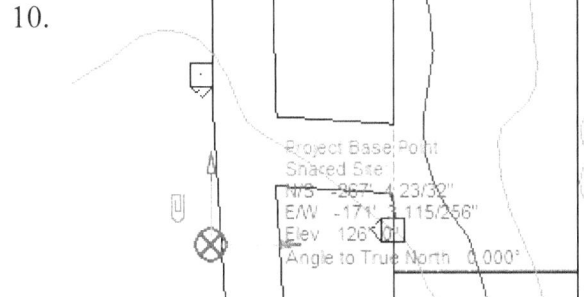

11. Change the Angle to Truth North to **90.00**.
    Press **Apply**.
    The site plan will rotate 90 degrees.

Collaboration

12. Move the import site plan so that the corner of the rectangle is coincident with the project base point.

13. Select **OK**.

    If you select Save Now, this will modify the linked file.

14. Activate the Insert ribbon.

    Select **Link Revit** from the Link panel.

15. Select the *i_shared_coords* file.
    Set the Positioning to: **Manual - Base point**.
    Press **Open**.

16. Place the Revit file to the left of the site plan.

17. Select the **Align** tool on the Modify panel on the Modify ribbon.

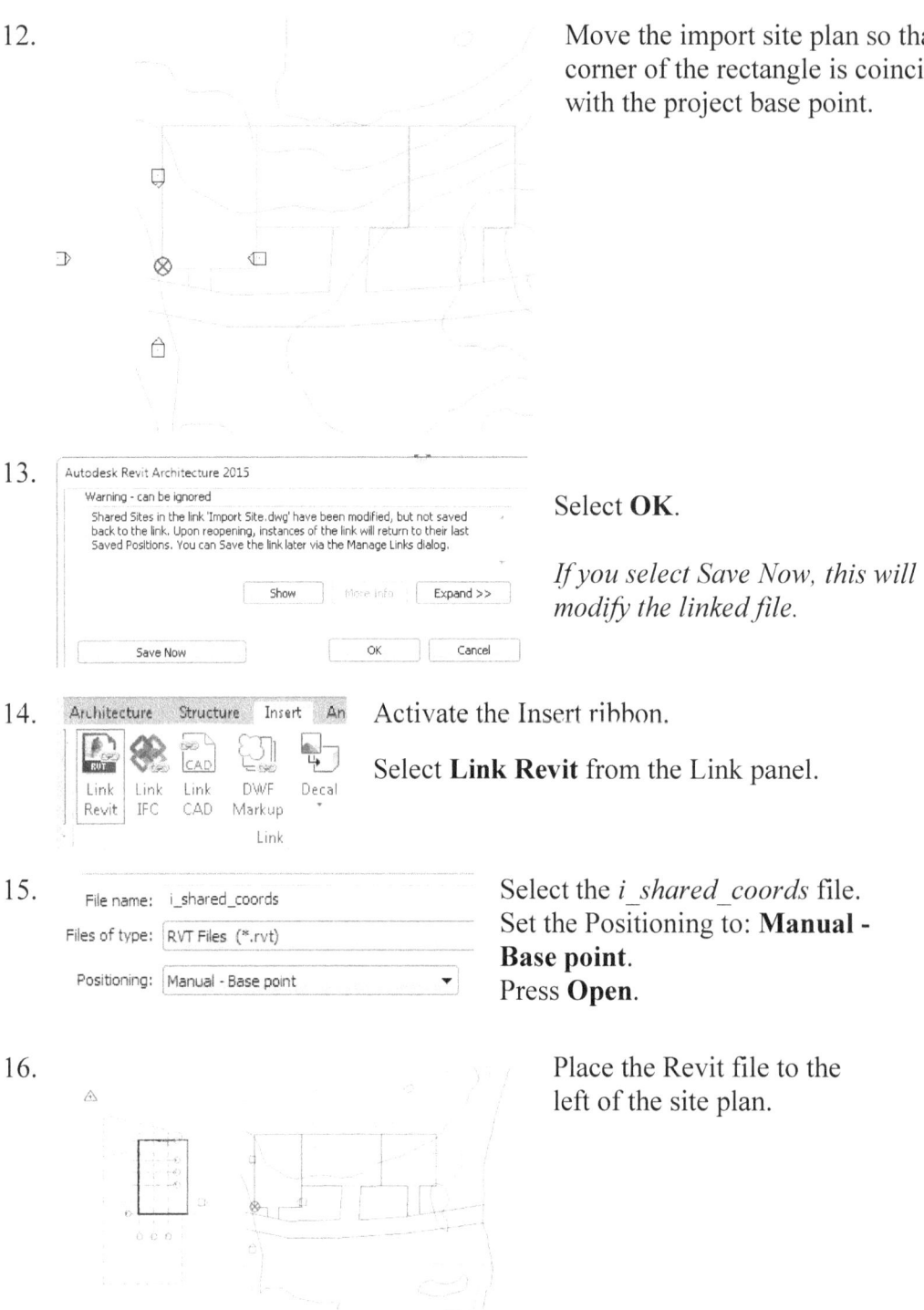

10-17

18. 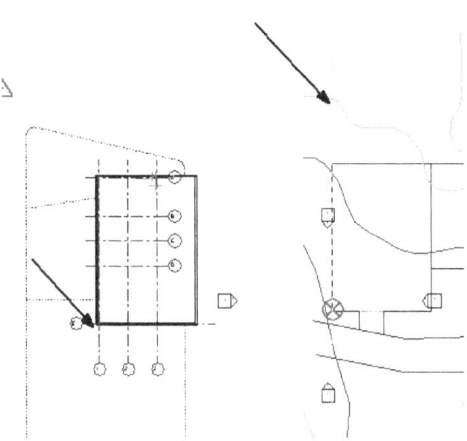 Select the vertical line above the project base point.
Then select Grid line 1 on the shared_coords imported file.

19. 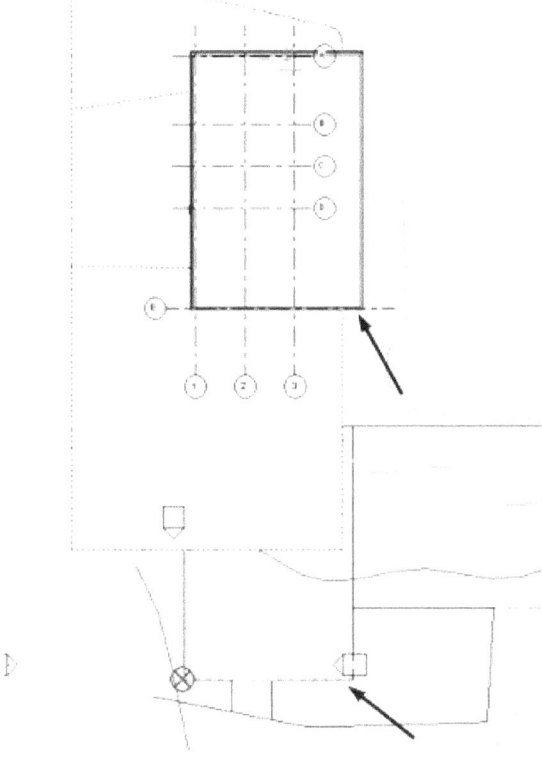 Change the display to wireframe, if necessary.

You can also move the linked Revit file up to locate the next alignment point.

Select the horizontal line on the project base point.
Then select Grid line E.

Cancel out of the ALIGN command.

20. Select the *i_shared_coords* Linked Revit Model.

In the Properties pane:
Click on the Shared Site button.

Collaboration

21. Enable **Record current position as *i_shared_coords.rvt***.

Press **Change**.

22. Press **OK**.

23. Elevations (Building Elevation)
 East
 North
 South
 West

Activate the North Elevation.

24.

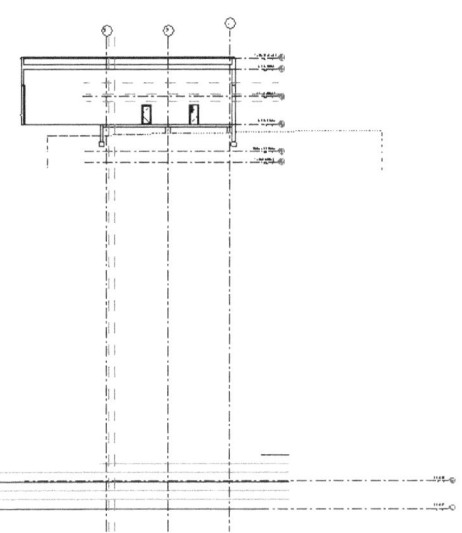

Level 1 and Level 2 are in the host project.

The other levels shown are in the Linked Revit model.

10-19

25.  Activate the **Site** plan.

26. 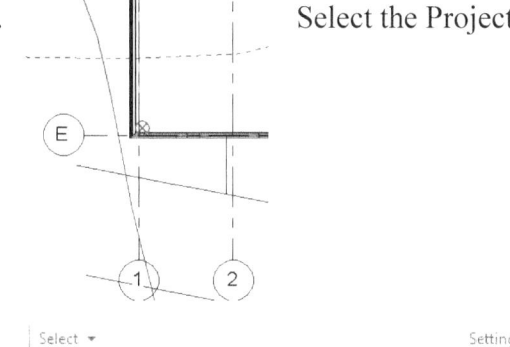 Activate the Manage ribbon.

    Select **Coordinates→Report Shared Coordinates** from the Project Location panel.

27. Select the Project Base point.

28. The coordinate information will display on the Options bar.

29. Close without saving.

Collaboration

Command Exercise
Exercise 10-4 – Worksets

Drawing Name: **i_Urban_house.rvt**
Estimated Time to Completion: 90 Minutes

Scope

Use of Worksets
Use of Revisions

Solution

1. Before you can use Worksets, you need to set Revit to use your name.

    Close any open projects.

    Go to **Options**.

2. Select the **General** tab.
   Enter your first name in the Username text field.
   Press **OK**.

   Username
   smossi@peralta.edu
   You are currently signed in to Autodesk 360. Your Autodesk ID is used as the username. If you need to change your username, you will need to sign out from Autodesk 360.
   Sign Out of Autodesk 360

   If you are using the Autodesk 360 account, you will see the user name assigned to that account.

10-21

3. Locate the *i_Urban_House* project.

   Press **Open**.

4. 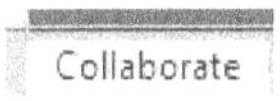 Select the **Collaborate** ribbon.

5. 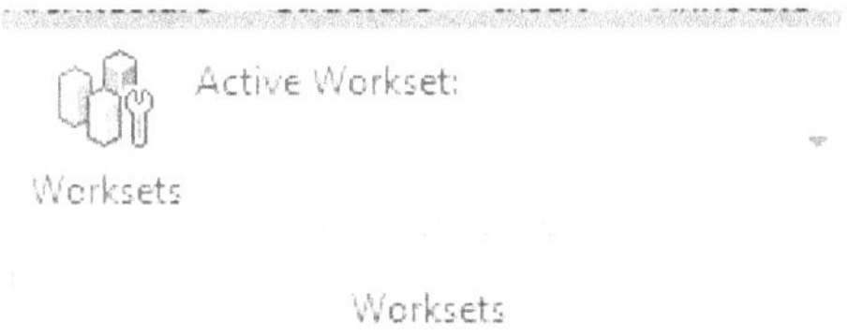 Select **Worksets** on the Worksets panel.

6. 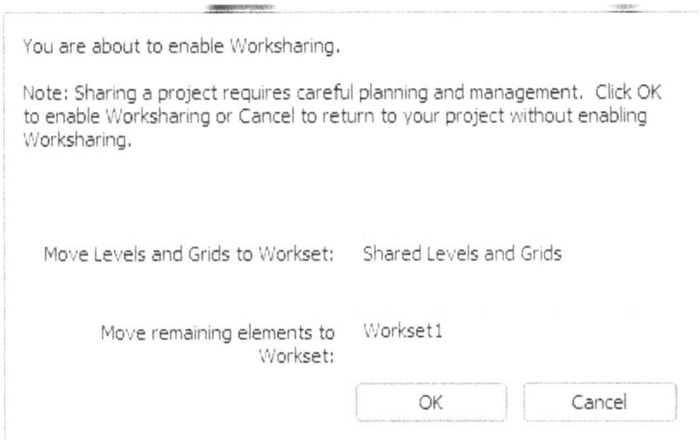 When you first enable Worksets, you will see this dialog.

   Press **OK**.

7. 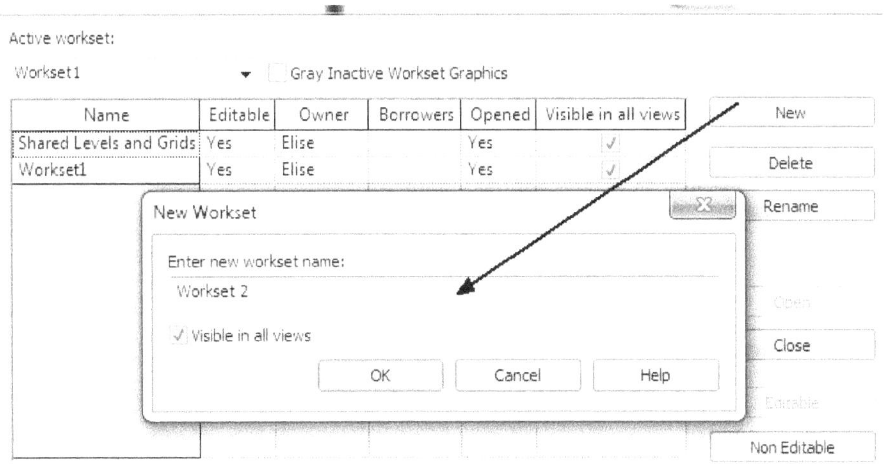 Create new worksets by pressing **New** and entering a name for the workset.

8. 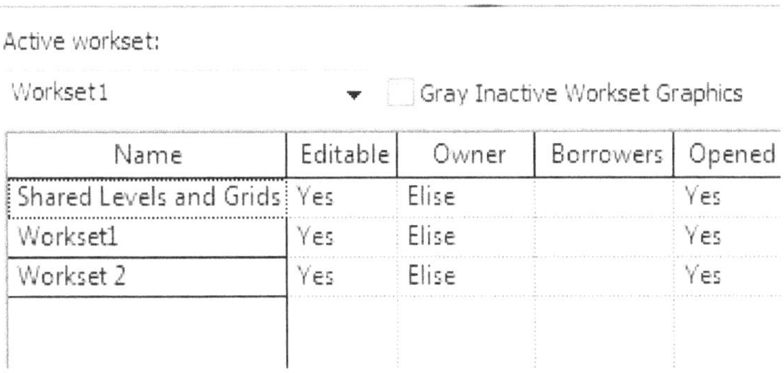 Identify the workset assigned to you.
   Set Editable to **Yes**.
   Your name should appear as the Owner of the workset.

Collaboration

9. 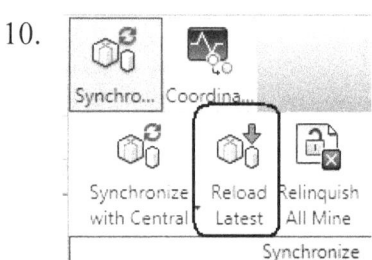 After each change, go to the Collaborate ribbon and select **Synchronize Now**.

This ensures that your work is saved and that other team members can see your changes.

10. If you select an element which has been modified, you may receive a prompt to load the latest version of the project.

Select **Reload Latest** from the Collaborate ribbon to ensure you are working on the latest saved version of the project.

*You also want to **Reload Latest** after any breaks or long periods away from the project.*

11. Other team members may need to borrow elements assigned to your workset in order to make changes. You can Grant or Deny those requests.

 Every 30 minutes or so, you should check your **Editing Requests**. Editing Requests also shows any pending requests you may have.

12. 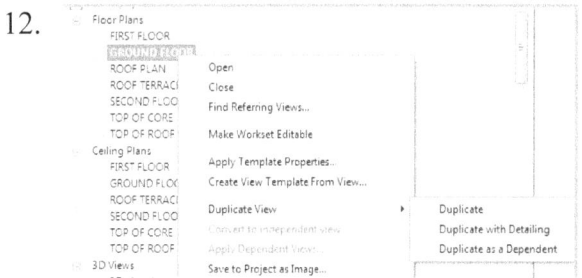 Highlight the view you are assigned to work on.
Right click and select **Duplicate View→ Duplicate**.

13. Rename the view with the [View Name]-Workset [Number].
Press **OK**.

14.  Go to the View ribbon.
Select the **Revisions** tool on the Sheet Composition panel.

10-23

15.

Sequence	Numbering	Date	Description	Issued	Issued to	Issued by	Show
1	Numeric	08.02	Move Door 2 in the kitchen -		Workset 1	Elise Moss	Cloud and Tag
2	Numeric	08.03	Change floor type to use gol		Workset 1	Elise Moss	Cloud and Tag
3	Numeric	08.02	Demo Wall next to staircase		Workset 1	Elise Moss	Cloud and Tag
4	Numeric	08.02	Change Opening to Staircas		Workset 1	Elise Moss	Cloud and Tag
5	Numeric	08.03	Add door tags to all the doo		Workset 2	Elise Moss	Cloud and Tag
6	Numeric	08.02	Change Door 17 Width to 4' -		Workset 2	Elise Moss	Cloud and Tag
7	Numeric	08.02	Demo Door 19		Workset 2	Elise Moss	Cloud and Tag
8	Numeric	08.02	Add a 36" x 84" Cased Openi		Workset 3	Elise Moss	Cloud and Tag
9	Numeric	08.02	Demo the short horizontal w		Workset 3	Elise Moss	Cloud and Tag
10	Numeric	08.03	Add a roof		Workset 4	Elise Moss	Cloud and Tag
11	Numeric	08.03	Add a shade awning		Workset 4	Elise Moss	Cloud and Tag
12	Numeric	08.03	Change Floor Type on Roof		Workset 5	Elise Moss	Cloud and Tag
13	Numeric	08.03	Change Railing Type on Roo		Workset 5	Elise Moss	Cloud and Tag

This is the table where revisions are listed as well as the team member assigned to make the change. *I have assigned changes to a Workset number. The instructor will assign each student a workset group to make changes.*

The Sequence Number is the same as the revision number.

16.  To place a revision cloud:
Go to the Annotate ribbon.
Select **Revision Cloud** from the Detail panel.

Revisions should be placed on the existing view.

17. Place a small cloud in the general area of the change.

18. In the Properties pane:
In the Revision field, select the Sequence Number that is used for your change.
Press **OK**.

19.  Select the **Green Check** on the Mode panel to **Finish Cloud**.

Collaboration

20.  Activate the **Annotate** ribbon.

    Select **Tag by Category** from the Tag panel.

21. Pick the revision cloud.

    The first time you pick a revision cloud, you will be prompted to load the revision cloud tag. Press **Yes**.

22. Browse to the *Annotations* folder.

    Scroll down to locate the *Revision Tag* family.

    Press **Open**.

23. 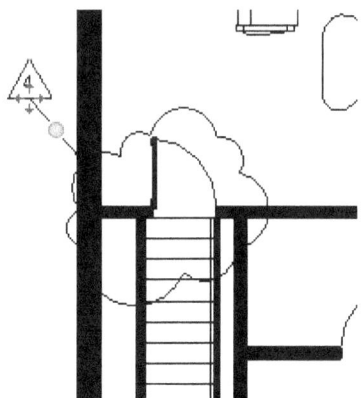 The tag will display the sequence number assigned to the revision cloud.

    To change the sequence number, select the rev cloud, right click, select Element Properties, and change the Revision value.

24. Create a sheet for the view and name the sheet.
    The sheet should be named using [View] - [Workset Number].

25. Set Drawn By to your name.

26. Make the changes assigned to your workset.

10-25

Other Hints:

- Name any sheets you create. That way you can distinguish between your sheet and other sheets.
- Create a view that is specific to your changes or workset, so that you have an area where you can keep track of your work.
- Create one view for the existing phase and one view for the new construction phase. That way you can see what has changed. Name each view appropriately
- Use View Properties to control the phase applied to each view.
- Check Editing Requests often
- Duplicate any families you need to modify, rename, and redefine. If you modify an existing family, it may cause problems with someone else's workset.

Workset Checklist

1. Before you open any files, make sure your name is set in Options.
2. The main file is stored on the server. The file you are working on should be saved locally to your flash drive or your folder. It should be named Urban House_[Your Name]. If you don't see your name on the file, you need to re-save or re-load the file.

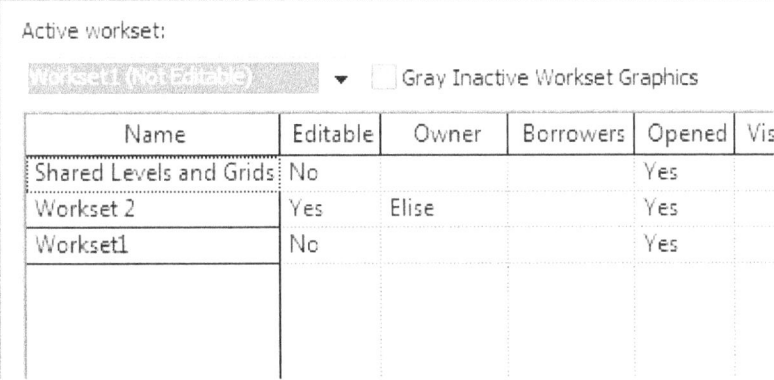

3. Go to Work sets and verify that your name is next to the Work set you are assigned.

4. Verify that the Active work set is the work set you are assigned.
5. Re-load the latest from Central so you can see all the updates done by other users.
5. When you are done working, save to Central – so other users can see your changes.
6. When you close, relinquish your work sets, so others can keep working.

Collaboration

The **Industry Foundation Classes (IFC)** data model is used to describe building and construction industry data.

It is a platform neutral, open file format specification that is not controlled by a single vendor or group of vendors. It is meant to promote data exchange between CAD vendors. It is a popular format in Europe and is promoted by several governments for use in their AEC projects. It is similar to the step/stp format used in mechanical design file exchanges. Revit 2015 has been improved for file import and export of IFC files.

Prior to importing or exporting with Revit, it is critical to set the IFC Options. The options allow you to map the different Revit elements appropriately.

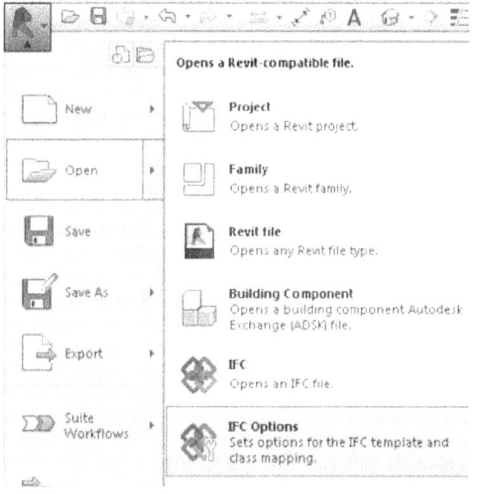

To set the mapping options, use the Application Menu to go to Open→IFC Options.

Import IFC Options

Default Template for IFC Import:

Import IFC Class Mapping: C:\Users\Elise\SkyDrive\Revit Certification Guide 2015\Certification Guide 2015 Exercises\StairimportIFCClassMapping

IFC Class Name	IFC Type	Revit Category	Revit Sub-Category
IfcAirTerminal		Air Terminals	
IfcAirTerminalType		Air Terminals	
IfcAnnotation		Generic Annotations	
IfcBeam		Structural Framing	
IfcBeamType		Structural Framing	
IfcBoiler		Mechanical Equipment	
IfcBoilerType		Mechanical Equipment	
IfcBuildingElementPart		Parts	
IfcBuildingElementPartType		Parts	
IfcBuildingElementProxy		Generic Models	

Then select which IFC Types to map to the appropriate Revit Category.

This ensures the geometry comes in properly. You can also save and restore IFC mapping files.

Command Exercise

Exercise 10-5 – Import an IFC File

Drawing Name: Duplex_A_20110505
Estimated Time to Completion: 5 Minutes

Scope

Use of IFC Mapping files
Import IFC File

Solution

1. 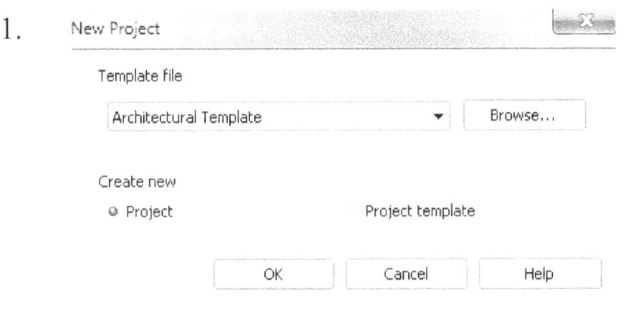 Start a new project using the Architectural template.

2. 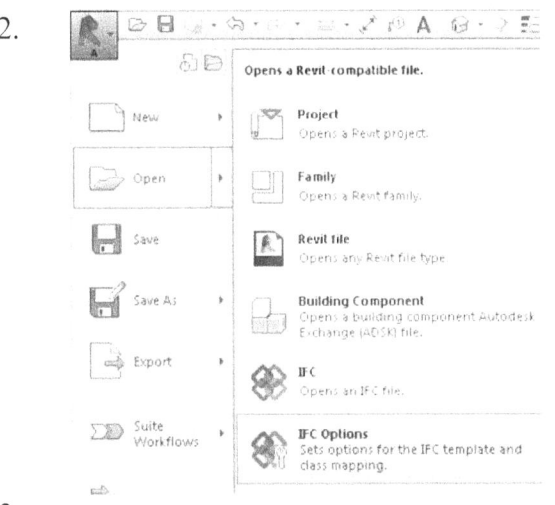 Use the Application Menu to go to **Open→IFC Options.**

   Note: you must have a project open in order to use the IFC Options.

3.  Select the **Load** button.

Collaboration

4.   Dds_BardNa.Ifc.sharedparameters
      Duplex_A_20110505.ifc.sharedparameters
      Staircase.ifc.sharedparameters

Select the *Duplex_A_20110505.ifc.sharedparameters* file.

Press **Open**.

5.  Select **Link IFC** on the Insert ribbon.

6. File name: Duplex_A_20110505
   Files of type: IFC Files (*.ifc)

   Select *Duplex_A_20110505*.

   Press **Open.**

7.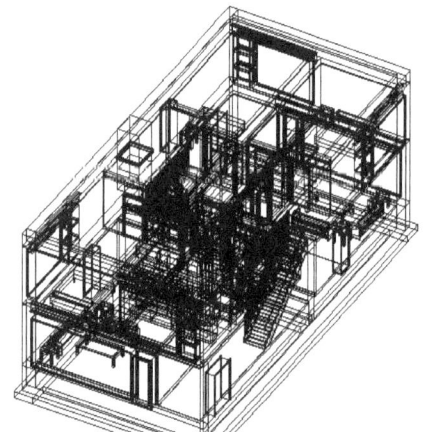

   The model is imported.

8. Close without saving.

Practice Exam

1. Worksharing allows team members to:
   - A. Share families from different projects
   - B. Share views from different projects
   - C. Work on the same parts of a project simultaneously
   - D. Work on different parts of the same project

2. Set Phase Filters for a view in the:
   - A. Properties pane
   - B. Design Options
   - C. Ribbon
   - D. Project Browser

3. A rectangular column family comes with three types: 18″ x 18″, 18″ x 24″, and 24″ x 24″. To add a 6″ x 6″ type, you:
   - A. Duplicate, rename and modify the instance parameters
   - B. Duplicate, rename and modify the type parameters
   - C. Rename and modify the type parameters
   - D. Duplicate, rename the family and add a type.

4. A door is _____ a wall.
   - A. hosts
   - B. defines
   - C. is hosted by
   - D. is defined by

5. The Coordination Review tool is used when you use:
   - A. Worksets
   - B. Linked Files
   - C. Interference Checking
   - D. Phase

Answers
1) D; 2) A; 3) D; 4) C; 5) B

About the Author

Autodesk
Certified Instructor

Elise Moss has worked for the past thirty years as a mechanical engineer in Silicon Valley, primarily creating sheet metal designs. She has written articles for Autodesk's Toplines magazine, AUGI's PaperSpace, DigitalCAD.com and Tenlinks.com. She is President of Moss Designs, creating custom applications and designs for corporate clients. She has taught CAD classes at DeAnza College, Silicon Valley College, SFSU, and for Autodesk resellers. She is currently teaching CAD at Laney College in Oakland. Autodesk has named her as a Faculty of Distinction for the curriculum she has developed for Autodesk products. She holds a baccalaureate degree in Mechanical Engineering from San Jose State.

She is married with three sons. Her older son, Benjamin, is an electrical engineer. Her middle son, Daniel, works with AutoCAD Architecture in the construction industry. His designs have been featured in architectural journals. Her youngest son, Isaiah, is now attending college. Her husband, Ari, is retired from a distinguished career in software development.

Elise is a third generation engineer. Her father, Robert Moss, was a metallurgical engineer in the aerospace industry. Her grandfather, Solomon Kupperman, was a civil engineer for the City of Chicago.

She can be contacted via email at elise_moss@mossdesigns.com.

More information about the author and her work can be found on her website at www.mossdesigns.com.

Other books by Elise Moss
AutoCAD Architecture 2015 Fundamentals
Autodesk Revit Architecture 2015 Basics

www.ingramcontent.com/pod-product-compliance
Lightning Source LLC
Chambersburg PA
CBHW081714170526
45167CB00009B/3572